International Conference on Intelligent Transportation and Vehicle Engineering (ICITVE 2019)

IOP Conference Series: Materials Science and Engineering
Volume 787

Harbin, China
6 - 8 December 2019

ISBN: 978-1-7138-2669-9
ISSN: 1757-8981

Printed from e-media with permission by:

Curran Associates, Inc.
57 Morehouse Lane
Red Hook, NY 12571

Some format issues inherent in the e-media version may also appear in this print version.

Institute of Physics
Dirac House, Temple Back
Bristol BS1 6BE UK

Phone: 44 1 17 929 7481
Fax: 44 1 17 920 0979

techtracking@iop.org

Additional copies of this publication are available from:

Curran Associates, Inc.
57 Morehouse Lane
Red Hook, NY 12571 USA
Phone: 845-758-0400
Fax: 845-758-2633
Email: curran@proceedings.com
Web: www.proceedings.com

TABLE OF CONTENTS

Preface

2019 International Conference on Intelligent Transportation and Vehicle Engineering (ICITVE 2019), hosted by School of Automotive and Transportation Engineering, Heilongjiang Institute of Technology, was successfully held in Chongqing, China during December 6-8, 2019. ICITVE 2019 aims to provide a platform for experts, scholars, and practitioners in the field of Intelligent Transportation and Vehicle Engineering to share the latest research results, discuss existing problems and challenges, explore cutting-edge technologies, and enhance their cooperation.

We are honored to have Prof. Baixue Fu, Dean of School of Automotive and Transportation Engineering, Heilongjiang Institute of Technology, China, and Prof. Traian Mazilu from University Politehnica of Bucharest, Romania to jointly chair this conference. And we invited over 30 excellent scholars form east and west to constitute the Academic Committee of ICITVE 2019. During the conference, five remarkable experts gave excellent keynote speeches related to the field of Intelligent Transportation and Vehicle Engineering on site. Prof. Devinder Yadav from University of Nottingham Ningbo, China discussed *emission control at airports and advances in alternate propulsion in civil aviation industry*. Assoc.Prof. Seyed Mohammadreza Ghadiri from Malaysia University of Science & Technology, shared his research on *Speed Limit Obedience Factor (SLOF) A New Countermeasure to Improve the Efficiency of The Advisory Intelligent Speed Adaptation*. Prof. Said Easa from Ryerson University Toronto, Canada held a speech on *Smart Mobility: Challenges and Solution*s. Prof. Fangwu Ma from Jilin University, China discussed the *Outlook of Ecosystem of New Mobility*. And Senior Engineer Chengyong Niu, Project Supervisor from Chongqing Vehicle Test & Research Institute Co., Ltd, China had a speech on *Test and application challenges for autonomous vehicle at level 3 and below*. Thanks to these wonderful speeches given by the distinguished experts, the conference has formed a lively and heated academic vibe on site.

We are happy to present the proceedings of ICITVE 2019 featured the new advances and research results in the fields of transportation and vehicle engineering research. All papers have been through rigid review process to comply with the requirements of international publication.

We would to like to express our sincere gratitude to all the participants attending the conference, who made this conference a great success. We want to thank our honorific chairmen for hosting this conference, the distinguished keynote speakers for giving insightful speeches, and all authors for submitting their study paper. We hope this conference and the proceedings of it could serve as a good reference for those who work in the field of Intelligent Transportation and Vehicle Engineering. We aim to hold this conference every year to make it an convenient platform for people to share views and experiences in intelligent transportation and vehicle engineering and related areas. We would like to see you next year. Thank you.

Committee of ICITVE 2019

IOP Conf. Series: Materials Science and Engineering **787** (2020) 011001 doi:10.1088/1757-899X/787/1/011001

Committee member

Academic Committee Chairs

Prof. Baixue Fu, Dean of School of Automotive and Transportation Engineering, Heilongjiang Institute of Technology, China

Prof. Traian Mazilu, Director of Department of Railway Vehicles, University Politehnica of Bucharest, Romania

Academic Committee Member

Assis. Prof.Hong Ki An, Huaiyin Institute of Technology, China
Assoc. Prof.Da-Jie Lin, Feng Chia University, Taiwan, China
Prof.Yuh-Chung Hu, National ILan University, Taiwan, China
Assis.Prof.Mohammed Abdo Hashem, Universiti Malaysia Pahang, Malaysia
Assoc. Prof.Muhammad Ali, HITEC University, Taxila, Pakistan
Prof. Peter Klaus, Technical University of Munich Asia, German
Dr. Andreas Rau, Technical University of Munich Asia, German
Assoc. Prof.Marta Gonzalez, University of California, Berkeley
Assis.Prof. Alexey Pozdnukhov, University of California, Berkeley
Dr. Kun An, Monash University, Australia
A.Prof. Qiang Wang, Heilongjiang Institute of Technology, China
Dr.Mehdi Moeinaddini, University of Liège, Belgium
Dr. Ali Rahman, Southwest Jiaotong University, China
Prof. Devinder Yadav, University of Nottingham Ningbo, China
Assis.Prof. Liu Yang, National University of Singapore, Singapore
Dr. Zohreh Asadi Shekari, Universiti Teknologi Malaysia, Malaysia
Assis.Prof. Chengpeng Wan, Wuhan University of Technology, China
Prof. Chi-Hua Chen, Fuzhou University, China
A.Prof. Felix Ling Ngee Leh, University Tun Hussein Onn Malaysia, Malaysia
A.Prof. Adnan Bin Zainorabidin, University Tun Hussein Onn Malaysia, Malaysia
Assis.Prof.Keping Yu, Waseda University, Japan
A. Prof. Madalina Dumitriu, University Politehnica of Bucharest, Romania
Dr.Huizhi Xu, Northeast Forestry University,China
Dr.Wei Ma, Carnegie Mellon University, USA
Assis.Prof. Xiaozheng He, Rensselaer Polytechnic Institute, USA
Dr.Qian Zhou, Chongqing Jiaotong University, China
Dr. Amila Jayasinghe, University of Moratuwa, Moratuwa, Sri Lanka
Dr.Behram Wali, Massachusetts Institute of Technology, USA
Prof. Said Easa, Ryerson University Toronto, Canada
Prof. Virginia P. Sisiopiku, University of Alabama at Birmingham, USA
Dr. Wen Long Yue, Department of Transport, Victoria, Australia
Niu Chengyong, Chongqing Vehicle Test & Research Institute Co., Ltd, China

IOP Conf. Series: Materials Science and Engineering **787** (2020) 011002 doi:10.1088/1757-899X/787/1/011002

Peer review statement

All papers published in this volume of *IOP Conference Series: Materials Science and Engineering* have been peer reviewed through processes administered by the proceedings Editors. Reviews were conducted by expert referees to the professional and scientific standards expected of a proceedings journal published by IOP Publishing.

IOP Conf. Series: Materials Science and Engineering **787** (2020) 012001 doi:10.1088/1757-899X/787/1/012001

An analysis of the calibration method of accelerometer bias and scale factor

Jiangwei Jia[1], Jiaxing Zheng[1*], Xingshu Wang[1]

[1]College of Advanced Interdisciplinary Studies, National University of Defense Technology, Changsha, Hunan, 410000, China

[*]Corresponding author's e-mail: zhengjiaxing@nudt.edu.cn

Abstract. Accelerometer bias and scale factor are important parameters of the accelerometer. The accuracy of calibration of bias and scale factor will directly affect navigation accuracy. This paper studies the calibration of accelerometers under out-door conditions. In some applications, the inertial measurement unit (IMU) cannot rotate with any angle due to the restriction of equipment and site. The pitch and roll rotation range is -45° to 45°, and the yaw is -180° to 180°. In this paper, the bias and scale factor identification equations under different positions are derived based on the inertial navigation error equation. The results show that when the IMU rotates around the pitch axis, as long as there are three pitch positions, the bias and scale factor of the y-direction and z-direction accelerometers can be calibrated. When the IMU rotates 45° around the roll axis and then rotates 90° around the yaw axis, the x-direction accelerometer scale factor can be calibrated. If the bias and scale factor of the y-direction and z-direction accelerometer are known, the bias of the x-direction accelerometer can also be calibrated. Then, several simulations are conducted which are consistent with the theoretical results. At last, optimized calibration path is designed. Within 20 min, the calibration accuracy of bias is better than 5 mGal, and the calibration accuracy of scale factor is better than 5 ppm.

1. Introduction

Accelerometer bias and scale factor are important error sources in inertial navigation systems[1-3]. Ideally, if the input of accelerometers is zero, the output should also be zero. But in actual applications, the output of accelerometers is not zero, which is the accelerometer bias. The output of the accelerometer is a digital electrical signal that needs to be converted to velocity increment. The ratio of the velocity increment to the digital electrical signal is the accelerometer scale factor[4-5]. The bias and scale factor errors can affect the navigation accuracy badly, so the bias and scale factor need to be accurately calibrated.

Calibration methods mainly include discrete calibration and systematic calibration. Systematic calibration can meet the need of out-door calibration and becomes a research hotspot[6-7]. Yang Xiaoxia proposed a systematic calibration model, and analyzed the identification of inertial measurement unit (IMU) error parameters[8]. Shi Wenfeng designed a ten-position systematic calibration method to improve calibration accuracy[9]. Wang Zihui[10] proposed an eight-position systematic calibration method based on the two-axis turntable, which is superior to the traditional calibration method for both calibration time and navigation precision. These studies are based on the premise that the IMU can rotate to any position. In some applications, the IMU cannot rotate with any angle due to the restriction of equipment and site. The pitch and roll rotation range is -45° to 45°, and the yaw is -180° to 180°. Due to the constraints of the rotational position, the rotational path needs to be redesigned for calibration.

This paper mainly studies the calibration of accelerometer bias and scale factor. The rest of the paper is organized as follows: section 2 conducts theoretical analysis based on inertial navigation error equation and proves the identification of accelerometer bias and scale factor; section 3 conducts simulations to verify the theoretical analysis and section 4 makes a conclusion of this paper.

2. Theoretical analysis

Ignoring the gravity error, the velocity error model of IMU can be written as follows[11-12]:

$$\delta\dot{\vec{V}}^n = \vec{f}^n \times \vec{\varphi} + C_b^n \delta\vec{f}^b - \left(2\vec{\omega}_{ie}^n + \vec{\omega}_{en}^n\right) \times \delta\vec{V}^n - \left(2\delta\vec{\omega}_{ie}^n + \delta\vec{\omega}_{en}^n\right) \times \vec{V}^n \tag{1}$$

Where $\vec{V}^n$ is the velocity of IMU; $\vec{\varphi}$ is the attitude error of IMU; $\vec{f}^n$ and $\vec{f}^b$ are the specific force in the navigation frame (*n*-frame) and body frame (*b*-frame) respectively; C_b^n is the direction cosine matrix from *b*-frame to *n*-frame; $\vec{\omega}_{ie}^n$ is the angular velocity of the earth rotation in *n*-frame; $\vec{\omega}_{en}^n$ is the angular velocity of *n*-frame with respect to Earth-centric fixed frame (*e*-frame); $\delta\vec{\omega}_{ie}^n$ and $\delta\vec{\omega}_{en}^n$ are the differential of $\vec{\omega}_{ie}^n$ and $\vec{\omega}_{en}^n$ respectively.

The accelerometer errors include accelerometer bias and scale factor error, so accelerometer errors can be written as follows:

$$\delta\vec{f}^b = \delta\vec{f}^b\left(\nabla^b\right) + \delta\vec{f}^b\left(\vec{S}_a\right) = \nabla^b + \begin{bmatrix} S_{ax} & & \\ & S_{ay} & \\ & & S_{az} \end{bmatrix} \begin{bmatrix} f_x^b \\ f_y^b \\ f_y^b \end{bmatrix} \tag{2}$$

Where $\delta\vec{f}^b\left(\nabla^b\right)$ is accelerometer bias; $\delta\vec{f}^b\left(\vec{S}_a\right)$ is accelerometer scale factor error. $\vec{\omega}_{ie}^n$, $\vec{\omega}_{en}^n$, $\delta\vec{\omega}_{ie}^n$ and $\delta\vec{\omega}_{en}^n$ can be expressed as follows:

$$\vec{\omega}_{ie}^n = \begin{bmatrix} 0 & \omega_{ie}\cos L & \omega_{ie}\sin L \end{bmatrix}^T \tag{3}$$

$$\vec{\omega}_{en}^n = \begin{bmatrix} \frac{-V_N}{R_N + h} & \frac{V_E}{R_E + h} & \frac{V_E}{R_E + h} tgL \end{bmatrix}^T \tag{4}$$

$$\delta\vec{\omega}_{ie}^n = \begin{bmatrix} 0 & -\omega_{ie}\sin L\delta L & \omega_{ie}\cos L\delta L \end{bmatrix}^T \tag{5}$$

$$\delta\vec{\omega}_{en}^n = \begin{bmatrix} 0 & 0 & \frac{V_N}{\left(R_N + h\right)^2} \\ 0 & 0 & -\frac{V_E}{\left(R_E + h\right)^2} \\ 0 & \frac{V_E \sec^2 L}{R_E + h} & -\frac{V_E tgL}{\left(R_E + h\right)^2} \end{bmatrix} \begin{bmatrix} \delta\lambda \\ \delta L \\ \delta h \end{bmatrix} + \begin{bmatrix} 0 & -\frac{1}{R_N + h} & 0 \\ \frac{1}{R_E + h} & 0 & 0 \\ \frac{tgL}{R_E + h} & 0 & 0 \end{bmatrix} \begin{bmatrix} \delta V_E \\ \delta V_N \\ \delta V_U \end{bmatrix} \tag{6}$$

Substituting equations (2)-(6) into equation (1), we can obtain:

$$\begin{bmatrix}\delta\dot{V}_E\\ \delta\dot{V}_N\\ \delta\dot{V}_U\end{bmatrix}=\begin{bmatrix}0 & 2\left(V_N\omega_{ie}\cos L+V_U\omega_{ie}\sin L\right)+\dfrac{V_EV_N}{(R_E+h)\cos^2 L} & \dfrac{-V_EV_NtgL}{(R_E+h)^2}+\dfrac{V_EV_U}{(R_E+h)^2}\\ 0 & -2V_E\omega_{ie}\cos L-\dfrac{V_E^{\ 2}}{(R_E+h)\cos^2 L} & \dfrac{V_E^{\ 2}tgL}{(R_E+h)^2}+\dfrac{V_NV_U}{(R_E+h)^2}\\ 0 & -2V_E\omega_{ie}\sin L & \dfrac{2g}{R_E}-\dfrac{V_E^{\ 2}}{(R_E+h)^2}-\dfrac{V_N^{\ 2}}{(R_E+h)^2}\end{bmatrix}\begin{bmatrix}\delta\lambda\\ \delta L\\ \delta h\end{bmatrix}$$
$$+\begin{bmatrix}\dfrac{V_NtgL-V_U}{R_E+h} & 2\omega_{ie}\sin L+\dfrac{V_EtgL}{R_E+h} & -2\omega_{ie}\cos L-\dfrac{V_E}{R_E+h}\\ -2\omega_{ie}\sin L-\dfrac{2V_EtgL}{R_E+h} & \dfrac{-V_U}{R_N+h} & \dfrac{-V_N}{R_N+h}\\ 2\omega_{ie}\cos L+\dfrac{2V_E}{R_E+h} & \dfrac{2V_N}{R_N+h} & 0\end{bmatrix}\begin{bmatrix}\delta V_E\\ \delta V_N\\ \delta V_U\end{bmatrix} \tag{7}$$
$$+\begin{bmatrix}0 & -f_U & f_N\\ f_U & 0 & -f_E\\ -f_N & f_E & 0\end{bmatrix}\begin{bmatrix}\varphi_E\\ \varphi_N\\ \varphi_U\end{bmatrix}+C_b^n\nabla^b+C_b^n\begin{bmatrix}S_{ax} & & \\ & S_{ay} & \\ & & S_{az}\end{bmatrix}\begin{bmatrix}f_x^b\\ f_y^b\\ f_y^b\end{bmatrix}$$

During the calibration, the time of rotation is short. Ignoring the variety of velocity and position, we can obtain:

$$\begin{bmatrix}\delta\dot{V}_E\\ \delta\dot{V}_N\\ \delta\dot{V}_U\end{bmatrix}=\begin{bmatrix}0 & -g & 0\\ g & 0 & 0\\ 0 & 0 & 0\end{bmatrix}\begin{bmatrix}\varphi_E\\ \varphi_N\\ \varphi_U\end{bmatrix}+C_b^n\nabla^b+C_b^n\begin{bmatrix}S_{ax} & & \\ & S_{ay} & \\ & & S_{az}\end{bmatrix}\begin{bmatrix}f_x^b\\ f_y^b\\ f_z^b\end{bmatrix}=\begin{bmatrix}-g\varphi_N\\ g\varphi_E\\ \varphi_U\end{bmatrix}+\delta\vec{f}^n \tag{8}$$

Where $\delta\vec{f}^n=C_b^n\delta\vec{f}^b$ is the error of accelerometer output in n-frame.

The difference between the differential of velocity error before and after rotation can be expressed as follows[13]:

$$\begin{cases}\delta\dot{V}_E(T)-\delta\dot{V}_E(0)=-g\Delta\varphi_N+\Delta\delta f_E\\ \delta\dot{V}_N(T)-\delta\dot{V}_N(0)=g\Delta\varphi_E+\Delta\delta f_N\\ \delta\dot{V}_U(T)-\delta\dot{V}_U(0)=\Delta\delta f_U\end{cases} \tag{9}$$

Where $\delta\dot{V}_E(T)$, $\delta\dot{V}_N(T)$ and $\delta\dot{V}_U(T)$ are the east, north and vertical components of $\delta\dot{\vec{V}}^n$ after rotation; $\delta\dot{V}_E(0)$, $\delta\dot{V}_N(0)$ and $\delta\dot{V}_U(0)$ are the east, north and vertical components of $\delta\dot{\vec{V}}^n$ before rotation; $\Delta\vec{\varphi}$ and $\Delta\delta\vec{f}^n$ are the variety of $\vec{\varphi}$ and $\delta\vec{f}^n$; $\Delta\varphi_E$ and $\Delta\varphi_N$ are the east and north components of $\Delta\vec{\varphi}$; δf_E, δf_N and δf_U are the east, north and vertical components of $\delta\vec{f}^n$; $\Delta\delta f_E$, $\Delta\delta f_N$ and $\Delta\delta f_U$ are the east, north and vertical components of $\Delta\delta\vec{f}^n$. In the case where the error of the gyro only includes bias, three components of $\Delta\vec{\varphi}$ are zero. $\Delta\delta\vec{f}^n$ can be expressed as follows:

$$\begin{aligned}\Delta\delta\vec{f}^n&=C_b^n(T)\left(\begin{bmatrix}\nabla_x^b\\ \nabla_y^b\\ \nabla_z^b\end{bmatrix}+\begin{bmatrix}S_{ax} & & \\ & S_{ay} & \\ & & S_{az}\end{bmatrix}\begin{bmatrix}f_x^b(T)\\ f_y^b(T)\\ f_z^b(T)\end{bmatrix}\right)-C_b^n(0)\left(\begin{bmatrix}\nabla_x^b\\ \nabla_y^b\\ \nabla_z^b\end{bmatrix}+\begin{bmatrix}S_{ax} & & \\ & S_{ay} & \\ & & S_{az}\end{bmatrix}\begin{bmatrix}f_x^b(0)\\ f_y^b(0)\\ f_z^b(0)\end{bmatrix}\right)\\ &=C_b^n(T)\begin{bmatrix}S_{ax}f_x^b(T)+\nabla_x^b\\ S_{ay}f_y^b(T)+\nabla_y^b\\ S_{az}f_z^b(T)+\nabla_z^b\end{bmatrix}-C_b^n(0)\begin{bmatrix}S_{ax}f_x^b(0)+\nabla_x^b\\ S_{ay}f_y^b(0)+\nabla_y^b\\ S_{az}f_z^b(0)+\nabla_z^b\end{bmatrix}\end{aligned} \tag{10}$$

Where $\vec{f}^b(T)$ is the output of accelerometer after rotation; $\vec{f}^b(0)$ is the output of accelerometer before rotation; $f_x^b(T)$, $f_y^b(T)$ and $f_z^b(T)$ are the *x*, *y* and *z* components of $\vec{f}^b(T)$; $f_x^b(0)$, $f_y^b(0)$ and $f_z^b(0)$ are the *x*, *y* and *z* components of $\vec{f}^b(0)$. During the calibration, velocity error can be obtained by Kalman filter. From equation (9), when three components of $\Delta\vec{\varphi}$ are zero, three components of $\Delta\delta\vec{f}^n$ can be regarded as known parameters. So we can establish equation at the different rotational positions, and analyze the identification of the accelerometer bias and scale factor.

2.1. Rotating around pitch-axis

The pitch of IMU is 45° at the initial moment and initial attitude is [45°, 0°, 0°]. Then, IMU rotates around pitch-axis with -45° and the attitude becomes [0°, 0°, 0°]. $C_b^n(T)$, $C_b^n(0)$, $\vec{f}^b(T)$ and $\vec{f}^b(0)$ can be expressed as follows:

$$C_b^n(T)=\begin{bmatrix}1&0&0\\0&1&0\\0&0&1\end{bmatrix} \tag{11}$$

$$C_b^n(0)=\begin{bmatrix}1&0&0\\0&1/\sqrt{2}&-1/\sqrt{2}\\0&1/\sqrt{2}&1/\sqrt{2}\end{bmatrix} \tag{12}$$

$$\vec{f}^b(T)=\begin{bmatrix}0\\0\\g\end{bmatrix} \tag{13}$$

$$\vec{f}^b(0)=\begin{bmatrix}0\\g/\sqrt{2}\\g/\sqrt{2}\end{bmatrix} \tag{14}$$

Substituting equations (11)-(14) into equation(10), we can obtain:

$$\begin{aligned}\Delta\delta\vec{f}^n&=\begin{bmatrix}1&0&0\\0&1&0\\0&0&1\end{bmatrix}\begin{bmatrix}\nabla_x^b\\\nabla_y^b\\S_{az}g+\nabla_z^b\end{bmatrix}-\begin{bmatrix}1&0&0\\0&1/\sqrt{2}&-1/\sqrt{2}\\0&1/\sqrt{2}&1/\sqrt{2}\end{bmatrix}\begin{bmatrix}\nabla_x^b\\S_{ay}g/\sqrt{2}+\nabla_y^b\\S_{az}g/\sqrt{2}+\nabla_z^b\end{bmatrix}\\&=\begin{bmatrix}0\\-S_{ay}g/2+S_{az}g/2+\left(1-1/\sqrt{2}\right)\nabla_y^b+1/\sqrt{2}\nabla_z^b\\-S_{ay}g/2+S_{az}g/2-1/\sqrt{2}\nabla_y^b+\left(1-1/\sqrt{2}\right)\nabla_z^b\end{bmatrix}\end{aligned} \tag{15}$$

$\Delta\delta\vec{f}^n$ can be obtained from the Kalman filter. From equation (15), ∇_z^b can be calculated. When the attitude is [0°, 0°, 0°], the error of the *z*-direction accelerometer is $S_{az}g+\nabla_z^b$, which can be obtained from the Kalman filter. So S_{az} can be calculated.

The pitch of IMU is 45° at the initial moment and initial attitude is [45°, 0°, 0°]. Then IMU rotates around pitch-axis with -90° and the attitude becomes [-45°, 0°, 0°]. $C_b^n(T)$, $C_b^n(0)$, $\vec{f}^b(T)$ and $\vec{f}^b(0)$ can be expressed as follows:

$$C_b^n(T)=\begin{bmatrix}1&0&0\\0&1/\sqrt{2}&1/\sqrt{2}\\0&-1/\sqrt{2}&1/\sqrt{2}\end{bmatrix} \tag{16}$$

$$C_b^n(0)=\begin{bmatrix}1 & 0 & 0\\ 0 & 1/\sqrt{2} & -1/\sqrt{2}\\ 0 & 1/\sqrt{2} & 1/\sqrt{2}\end{bmatrix} \tag{17}$$

$$\vec{f}^b(T)=\begin{bmatrix}0\\ -g/\sqrt{2}\\ g/\sqrt{2}\end{bmatrix} \tag{18}$$

$$\vec{f}^b(0)=\begin{bmatrix}0\\ g/\sqrt{2}\\ g/\sqrt{2}\end{bmatrix} \tag{19}$$

Substituting equations (16)-(19) into equation (10), we can obtain:

$$\begin{aligned}\Delta\delta\vec{f}^n &=\begin{bmatrix}1 & 0 & 0\\ 0 & 1/\sqrt{2} & 1/\sqrt{2}\\ 0 & -1/\sqrt{2} & 1/\sqrt{2}\end{bmatrix}\begin{bmatrix}\nabla_x^b\\ -S_{ay}g/\sqrt{2}+\nabla_y^b\\ S_{az}g/\sqrt{2}+\nabla_z^b\end{bmatrix}-\begin{bmatrix}1 & 0 & 0\\ 0 & 1/\sqrt{2} & -1/\sqrt{2}\\ 0 & 1/\sqrt{2} & 1/\sqrt{2}\end{bmatrix}\begin{bmatrix}\nabla_x^b\\ S_{ay}g/\sqrt{2}+\nabla_y^b\\ S_{az}g/\sqrt{2}+\nabla_z^b\end{bmatrix}\\ &=\begin{bmatrix}0\\ -S_{ay}g+\left(S_{az}g+\sqrt{2}\nabla_z^b\right)\\ -\sqrt{2}\nabla_y^b\end{bmatrix}\end{aligned} \tag{20}$$

From equation (20), ∇_y^b can be calculated. According to equation (15), if ∇_z^b and S_{az} are known, S_{ay} can be calculated. Therefore, when the IMU rotates around the pitch axis, as long as there are three different pitch positions, ∇_y^b, ∇_z^b, S_{ay} and S_{az} can be calculated. In these cases, the bias and scale factor of the *y*-direction and *z*-direction accelerometers can be calibrated.

2.2. Rotating around roll-axis and yaw-axis

The identification of the bias and scale factor of the y-direction and z-direction accelerometer has been analyzed in Section 2.1. The rotational path is designed to calibrate the bias and scale factor of the x-direction accelerometer in this section.

The roll of IMU is 45° at the initial moment and initial attitude is [0°, 45°, 0°]. Then IMU rotates around yaw-axis with 90° and the attitude becomes [0°, 45°, 90°]. $C_b^n(T)$, $C_b^n(0)$, $\vec{f}^b(T)$ and $\vec{f}^b(0)$ can be expressed as follows:

$$C_b^n(T)=\begin{bmatrix}0 & -1 & 0\\ 1/\sqrt{2} & 0 & -1/\sqrt{2}\\ 1/\sqrt{2} & 0 & 1/\sqrt{2}\end{bmatrix} \tag{21}$$

$$C_b^n(0)=\begin{bmatrix}1/\sqrt{2} & 0 & 1/\sqrt{2}\\ 0 & 1 & 0\\ -1/\sqrt{2} & 0 & 1/\sqrt{2}\end{bmatrix} \tag{22}$$

$$\vec{f}^b(T)=\begin{bmatrix}0\\ g/\sqrt{2}\\ g/\sqrt{2}\end{bmatrix} \tag{23}$$

$$\vec{f}^b(0)=\begin{bmatrix} -g/\sqrt{2} \\ 0 \\ g/\sqrt{2} \end{bmatrix} \tag{24}$$

Substituting equations (21)-(24) into equation (10), we can obtain:

$$\begin{aligned}\Delta\delta\vec{f}^n &= \begin{bmatrix} 0 & -1 & 0 \\ 1/\sqrt{2} & 0 & -1/\sqrt{2} \\ 1/\sqrt{2} & 0 & 1/\sqrt{2} \end{bmatrix}\begin{bmatrix} \nabla_x^b \\ S_{ay}g/\sqrt{2}+\nabla_y^b \\ S_{az}g/\sqrt{2}+\nabla_z^b \end{bmatrix} - \begin{bmatrix} 1/\sqrt{2} & 0 & 1/\sqrt{2} \\ 0 & 1 & 0 \\ -1/\sqrt{2} & 0 & 1/\sqrt{2} \end{bmatrix}\begin{bmatrix} -S_{ax}g/\sqrt{2}+\nabla_x^b \\ \nabla_y^b \\ S_{az}g/\sqrt{2}+\nabla_z^b \end{bmatrix} \\ &= \begin{bmatrix} S_{ax}g/2-S_{ay}g/\sqrt{2}-S_{az}g/2-\nabla_x^b/\sqrt{2}-\nabla_y^b-\nabla_z^b/\sqrt{2} \\ -S_{az}g/2+\nabla_x^b/\sqrt{2}-\nabla_y^b-\nabla_z^b/\sqrt{2} \\ -S_{ax}g/2 \end{bmatrix}\end{aligned} \tag{25}$$

From equation (25), S_{ax} can be calculated. When ∇_y^b, ∇_z^b, S_{ay} and S_{az} are known, ∇_x^b can also be calculated.

3. Simulation

Section 2 conducts theoretical analysis and demonstrates the effect of the different rotational paths on accelerometer calibration. This section conducts simulation to verify the theoretical analysis. The accelerometer scale factor is set to 100 ppm and other specific parameters are set in Table 1:

Table 1. IMU specifications.

	Error items	Bias instability	Random walk
IMU	Gyros	0.003°/h	0.0005° / √s
	Accelerometers	20 mGal	20 mGal/ √s

3.1. Rotating around pitch-axis

It can be seen from section 2.1 that when IMU rotates around the pitch-axis, as long as there are three pitch positions, the bias and scale factor of the y-direction and z-direction accelerometers can be calibrated. This section conducts simulation to verify the theoretical result. The rotational path and the results of calibration are shown in Figures. 1-3.

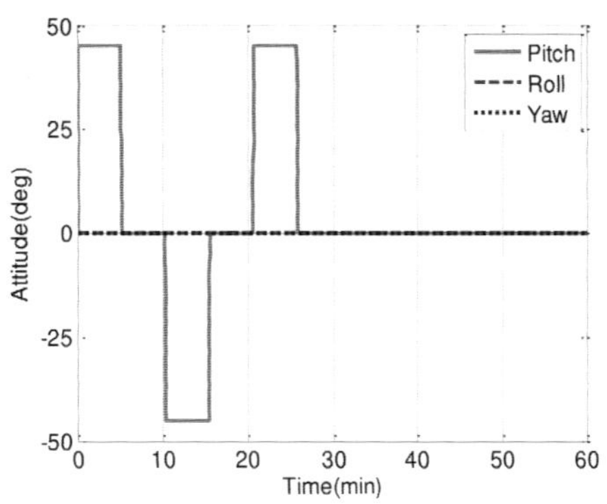

Figure 1. Attitude of IMU

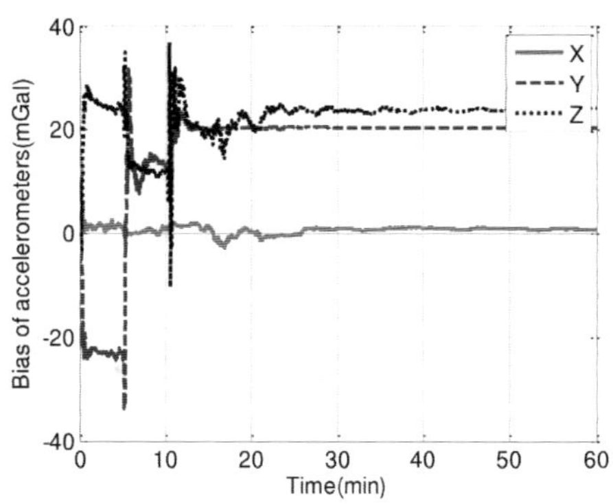

Figure 2. Bias of accelerometers

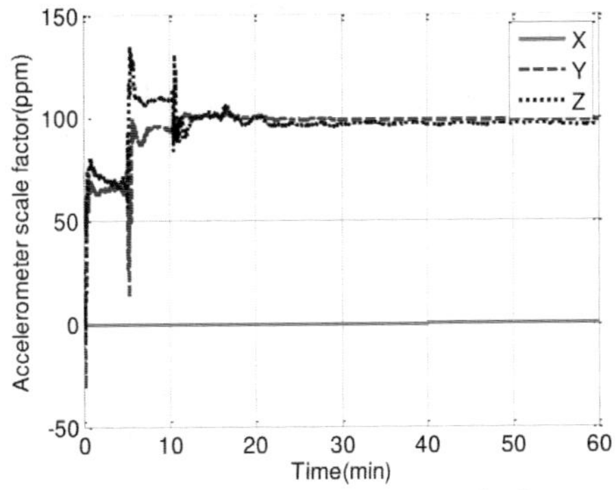

Figure 3. Accelerometer scale factor

It can be seen from Figures. 2-3 that the bias and scale factor of the x-direction and y-direction accelerometers are calibrated within 20 min, which is consistent with the theoretical analysis. The calibration error of the bias is less than 5 mGal, and the calibration error of the scale factor is less than 5 ppm.

3.2. Rotating around roll-axis and yaw-axis

As known in section 2.2, when the IMU rotates 45° around the roll axis and then rotates 90° around the yaw axis, the x-direction accelerometer scale factor can be calibrated. This section conducts simulation to verify the theoretical result. The rotational path and the results of calibration are shown in Figures. 4-6.

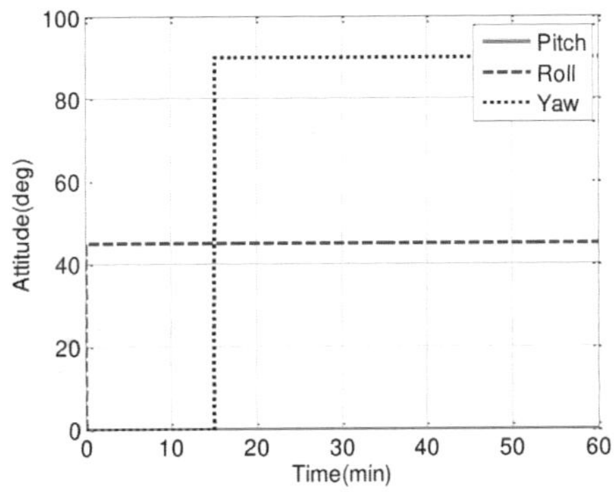

Figure 4. Attitude of IMU

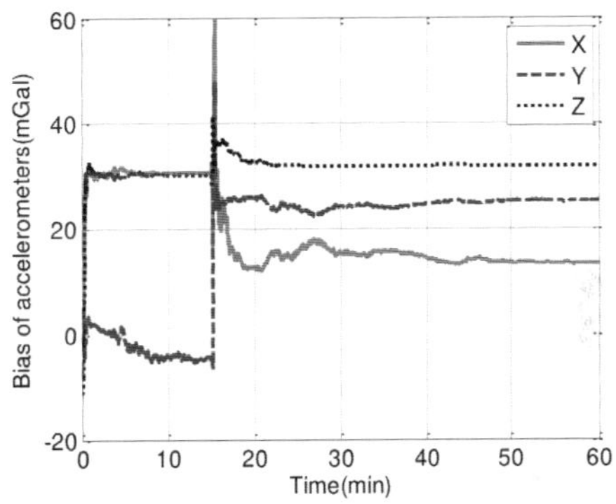

Figure 5. Bias of accelerometers

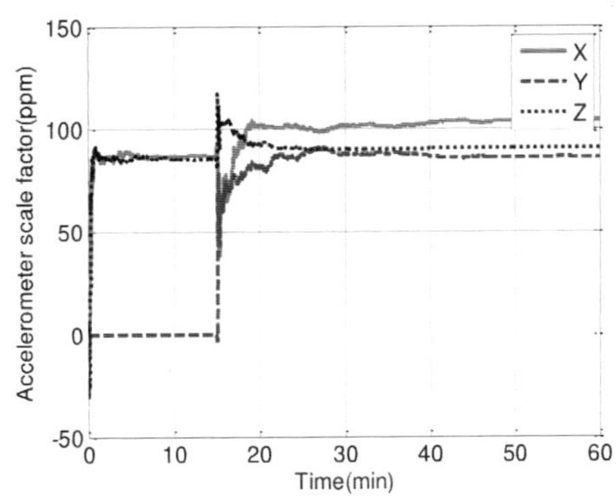

Figure 6. Accelerometer scale factor

It can be seen from Figure. 6 that the x-direction accelerometer scale factor can be calibrated, which is consistent with the theoretical analysis. The calibration error of the x-direction accelerometer scale factor is less than 5 ppm, while the calibration error of the x-direction accelerometer bias is large. The bias needs to be calibrated in conjunction with the rotational path in section 3.1.

3.3. Optimizing rotational path

In sections 3.1 and 3.2, we conduct simulations to verify previous theoretical analysis. In this section, we combine the calibration methods of sections 3.1 and 3.2 and optimize them to calibrate the accelerometers. The rotational path and the results of calibration are shown in Figures. 7-9.

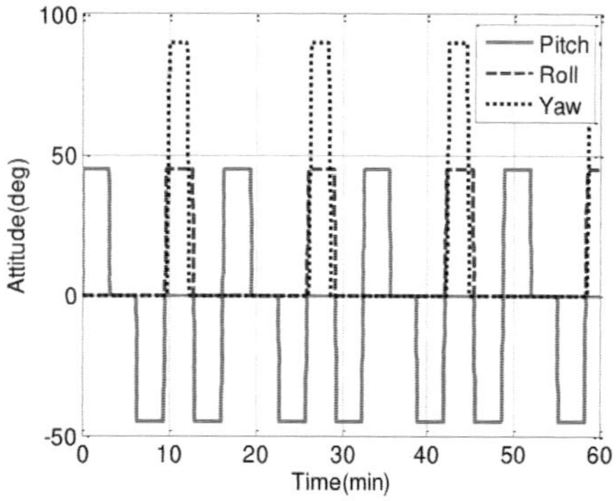

Figure 7. Attitude of IMU

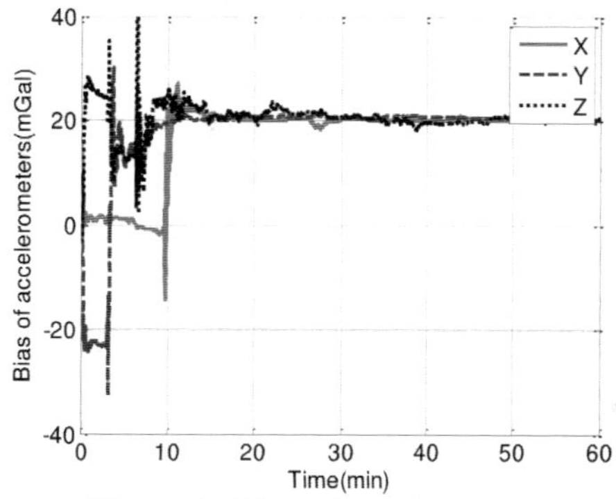

Figure 8. Bias of accelerometers

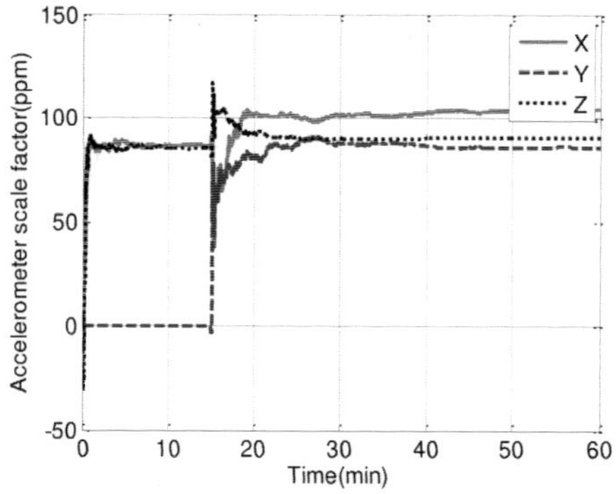

Figure 9. Accelerometer scale factor

It can be seen from Figures. 8-9 that the bias and scale factor of the accelerometers are calibrated within 20 min. The calibration error of the bias is less than 5 mGal, and the calibration error of the scale factor is less than 5 ppm. Therefore, the rotational path is effective.

4. Conclusion

This paper mainly studies the calibration of accelerometer bias and scale factor in the case that the rotation is limited. The analysis shows: when the IMU rotates around the pitch axis, as long as there are three different pitch positions, the bias and scale factor of the y-direction and z-direction accelerometers can be calibrated. When the IMU rotates 45° around the roll axis and then rotates 90° around the yaw axis, the x-direction accelerometer scale factor can be calibrated. Then, several simulations are conducted to verify the theoretical results. The simulation results agree well with theoretical results. At last, optimized rotational path is designed. And the rotational path is effective for calibration.

References

[1] Bai S. The development of accelerometers [J]. Process automation instrumentation, 1999, 20(9):1-5.
[2] Zhang X, Ning K. A Preferable Airborne Integrated Navigation Method Based on INS and GPS [J]. Journal of Sensors, 2018, 2018:1-14.

[3] Peng H, Xiong Z, Wang R, Liu J, and Wang J. Dynamic parameter identification research on the installation errors and the scale factor errors of the IMU [J]. Chinese Space Science and Technology, 2014, 34(1):42–49. 15.

[4] Wang K. Quartz flexible accelerometer partial value and scale factor stability study [D]. Harbin Engineering University, 2012.

[5] Guo J. Research on laser gyro sins dynamic error test and calibration[D]. Harbin: Harbin Institute of Technology, 2015.

[6] Chen J, Sun J, Li J, et al. The research of calibration of accelerator in inertial navigation system [J]. Microelectronics and computer, 2012,29(8):130-133.

[7] Sipos M, Paces P, Rohac J, et al. Analyses of Triaxial Accelerometer Calibration Algorithms [J]. IEEE Sensors Journal, 2012, 12(5):1157-1165.

[8] Yang X, Huang Y. Systematic calibration method for laser gyro SINS [J]. Journal of Chinese Inertial Technology 2008, 16(6):657-664.

[9] Shi W, Wang X, Zheng J, Zhan D, Wang Y. Multi-position systematic calibration method for RLG-SINS [J]. Infrared and Laser Engineering, 2016, 45(11).

[10] Wang Z, Cheng X, Fan S. Eight-position systematic calibration method for SINS based on two-axis turntable with temperature compensation [J]. Journal of Chinese Inertial Technology, 2019,27(1):23-31.

[11] Titterton D, Weston J L. Strapdown inertial navigation technology [M]. IET, 2004.

[12] Gao Z. Inertial navigation system technology [M]. Tsinghua University Press, 2012.

[13] Wei G. Research on Some Key Technologies for Double-axis Rotation Inertial Navigation System with Mechanically Dithered Ring Laser Gyroscope [D]. National University of Defense Technology.

IOP Conf. Series: Materials Science and Engineering **787** (2020) 012002 doi:10.1088/1757-899X/787/1/012002

Vehicle path tracking Based on Strong Tracking Cubature Kalman Filter While Encountering An Emergency Collision Avoidance

Bi Wei[1,2], Wang Wei[3], Bei Shaoyi[3], Jin Ben[3], Li Bo[3]

[1] School of Mechanical Engineering, Jiangsu University of Technology, Changzhou, 213001

[2] School of Mechanical Engineering and Automation, Zhejiang Sci-Tech University, Hangzhou, 310018

[3] School of Automotive and Traffic Engineering, Jiangsu University of Technology, Changzhou, 213001

Abstract: Considering the high speed emergency avoidance situations, the vehicle path tracking was studied in this paper. The three degrees of freedom nonlinear vehicle model and the cubature Kalman filter are applied to tracking vehicle emergency avoidance path. For the degradation of adaptive tracking performance in the vehicle emergency avoidance, an improved strong tracking cubature Kalman filter algorithm is proposed by introducing the fading factor into filtering process which is learnt from strong tracking filter. The algorithm has a simple implementation, high estimation accuracy and good robustness. Different target paths are selected and the cubature Kalman filter is applied to simulate path tracking in the vehicle emergency avoidance. The simulation results show that the proposed path tracking method can control the vehicle tracking the ideal collision avoidance path rapidly without collisions.

1. Introduction

In recent years, the highway traffic accident remains high, including rear-end collision caused by accidents accounted for more than 75%. High-speed emergency avoidance technology can reduce the incidence of accident, increasing the active safety of vehicle. Vehicle high-speed emergency avoidance system uses modern information technology and sensor technology to transmit information to the control unit, and then the vehicle actively avoids obstacles ahead to ensure the safety of vehicle.

Path tracking is to make the vehicles driving along the expected path, and ensure driving safety and comfort of the vehicle at the same time [1-3]. The operation control layer in the path tracking controller includes path following control, the speed tracking control, logical component control, emergency management and control of the actuator modules and components, among which the path tracking control is the core, and the most studied content at present. When vehicles in the process of emergency avoidance, high speed, strong nonlinear and uncertainty, high speed vehicle path tracking is intelligent vehicle development the key technical problems to be solved in the process of [4-5].

This article uses the idea of the strong tracking filtering algorithm into CKF, using the fading factor for real-time adjustment of gain matrix, thus enhances the system ability of adaptive tracking [6-9].In this paper, the nonlinear vehicle dynamics model is established, and then the strong tracking kalman filter algorithm and kalman filter volume algorithm are used for the expected avoiding path of emergency for adaptive tracking respectively. The results show that the strong tracking measurement kalman filter algorithm has higher tracking precision than volume of kalman filter algorithm .

2. Nonlinear 3-DOF vehicle dynamic model

Nonlinear 3-DOF model takes into account three degrees of freedom which are longitudinal movement along the x-axis of , lateral movement along the y-axis and yaw motion along z-axis to establish three degrees of freedom nonlinear model. Three state variables which are longitudinal, lateral yaw variables are mutually coupling in this model.

Limited vehicle lateral acceleration at 0.4g or less, in this situation cornering characteristics of tire is in linear range, and assuming the value of driving force is not great ,without regrading to the influence tangential force from the ground to tire's cornering characteristics, ignoring the influence to tire's concerning characteristics and aligning torque caused by left and right tire. There is no aerodynamic effect [12], the resulting model is shown in figure 1.

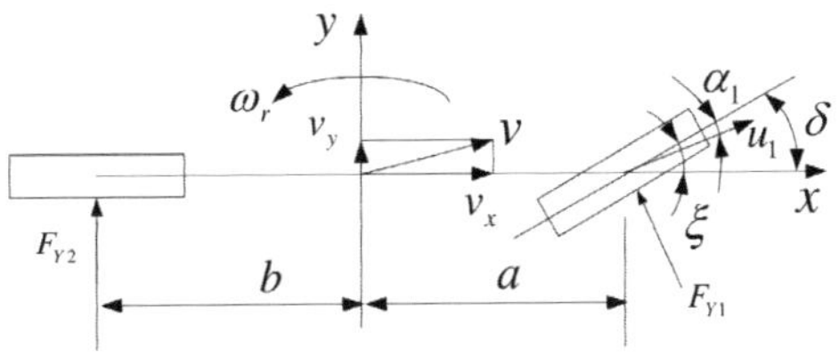

Figure 1. 3-DOF vehicle dynamics model.

Vehicle's movement differential equations show as following:

$$I\dot{\omega}_r = (ak_1 - bk_2)\beta + \frac{(a^2k_1 + b^2k_2)}{v_x}\omega_r - ak_1\delta \tag{1}$$

$$m(\dot{v}_y + v_x\omega_r) = (k_1 + k_2)\beta + \frac{(ak_1 - bk_2)}{v_x}\omega_r - k_1\delta \tag{2}$$

$$a_x = \dot{v}_x - v_y\omega_r \tag{3}$$

$$a_y = \dot{v}_y + v_x\omega_r \tag{4}$$

The state equation can be described by equation (5), and the observation equation is shown in formula (6).

$$\begin{cases} \dot{\omega}_r = \dfrac{(a^2k_1 + b^2k_2)}{I_z}\dfrac{\omega_r}{v_x} + \dfrac{(ak_1 - bk_2)}{I_z}\beta - \dfrac{ak_1}{I_z}\delta \\ \dot{\beta} = (\dfrac{ak_1 - bk_2}{m} - u^2)\dfrac{\omega_r}{v_x^2} + \dfrac{k_1 + k_2}{m}\dfrac{\beta}{v_x} - \dfrac{k_1}{m}\dfrac{\delta}{v_x} \\ \dot{v}_x = a_x + \beta v_x\omega_r \end{cases} \tag{5}$$

$$a_y = \frac{ak_1 - bk_2}{m}\frac{\omega_r}{v_x} + \frac{k_1 + k_2}{m}\beta - \frac{k_1}{m}\delta \tag{6}$$

Where m is the vehicle mass, I_z is the moment of inertia about z-axis, a_x, a_y are the longitudinal acceleration and lateral acceleration measured of center mass, β is the side slip angle of center mass, ω_r is the yaw rate, a, b is the distance from center of mass to front axle and rear axle, v_x, v_y is the longitudinal velocity and lateral velocity of center mass, k_1, k_2 are the cornering stiffness of the front axle and rear axle.

3. Strong Tracking CKF

CKF has a high estimated accuracy and it's easy to implement. However, to the path of adaptive tracking ability, CKF is poor. The reason is that when the system is stable state, the algorithm of gain matrix will be minimum, and when mutations in the state prediction residual, great changes have taken place. However, the system remained for the minimum gain matrix, it doesn't vary with the change of residual real-time adaptive adjustment, resulting in the unapproachable to the optimal gain matrix. Using the ideas of strong tracking, the fading factor is introduced into the algorithm of one step prediction covariance matrix to reduce the data for the current filter value, thus the adaptive adjustment of gain matrix, achieving the goal of stable tracking.

Calculating the fading factor needs to calculate jacobian matrix Φ and H_{k+1}. But both of them cannot directly getting from algorithm, they need to be improved, the literature [11] gives the improved fading factor of this, as follows:

$$N_{k+1} = V_{k+1} - R_{k+1} - H_{k+1} O_k H_{k+1}^T \quad (7)$$

$$M_{k+1} = P_{zz,\ k+1|k}^{(l)} - V_{k+1} + N_{k+1} \quad (8)$$

Wherein: $H_{k+1} = \left[P_{xz,\ k+1|k}^{(l)}\right]^T \left[P_{k+1|k}^{(l)}\right]^{-1}$, (l) said the introduction of fading factor variable.

The literature[12] introducing weakening factor β in the improved strong tracking algorithm is fading factor, increasing determine filtering abnormal threshold value to $\beta C_{0,\ k+1}$ ($C_{0,\ k+1}$ is filtering residual), it greatly reduced the probability of misjudgment filtering divergence in the normal working conditions , and also improved the accuracy of the filter at the same time. The calculation process is as follows:

$$N_{k+1} = V_{0,\ k+1} - \beta\left(H_{k+1} O_{k+1} H_{k+1}^T + R_{k+1}\right) \quad (9)$$

$$M_{k+1} = \beta H_{k+1} F P_k F^T H_{k+1}^T \quad (10)$$

4. Road model

Double lane and serpentine lines are typical road models which are commonly proposed in vehicle handling stability evaluation. To simulate vehicle state estimation, double lane and serpentine lines are used as the ideal road input. Meanwhile, the road trace takes the actual road centerline trajectory.

4.1. Double lane road input model

Double lane road test model dimensions are shown in figure 2. The actual road segment size parameters in figure 4 are, $s_0 = s_1 = s_2 = s_4 = 2u$, $s_3 = u$, $s_5 = 5u$, $s_6 = 3u$.

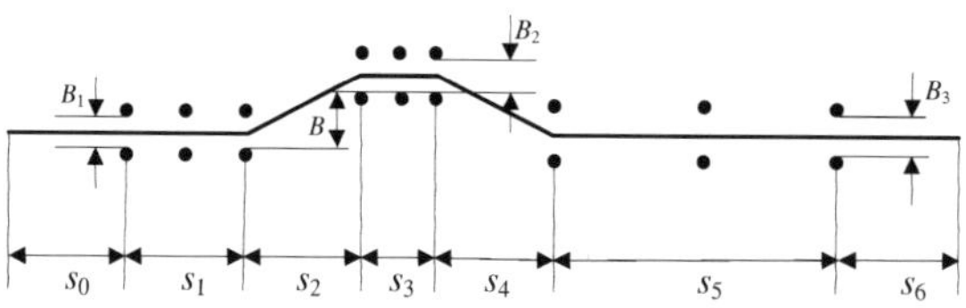

Figure 2. Double lane road test model diagram.

Where, u is the car's driving speed, and the changing lanes distance width B = 3.5m (assumed to be a standard motor vehicle lane width).

Take the centerline of the test road for the ideal way trajectory to track in the driver mind. Because the way in the corner points has a mutation, the car's actual travel path don't have a mutation, so smooth handling is conducted in the mutation department to make the road close to the ideal trajectory. The most simple and effective treatment is carring out third-order curve to fit to this polyline, so after fitting the road functions and a derivative at break point are continuous. The ideal path input after third-order curve fit is shown in figure 3.

The abscissa x is the longitudinal displacement of the car, and the ordinate y is the lateral displacement of the car.

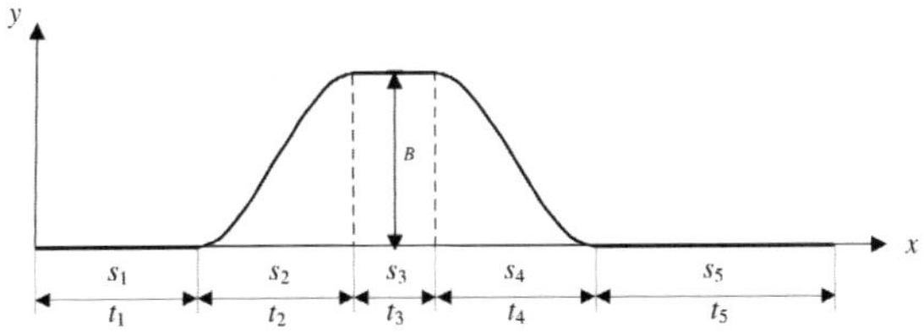

Figure 3. A double shift line graph after the third-order curve fit.

The third-order curve expression after fitting is :

$$f(x)=\begin{cases} 0 & x\in s_1 \\ e_0+e_1x+e_2x^2+e_3x^3 & x\in s_2 \\ B & x\in s_3 \\ e_0'+e_1'x+e_2'x^2+e_3'x^3 & x\in s_4 \\ 0 & x\in s_5 \end{cases} \tag{11}$$

Where, parameters are as follows:

$a_0=s_1$, $a_1=s_1+s_2$, $a_2=s_1+s_2+s_3$, $a_3=s_1+s_2+s_3+s_4$, $d=a_1-a_0, d'=a_2-a_3$,
$e_0=a_0^2(3a_1-a_0)B/d^3$, $e_1=-6a_0a_1B/d^3$, $e_2=3(a_0+a_1)B/d^3$, $e_3=-2B/d^3$,
$e_0'=a_3^2(3a_2-a_3)B/d'^3$, $e_1'=-6a_3a_2B/d'^3$, $e_2'=3(a_3+a_2)B/d'^3$, $e_3'=-2B/d'^3$.

Longitudinal displacement of the car can be obtained by $x=ut$, and the function $f(x)$ in formula (5) can be converted into a function of time $f(t)$. The specific expressions are:

$$f(t)=\begin{cases} 0 & t\in t_1 \\ g_0+g_1t+g_2t^2+g_3t^3 & t\in t_2 \\ b & t\in t_3 \\ h_0+h_1t+h_2t^2+h_3t^3 & t\in t_4 \\ 0 & t\in t_5 \end{cases} \tag{12}$$

Where, $g_0=e_0$, $g_i=e_iu^j\,(j=1,2,3)$, $h_0=e_0'$, $h_i=e_i'u^j\,(j=1,2,3)$.

4.2. Serpentine road input model

Serpentine test road model dimensions are shown in figure 4.

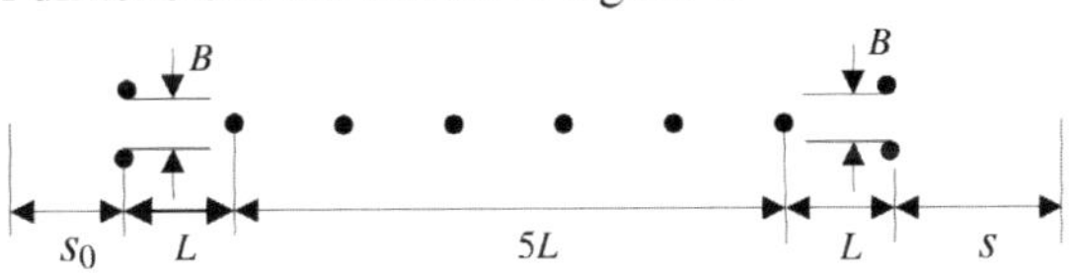

Figure 4. Serpentine road input model diagram.

The actual road segment size parameters in figure 4 are: $s_0=L=2u$, $5L=10u$, $s=3u$.

u is vehicle speed, pole width $B=2.46\text{m}$ (the width of the text is designed according to the car models in the test).

The ideal tracking trajectory by an actual driver shall be shown in figure 5. After a cubic fitting the road, the third-order curve with continuous derivative is:

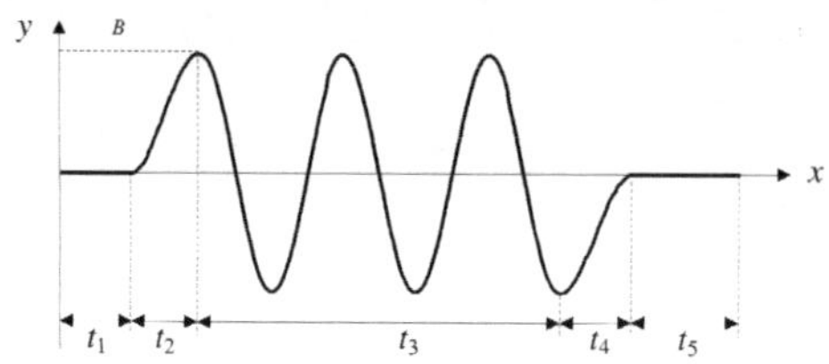

Figure 5. A serpentine line graph after the third-order curve fit.

The serpentine road function of time after fitting is as follows: (a similar derivation with double lane, so will not repeat them, but only give results).

$$f(t)=\begin{cases} 0 & t\in t_1 \\ H(1.0+\sin(\pi(t+1)/2))/2.0 & t\in t_2 \\ H\cos(\pi t/2) & t\in t_3 \\ H(-1.0+\sin(\pi(t+1)/2))/2.0 & t\in t_4 \\ 0 & t\in t_5 \end{cases} \tag{13}$$

Where, $H = B/2$.

5. The results of simulation and analysis

This paper is to simulate vehicle emergency avoidance path tracking simulation experiment, computer simulation model is established based on Matlab, and selects the double line and serpentine road model, the strong tracking kalman filter for a given volume path for precise tracking, kalman filtering and strong tracking filtering respectively with volume contrast research its tracking precision. The model parameters are as follows: the mass M=1818.2 kg, suspension quality Ms=1418 kg, center of mass to the front axle distance a=1.463m, the center of mass to the rear axle distance b=1.585m, the rotational inertia of the vehicle I_Z=3885 $kg\cdot m^2$, suspension mass moment of inertia of axis of I_X=980 $kg\cdot m^2$, suspension winding quality and product of inertia of I_{XZ}=0, the front wheel cornering stiffness （$2k_f$） k_f=62618 N/rad, and rear wheel cornering stiffness （$2k_r$） k_r=110185 N/rad, front suspension damping D_f=3430 $N\cdot m\cdot s/rad$, roll Angle after suspension damping D_r=3430 $N\cdot m\cdot s/rad$ roll Angle, roll lever h=0.488 m, C_f=100548 $N\cdot m/rad$, dip Angle of the side back rigid C_r=32732 $N\cdot m/rad$ Angle, steering system transmission ratio i=24.9, the gravitational acceleration g=9.8 $m\cdot s^{-2}$.

5.1. Double line simulation results

Two algorithms are adopted to simulate the vehicle tracking double line avoiding path, taking speed u= 108 km/h and 90 km/h respectively.

5.1.1. The simulation results of u=108km/h. Figure 6 (a) is a model's simulation results of lateral displacement of using ISTCKF and CKF algorithm tracking collision avoidance path for double line, compared with using the algorithm of ISTCKF in the figure 6 (a), CKF algorithm with the desired avoidance path is more consistent. In addition, from the figure 6 (b) the actual lateral displacement and tracking collision avoidance path absolute error, using ISTCKF algorithm has higher tracking precision than CKF. So the ISTCKF algorithm that adopted in this paper can make the car avoid better at a high speed.

IOP Conf. Series: Materials Science and Engineering **787** (2020) 012002 doi:10.1088/1757-899X/787/1/012002

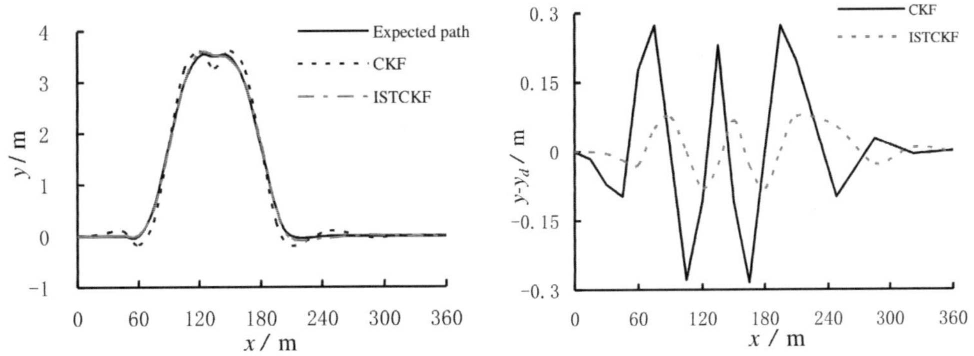

（a） the lateral displacement simulation results

（b） simulation results of the absolute error of lateral displacement

Figure 6. Double lane path tracking simulation results（u=108km/h）.

5.1.2. The simulation results of u =90km/h. Figure 7 is the simulation results of two kinds of algorithm in u= 90 km/h, it can be seen that the entire movement process, the response of the vehicle and the trend of u= 108 km/h, verifying again that compared with using the algorithm of ISTCKF, CKF algorithm has higher tracking precision.

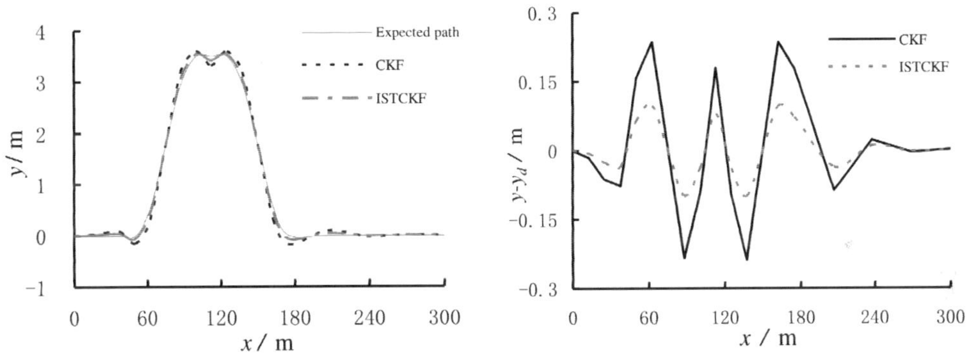

（a） the lateral displacement simulation results

（b） simulation results of the absolute error of lateral displacement

Figure 7. Double lane path tracking simulation results（u=90km/h）.

5.2. Serpentine line simulation results

Two algorithms are adopted to simulate the vehicle tracking serpentine avoidance path, take speed u= 90 km/h and u=72 km/h respectively.

5.2.1. The simulation results of u =90km/h. Figure 8 (a) a model's simulation results of lateral displacement of using ISTCKF and CKF algorithm tracking collision avoidance path for serpentine, using the algorithm of ISTCKF compared with CKF algorithm and the desired avoidance path is more consistent. In addition, from the figure 6 (b) the actual lateral displacement and tracking collision avoidance path absolute error, using ISTCKF algorithm has higher tracking precision than CKF. So the ISTCKF algorithm that adopted in this paper can make the car avoid better at a high speed.

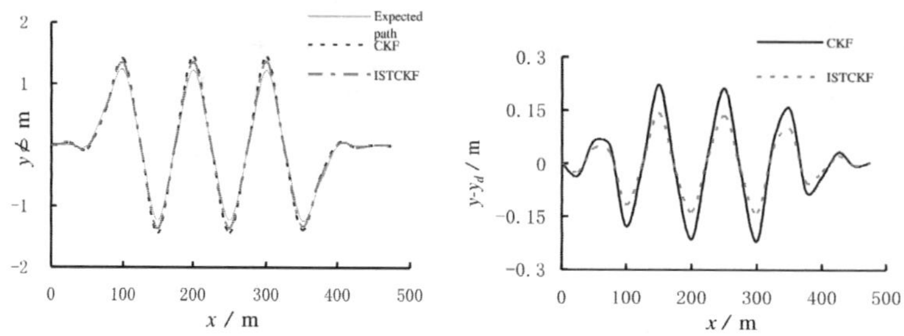

（a） the lateral displacement simulation results

（b） simulation results of the absolute error of lateral displacement

Figure 8. Serpentine path Tracking simulation results （u=90km/h）.

5.2.2. The simulation results of u =72km/h. Figure 9 is the simulation results of two kinds of algorithm in u= 72 km/h , it can be seen that the entire movement process, the response of the vehicle and the trend of u= 90 km/h, verifying again that compared with using the algorithm of ISTCKF, CKF algorithm tracking precision is higher.

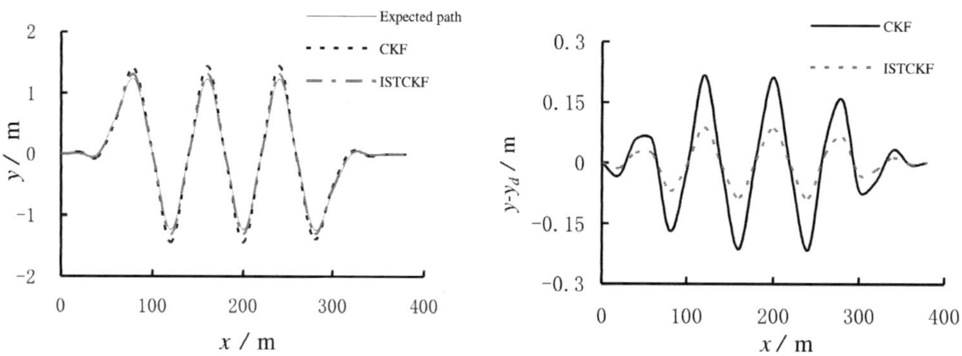

（a） the lateral displacement simulation results

（b） simulation results of the absolute error of lateral displacement

Figure 9. Serpentine path Tracking simulation results （u=72km/h）.

6. Conclusion

Since the low volume precision of traditional kalman filtering algorithm of path tracking, drawing lessons from the theory of strong tracking filter, an improved kalman filtering algorithm of strong tracking volume is put forward. Nonlinear vehicle dynamics model is established, and then use the strong tracking kalman filter algorithm and kalman filter volume algorithm to the expected avoiding path of emergency for adaptive tracking respectively. The results show that the Strong tracking measurement kalman filter algorithm has higher tracking precision than Volume of kalman filter algorithm.

Therefore, vehicle emergency avoidance path tracking problem, that uses strong tracking volume kalman filtering algorithm, can provide certain theoretical guidance to vehicle dynamic control system software design.

Acknowledgments

This work is partially supported by the National Science Foundation of China (Grant no. 51305175 and 51705220) , the Jiangsu Province Higher Education Natural Science Research Project (grant no. 17KJD580001), the Jiangsu Provincial Higher Education Natural Science Research Major Project (grant no. 17KJA580003), the National Science Foundation of JiangSu Province (Grant no.

BK2012586), the National Science Foundation of Jiangsu University of Technology (Grant no. KYY14041), the 333 Project of Jiangsu Province (BRA2015365) and Jiangsu province "six personnel peak" fund projects (Grant no. 2012-ZBZZ-023) and (Grant no. 2013-ZBZZ-039）, Policy Guidance Program (Collaboration)-Prospective Joint Research Project (BY2015028-02). The authors also gratefully acknowledge the helpful comments and suggestions of the reviewers, which have improved the presentation.

References

[1] Lee K., Kim D., Chung W. Car Parking Control Using a Trajectory Tracking Controller[J].IEEE. International Joint Conference.Washington DC:IEEE, 2006:2058-2063.

[2] Hatwal H, Mikulcik E C. Some Inverse Solutions to an Automobile Path-tracking Problem with Input Control of Steering and Braks[J], Vehicle System Dynamics, 1986, 15(2)：61-71.

[3] Aracil J，Heredia G，Ollero A.GlobalStability Analysis of Fuzzy Path Tracking Using Frequency Response[J]. Engineering Applications of Artificial Intelligence，2000(13)：109-119.

[4] Hsu T., Liu J F.,Yu P N. Development of an Automatic Parking System for vehicle[C].IEEE, Vehicle Power and Propulsion conference, Washington DC:IEEE, 2008:1-6.

[5] Muller B, Deutscher J, Grodde S. Trajectory Generation and Feedforward Control for Parking a Car[C].IEEE. International Conference on Control Application. Washington DC:IEEE, 2006:163-168.

[6] Julier S.J., Uhlmann J.K. A new method for the nonlinear transformation of means and covariances in filters and estimators[J]. IEEE Trans. on AC, 45(3), 2000:477-482.

[7] Arasaratnam I, Haykin S. Cubature Kalman filters[J]. IEEE Trans. on Automatic Control, 2009, 54(6):1254-1269.

[8] Hiroshi Mouri.Investigation of automatic path tracking using an extended Kalman filter [J]. Society of Automotive Engineers of Japan, 2002, 23:61-67.

[9] Zhang Xinchun, Guo Chengjun. Square-root imbedded cubature Kalman filtering[J]. Control Theory & Applications, 2013, 30(9):1116-1121.

[10] Bai Yan, Jia Xin, Zong Changfu, Guan Xin. Experimental Identification of Transfer Function for Vehicle Handling Stability Evaluation[J]. Automotive Engineering, 2014, 36(9):1074-1079.

[11] Wang Xiaoxu, Zhao Lin, Xia Quanxi, Hao Yong. Strong tracking filter based on unscented transformation[J]. Control and Decision, 2010, 25(7):1063-1068.

[12] Xu Shusheng, Lin Xiaogong, Li Xinfei. Strong tracking adaptive square-root cubature kalman Filter algorithm[J]. Acta Electronica Sinica, 2014, 42(12):2394-2400.

IOP Conf. Series: Materials Science and Engineering **787** (2020) 012003 doi:10.1088/1757-899X/787/1/012003

Architecture Design of Resource Support System for the Rail-Water Intermodal Transportation Cloud Platform

Jiaxin Yuan[1], Qiang Huang[1], Mantao Wang[1*] and Qianxi Feng[1]

[1] School of Information Engineering, Sichuan Agricultural University, Yaan, Sichuan, 625000, China

[*]Corresponding author: wangmantao@sicau.edu.cn

Abstract. Due to the unbalanced development of intermodal information development in different regions of China, the rigid structure of resource system and the low degree of information sharing, the development of intermodal transportation informationization is restricted to a large extent. According to the architecture design of the rail-water intermodal transportation cloud platform, this paper reconstructs the traditional IaaS and PaaS architectures, and establishes a multi-area redundant architecture, so that these virtual resources can form an organic whole in the cloud and interact with the intermodal organizations. At the same time, the direct coupling between the hardware infrastructure and software of the "chimneys" system architecture are eliminated by using the layered structure based on the virtualization technology. The resource support system can reduce the use cost of information, strengthen the management efficiency of software and hardware resources of cloud platform and a large number of heterogeneous business applications.

1. Introduction

With the proposal of The Belt And Road Initiative in China, the role of rail-water intermodal transportation in the field of logistics is becoming more and more important. Although the current rail-water intermodal cloud platform has realized the common facility virtualization and application management architecture, it does not consider the dispatch of server resources on demand allocation, resulting in a large amount of waste of resources, which makes poor overall performance of the server. Therefore, many scholars focus on how to design an overall structure of the rail-water intermodal transportation cloud platform correctly. As the basic service layer, the resource support system plays an important role in the architecture design of the cloud platform of rail-water combined transport.

Therefore, through infrastructure virtualization technology, we improve the overall utilization of IT resources. Based on the mechanism of on-demand allocation of computing resources and the core concept of "maximum utilization of resources" of cloud computing, the traditional IaaS and PaaS architectures are reconstructed with container cloud being as the center. A three-tier system including infrastructure layer, core layer and operation and maintenance management layer is designed. The hardware resources and software applications are virtualized by infrastructure layer and core layer respectively. We propose a redundant mechanism which bring the offsite disaster tolerance center in the rail-water intermodal transportation cloud platform and isolate the different types of intermodal transportation applications logically, so as to form a unified cloud information interaction mode. The establishment of a flat operation and maintenance system which adapted to the horizontal division of virtual resource management model improves the degree of automation, effectively improves the operational efficiency of intermodal transportation applications and provides intermodal transportation application and data infrastructure services for the business layer so as to solve the problems in the process of rail-water intermodal informatization by using single virtualization method, improve the management efficiency of business applications and adapt to the environment of heterogeneous seriously intermodal transportation applications. [1,2]

2. The architecture design of the resource support system

Since most of the current cloud platforms only provide users with a single virtualization medium, the management efficiency of IaaS service based on virtual machine is relatively low. Besides, PaaS service, which is completely encapsulated in the development and deployment environment, can only support a limited software architecture and cannot adapt to the heterogeneous multimodal application environment. Resource support system, as the basic service layer, needs to manage hardware resources such as servers and network facilities for the lower service layer and heterogeneous combined transport applications for the upper service layer. [3,4]

2.1. Resource support system

We propose to reconstruct the traditional IaaS and PaaS architectures with using cloud container as the center. Firstly, IaaS virtualization technology is adopted to integrate the physical nodes scattered in multiple data centers into a huge virtual resource pool, and build a virtual cluster environment based on that. Then, hybrid resource virtualization is adopted to standardize and package resources, and a unified resource management environment is established on it. It can provide cross-regional continuous integration and automatic operation and maintenance for a large number of heterogeneous multimodal transport business applications, and provide multimodal transport applications and data infrastructure services for the business layer. The architecture is designed as Figure 1:

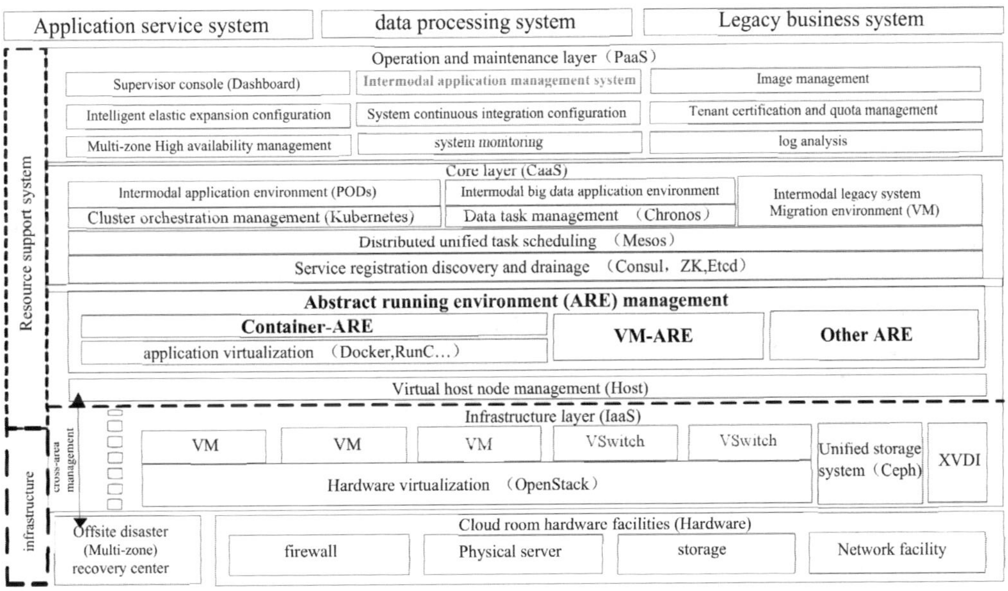

Figure 1. Resource Support System Framework of Intermodal Cloud Platform.

2.2 Layers of resource support system

As shown in the figure1, the resource support system is mainly divided into three layers, including the infrastructure layer, the core layer and the operation and maintenance management layer, which are described in detail below.

(1) Infrastructure Layer (IL): Provides virtual machine services. Hardware facilities of cloud room are abstracted to form a virtual resource pool, providing shared resources allocated according to requirements such as computing, storage, network and security.

(2) Core Layer (CL): Provides unified application virtualization services for the business layer, adopts standard image to encapsulate the intermodal application and dependencies, and hides the complexity of hardware and software virtualization by shielding the application heterogeneity through the Abstract Runtime Environment (ARE). The CL layer provides general IaaS/CaaS services upward encapsulating the IL layer.

(3) Operation and Management Layer (OML): Provides application management services and integrated PaaS services with IaaS and CaaS to the business layer through service interfaces such as unified authentication, resource allocation, system continuous integration, capacity scalability and mirroring management.

From the perspective of component composition, the IL layer adopts mature OpenStack to build IaaS services, and virtualizes server hardware into virtual machines to realize infrastructure sharing and on-demand distribution. By further abstracting and standardizing IaaS computing resources, computing resources at the CL layer ARE divided into virtual machine (VM) and Container (Container) to manage three kinds of combined applications: application service system, data processing system and legacy business system. Among them, Kubernetes realizes the arrangement management of the application service system, Chronos realizes the scheduling management of the data processing system, and Mesos completes the related resource allocation and distributed task scheduling. For legacy business systems that cannot be containerized, we adopt virtual machine encapsulation. These three kinds of system by the Consul framework for service registration and discovery and consumption. Through the intermodal application management system, continuous integration, monitoring and authentication components, the PaaS management environment in OML is formed to meet the requirements of the intermodal application.

3. Virtual resource pool of resource support system

Virtual resource pool is the abstraction of the resource support system for the management resources. We use the infrastructure layer and the core layer to virtualize the hardware resources and software applications respectively and finally forms a hybrid resource pool environment combining virtual machines and containers to meet the differentiated demand of heterogeneous applications for resources, as shown in figure 2:

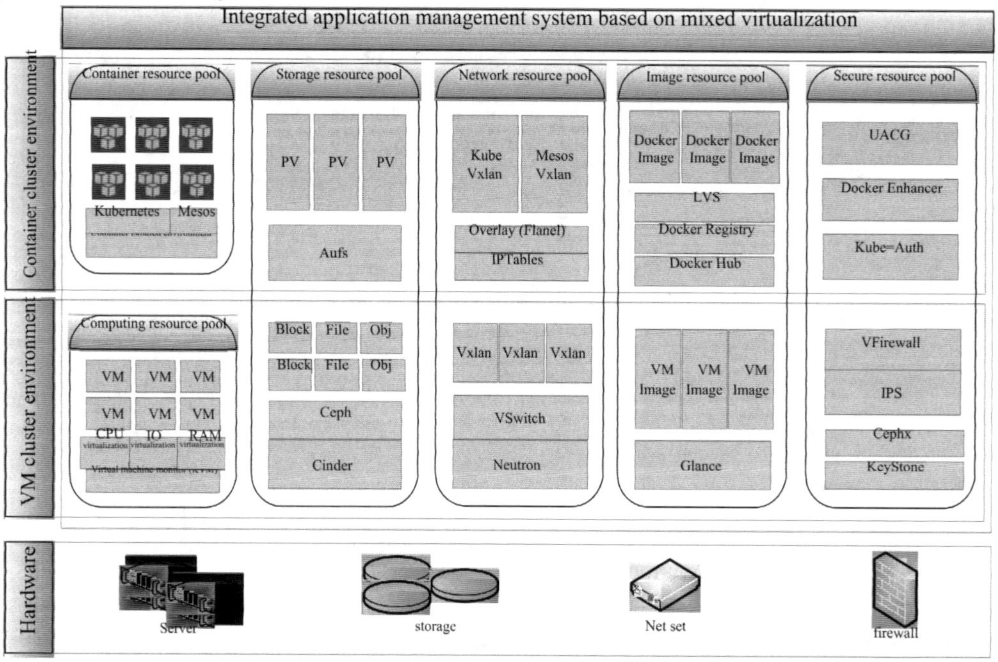

Figure 2. Virtual Resource Pool Design for Resource Support System.

As shown in figure 2, virtual resource pool is divided into six parts, including storage resource pool, security resource pool and mirror resource pool, which are described respectively as follows:

(1) Calculating resource pools

The computing resource pool is not only responsible for the allocation and sharing of computing resources, but also provides facility protection for the migration of legacy intermodal business applications. Its core is Nova component of OpenStack and key technology is based on server virtualization. Through the establishment of virtual hardware environment, it can be distinguished from physical hardware, and achieve the compatibility of multiple operating system instances in unified hardware.

(2) Container resource pool

Based on Docker, container resource pool provides packaging standard and operating environment for packaging intermodal information system, stores and uses the business system in the standard container image mode. These images form the container cloud environment under the scheduling of scheduling system, provide lightweight cloud service which is more suitable for service system than VM.

(3) Storage Resource Pool

With the help of the storage resource pool, the physical storage medium can be reasonably controlled and data block scheduling can be carried out under the condition of equipment abstraction. The logical attributes of the equipment are preserved, while the physical attributes no longer exist, which finally enables the storage system to achieve unified and centralized management.

(4) Network Resource Pool

Network resource pool is a software-defined network exchange architecture formed by using virtualization technology, including virtual machine network layer and container network. The construction of the virtual network is divided into two kinds of patterns which are vertical cutting and horizontal network integration. Separation of business, data, storage network and isolation among tenants belong to the vertical network cutting pattern, while the communication between application systems belongs to the horizontal network integration mode. The network resource pool adopts the Neutron VxLAN network mode to realize the separation of IaaS functional networks, and adopts the Overlay network of Mesos and Kuberntes to realize the tenant network separation and data communication between application systems.

(5) Secure Resource Pool

It needs to meet the three requirements of network communication security, computing resource security and access control at the same time. In the aspect of network communication security, the category of virtual firewall is defined, so that the physical firewall device can realize the transformation of logical firewall. In the aspect of computing resource security, Docker uses CGroup namespace to achieve resource isolation between containers on the same virtual machine. In the aspect of access control, we adopt the existing cloud facility access control scheme.

(6) Image resource pool

An image pool is a repository of virtual machines, components, and applications. According to the requirements on the environment of heterogeneous applications, it is divided into two parts: virtual machine image pool and container image pool. The virtual machine image is managed by Glance and stored by Cinder. Container image can be divided into public image and private image. The public image repository provides the basic image of operating system and public service components, and the private one manages all application images.

4. Redundant architecture of the resource support system

With the increasing complexity of software system and the extensive deployment of distributed applications, the operating environment of software becomes more and more complex, and the reliability of operating environment must be achieved through the redundancy of resources. Whether application cannot be accessed due to software and hardware failure, or due to a large number of concurrent business resulting in slow response, will affect the normal intermodal business. [5,6]

Most of the current "chimney " architecture adopts the hardware-based redundancy architecture of primary and standby servers. In case of failure, the availability of the system can be solved to some extent through routing switch. However, the hardware-based redundancy method not only consumes a lot of resources, but also cannot solve the problem of service unavailability when the equipment fails in the machine room. In addition, when the server performance bottleneck occurs, the hardware facilities must be extended. The performance bottleneck cannot be handled in a short time, leading to the decline of service quality [7,8]. Even in the distributed environment, the redundancy of hardware facilities in a single computer room or even a single area of multi-room hardware cannot guarantee the reliability of application services under the force majeure.

In order to ensure high availability of applications, it is necessary to establish a multiregional redundant architecture. We put forward introducing Remote Center that allows disaster to occur into redundant mechanism of rail-water transportation cloud platform, in order to make full use of the flexibility of virtual resource management, and improve the reliability of application service and fault tolerance. Therefore, the redundant architecture of the resource support system needs to simultaneously manage the virtual pool resources built by the physical server cluster based on multiple cloud computer rooms and the disaster recovery resources of the remote disaster tolerance cloud. The design is shown in figure 3:

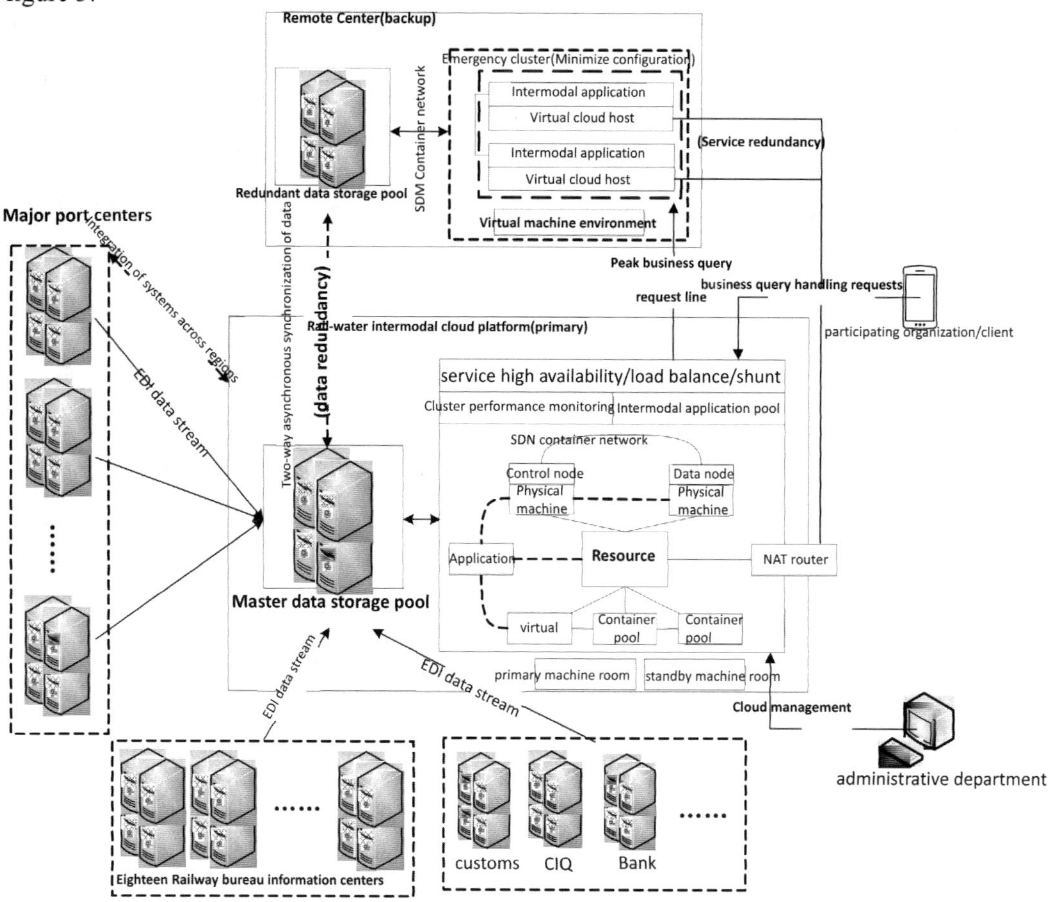

Figure 3. Redundant architecture of the resource support system.

The redundant structure of the resource support system can be divided into two logic zones: primary and backup. The primary zone manages multi-room mutual master backup mode. The Remote Center, are generally commercial cloud platforms with high reliability. When the main area fails or the service pressure is too high, resulting in the unavailability of the service or the request response time is too long, the resource support system can be rapidly configured in different places, start or add service nodes, and at the same time, the requests in the main area will be shunted and redirected to the standby area, so as to ensure the high availability of the service.

5. Conclusions

Based on the analysis of current existing problems in the architecture of the resource support system for the rail-water intermodal cloud platform, this paper proposes the new idea of resource support system architecture. Combining the core technologies and concepts of virtualization technology and multi-area redundant architecture, the resource support system is optimized by using container cloud technology. The reconstructed resource support system can improve the application environment of intermodal transportation with serious heterogeneity, solve the problem of service unavailability, and effectively improve the application efficiency of intermodal transportation.

Acknowledgements:

This work was supported by Sichuan Agricultural University education reform project X2013039, X2014025 "Agricultural Information Engineering" Sichuan key laboratory of higher education.

References:

[1] Zheng, P. (2012) Bottlenecks and key points of China's container sea and rail intermodal development. value engineering., 05 : 17-18.

[2] Zhang, Y. (2013) Design and implementation of virtual desktop cloud system server based on OpenStack. South China University of Technology, Guangzhou.

[3] Chen, L. (2007) The true meaning of the REST. Software World., 17: 38-39.

[4] Niu, L. (2010) Research and implementation of server framework based on REST. Beijing University of Posts and Telecommunications, Beijing.

[5] Hu, P. (2011) Research on operation and maintenance system of operators based on cloud computing technology. Telecom science., 12: 97-102.

[6] Ma, J., Zhou J., etc. (2015) Research on operation and maintenance mode of information system in cloud computing environment. Electric power information and communication technology., 08: 140-144.

[7] Tong, S. (2012) How the traditional IT architecture has shifted to the cloud mode. electronic finance., 04: 66.

[8] Brandt R.A., Andrew C.B. (1991) Information architecture: in search of efficient flexibility. MIS Quarterly., 15： 435-445.

IOP Conf. Series: Materials Science and Engineering **787** (2020) 012004 doi:10.1088/1757-899X/787/1/012004

A Realization Method of Video Information Exchange for Network Connected Vehicles

Deyou Zhan[1], Qingchun Jiao[2], Binghui Song[3]

[1]Zhejiang Fang Yuan test group

[2]Zhejiang University of science & technology

[3]Zhejiang Uniview Technology Co., Ltd.

*Corresponding author's e-mail: [1]24855693@qq.com, [2]jiaoqch@126.com, [3]song-bing-hui@163.com.

Abstract. This article through the design which has the function of transcoding, distribution and authentication exchange unit, achieve a face made car video information exchange method. REST method is used to realize the interface call of video information exchange, which includes control flow, media flow and structured data flow of video content analysis results. It can be used in video exchange scenarios of vehicle and vehicle, vehicle and infrastructure, vehicle and remote management platform.

1. Introduction

Intelligent transportation with modern communication technology, sensor technology and the rapid development of information processing technology has obtained the approval and carried out within a certain scope. The development of intelligent long-distance traffic scheduling, driverless vehicle interconnection, vehicle-road collaborative adaptive path planning and other technologies will greatly improve the efficiency of traffic at the same time, greatly reduce the loss. According to U.S. department of transportation estimates, the use of vehicle-to-infrastructure communications can effectively prevent 82% of accidents [1].

With the application of car video more and more abundant in recent years, the video information service in the vehicle provides a lot of content for drivers and passengers. For example multiple view road surveillance video from multiple cameras mounted on the vehicle, video shots highlighting traffic conditions on the planned route and on the highway, etc [2]. The exchange of information between the workshop video and video between the vehicle and the infrastructure and remote management platform greatly expands the perceived range. The scene model is shown in figure 1. This kind of application for multiple channel video transmission needs the support of high-speed broadband transmission. Currently, the global 5G network deployment is actively promoted, which provides a good environment for the integration of vehicle-mounted network and mobile cellular network [3], Jie Y [4] and others think that using cellular mobile network communication as auxiliary communication architecture is the future development direction of vehicular network.

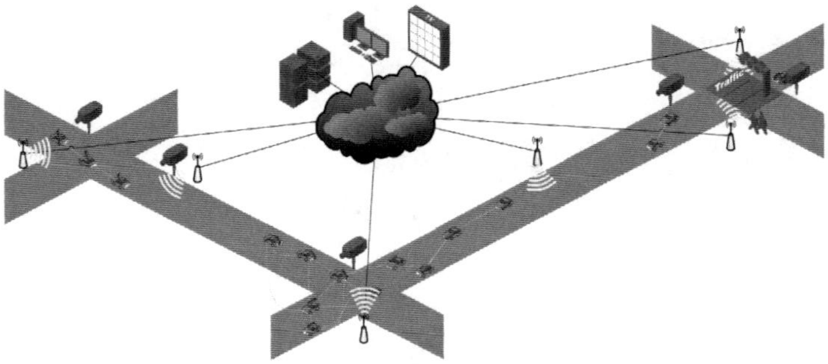

Figure 1. Scene model of network connected vehicles

Studies have shown that video data exchanged between vehicles, between vehicles and infrastructure provides a unique opportunity to advance driver assistance applications and intelligent transportation systems (ITS) [5]. Ikeda M [6], Zaidi S [7], Wu H [8] et al studied transmission characteristics and protocols of networked connected cars (video). Video source involved in the network connected vehicle is spread out from scattered vehicles and road infrastructure, which has the characteristics of large flow of video and high temporary combination. Liu Z and Abdi L studied the outward broadcast mechanism of vehicle-mounted video facing the scene of expressway [9, 10]. The data stream transmission of networked vehicles has obvious bidirectional characteristics, and video of each mobile vehicle transmits video resources outwards relatively fixed; however, the video streams received from interested parties are uncertain. Remote management platform has obvious convergence characteristics. The realization of video exchange for network connected vehicle needs universality, flexible access, and can meet the requirements of elastic expansion and dynamic selection.

2. Video exchange structure

In order to realize establishes the video exchange system between vehicles and vehicles, road infrastructure, remote management desks. By constructing a structure based on two-way network transmission, to complete the function of real-time exchange of all parties control stream, video stream with video intelligent structured data stream. Taking minimum exchange interconnection as the research object, two nodes are designed to conduct video exchange logic structure, as shown in Figure 2.

Combine the core functional components of the node logical structure participating in the exchange to form a video exchange unit, which mainly implements functions such as authorization authentication, access and output, transcoding and distribution of video streams. Each participating exchange node includes exchange unit and management terminal of video, video resource for exchange, and security isolation device for video and data. When the two nodes are connected to each other for video exchange, the video streams, video attributes and corresponding video content intelligent analysis results structured data of their respective are sent to each other.

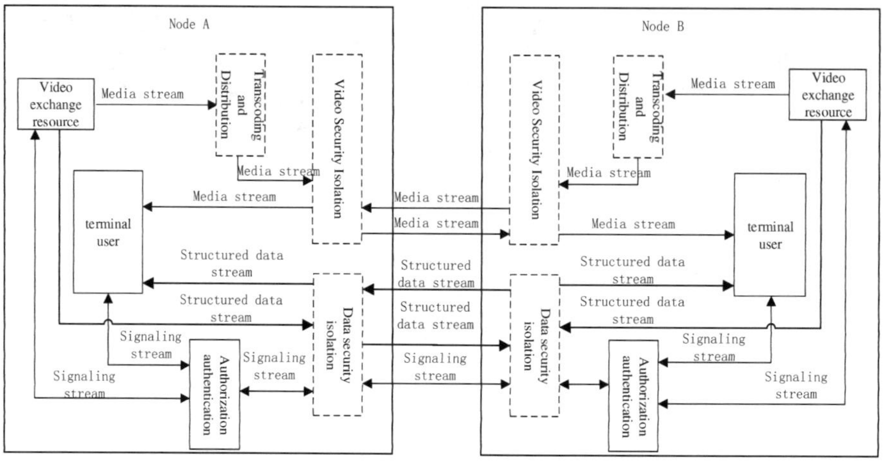

Figure 2.Two-node video information exchange logic diagram

Node A and node B can send video and data to each other and receive video and data sent by each other. The key step in the exchange process:

The 1st step: Nodes A and B respectively access the required camera video and structured data as video resources for exchange to form video exchange resources. Through negotiation with video resource management module, confirm the output mode to the other node; The 2nd step: Authorized video streaming and data transmission through the security isolation module, Video and data security isolation can be achieved by universal firewall and video network gate; The 3rd step (optional): If the video formats of node A and node B are not consistent, if H.265 and MPEG2 are adopted respectively, transcoding shall be carried out in their respective switching units; The 4th step (optional): When the received video stream needs to provide services to multiple video applications within the node, the service implementation is distributed through the video stream of the Video Exchange Unit (VEU).

3. Exchange protocol

The network connected vehicles involves a large number of heterogeneous devices. It is ideal to implement the control signaling stream, video stream and structured data stream all over on HTTP by using the universality features of TCP/IP and HTTP.

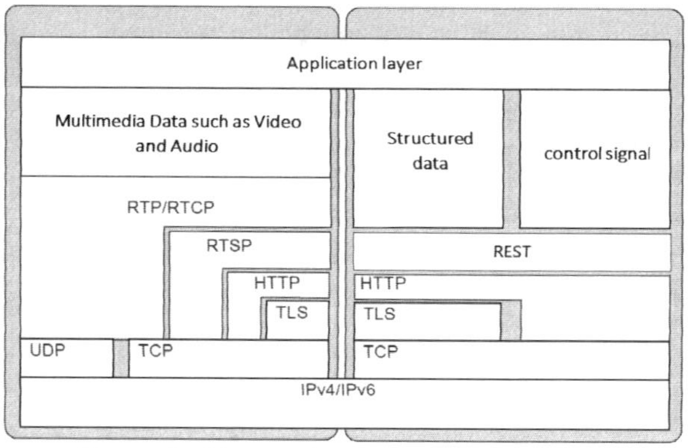

Figure 3.video exchange protocol structure

The protocol structure is shown in Figure 3. The three transmission channels supporting the application layer are video channel, control signal channel and structured data channel. Control signal channel for system control command transmission between exchange units. The video channel is used to transmit video streams. Video transmission must be coordinated through two channels: signaling channel and media stream. The structured data channel is used to transmit the analysis result data of the corresponding video stream. The control signaling and data interface are defined by a REST-style method, and the video channel is based on RTP/RTCP and RTSP.

RESTful is a platform-independent communication protocol that enables information exchange between heterogeneous systems [11], also be used as the interface of Internet of Things and M2M [12]. It is a way to create services by abstract all information to be resources in a unified way and any information that can be named can be used as a resource. Each resource defines a required or optional query string parameter, and the query string parameter appears as a name/value pair. For example, to query the resource list, the URI is: /VRS/Query/ResList.

4. Implementation operation

4.1. General Interaction Process

According to the function role of each video exchange node is set as the resource requester and the resource provider. The general process of interaction between the nodes is the method in which the resource requester first sends Request, and then the resource provider responds to the Response message. The process is shown in Figure 3:

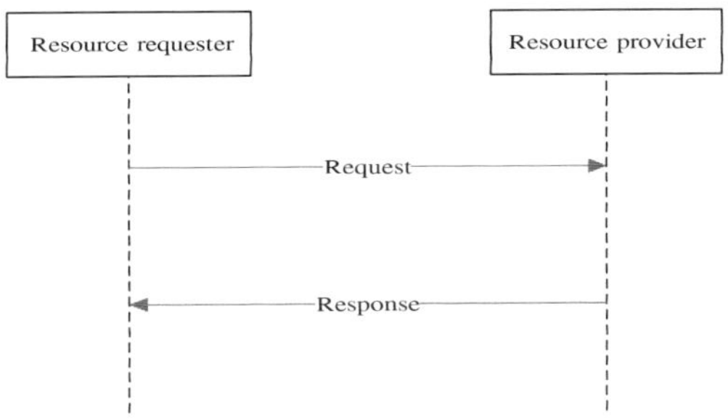

Figure 4.Universal interaction process

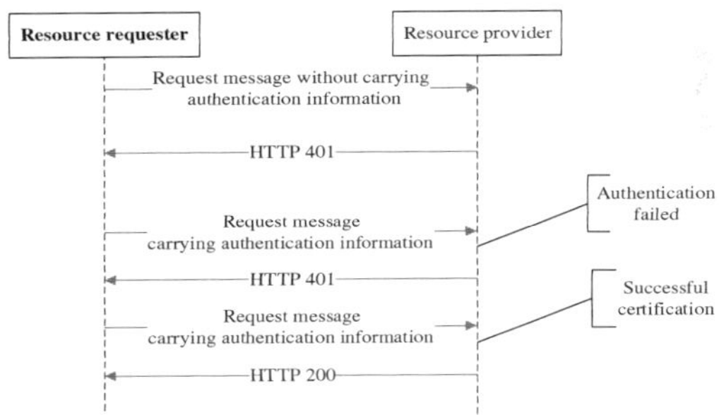

Figure 5. Authorization authentication process

4.2. Authorization Authentication implementation

The resource requester needs to pass the authorization authentication of the resource provider to obtain the corresponding service. Authorization authentication needs to carry authentication information in every request message; otherwise the request will be rejected. Carry out information authentication in the case of carrying authentication information. When the authentication is passed, the request will be responded; when the authentication fails, the request returns an error. The process is shown in Figure 5.

Each time the resource requester requests the service, the header field of the HTTP request message shall contain the Authorization field, described as Authorization: Digest-SHA1, username="username", timestamp="create message time", nonce="random number", Response= "password summary". Among them:

a) Digest-SHA1 is the authentication mode, and SHA1 is the digest algorithm, which is different from the Basic and Digest modes of the HTTP protocol;
b) Timestamp is the timestamp when creating this HTTP message. The format is as follows: YYYY-MM-DD Thh:mm:ss; Nonce is a 16-byte random number, represented by a hexadecimal string;
c) Response is a user password digest. The algorithm is response=SHA1 (timestamp+nonce+user+ password). The result of the digest calculation is represented by a hexadecimal string, and "+" means a string concatenation.

After receiving the Request, the resource provider sends an HTTP/1.1 401 Unauthorized response if the HTTP header field does not carry the Authorization field, or the Authorization field does not carry the complete username, timestamp, nonce, and response fields. If the header field is legal, the response value is calculated as described above, and compared with the value in the request message.

IOP Conf. Series: Materials Science and Engineering **787** (2020) 012004 doi:10.1088/1757-899X/787/1/012004

If they are the same, the authorization authentication is passed, and subsequent operations are performed; if not, the error response is sent.

In order to effectively prevent an attacker from sending a packet that the destination host has received to spoof the system, the authentication process may include the following operations:

a) The server verifies the timestamp. The difference between the time when the message is received and timestamp set an upper limit. It is recommended not to exceed 5 minutes. Otherwise, it is considered illegal and the authentication not pass.
b) Within the allowable time limit (for example, 5 minutes), the server caches the nonce value. The nonce value carried in the new message is not allowed to be duplicated with the nonce value in the cache. Otherwise, the authentications not pass.

4.3. Real-time video and formatted data call implementation

The real-time video stream call interface implements the introduction of a specified real-time video from the target node, and the request response process is as shown in Figure 6. The request URI is described as /VRS/Subscribe/RT Stream, and the signaling process is described as follows:

a) The resource request sends a call to the real-time video stream request to the resource provider;
b) The resource provider returns the live negotiation result to the resource requester according to the request and permission of the resource requester;
c) In case the request is successful, the resource provides a real-time video stream to the resource requester.

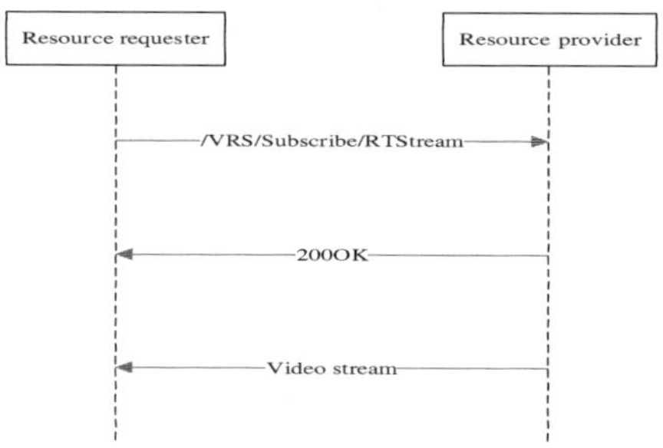

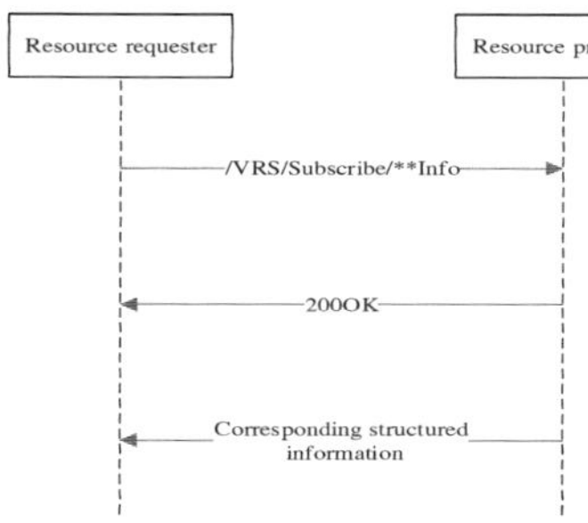

Figure 6. Real-time video stream call flow Figure 7. Video structured data subscription process

When the video content is recognized by the intelligent analysis device, and the structured feature information of the related vehicle target and other content are saved, the structured information can be reported to the resource requester by using a specified protocol interface, and the request URI is /VRS/Subscribe/ **info, **info means that you can replace it with different video structured features to subscribe to. The process is shown in Figure 7.

5. Expectation

The video exchange requirements of the network connected vehicles are clear, and the exchange scenarios are rich. The implementation method of the video control flow and the video structured data stream call as a resource by using the REST method has good adaptability for the combination of the network connected vehicles and the internet application. On the basis of the future 5G mobile communication transmission, combined with the front-end application of artificial intelligence video analysis, through the mutual exchange of information between the rich video subjects, It will produce positive impetus effect to the deep data analysis.

Acknowledgments
The article is funded by the scientific research project "Development and Demonstration of Common Technical Standards for Intelligent Internet of Vehicles" (Program No. 20180107) of Zhejiang Provincial Market Supervision Bureau.

References:

[1] ML Sichitiu, M Kihl, Inter-vehicle communication systems: a survey. IEEE Communication Surveys and Tutorials 10(2), 88–105 (2008)
[2] Li Q , Andreopoulos Y , Mihaela V D S (2007) Streaming-viability analysis and packet scheduling for video over in-vehicle wireless networks[J]. IEEE Transactions on Vehicular Technology, 56(6):3533-3549.
[3] A Benslimane, T Taleb, R Sivaraj, (2011) Dynamic clustering-based adaptive mobilegateway management in integrated VANET-3G heterogeneous wireless networks. IEEE Journal on Selected Areas in Communications 29(3), 559–570.
[4] Jie Y, Jin W, Li W, et al. (2017) Vehicle networking data-upload strategy based on mobile cloud services [J]. Eurasip Journal on Embedded Systems, 2016(1):22.
[5] HoraniM , Al-Refai G , Rawashdeh O . SAE (2017) Technical Paper Series [SAE International WCX™17: SAE World Congress Experience -, (APR. 04, 2017)] SAE Technical Paper Series - Towards Video Sharing in Vehicle-to-Vehicle and Vehicle-to-Infrastructure for Road Safety[J]. 1.
[6] Ikeda M , Honda T , Barolli L . (2015) Performance of optimized link state routing protocol for video streaming application in vehicular ad-hoc networks cloud computing.[M]. John Wiley and Sons Ltd.
[7] Zaidi S ,Bitam S , Mellouk A .(2018) Hybrid error recovery protocol for video streaming in vehicle ad hoc networks[J]. Vehicular Communications, S221420961730181X.
[8] Wu H T ,Horng G J . (2017) Establishing an Intelligent Transportation System with a Network Security Mechanism in an Internet of Vehicle Environment[J]. IEEE Access, PP(99):1-1.
[9] Liu Z , Dong M , Zhang B , et al. (2016) RMV: Real-Time Multi-View Video Streaming in Highway Vehicle Ad-Hoc Networks (VANETs)[C]// GLOBECOM 2016 - 2016 IEEE Global Communications Conference. IEEE.
[10] Abdi L ,Takrouni W , Meddeb A . (2017) Inter-vehicle video communications over Vehicular Ad Hoc Networks[C]// 2017 13th International Wireless Communications and Mobile Computing Conference (IWCMC). IEEE.
[11] He R Y . (2014) Comparison Analysis of RESTful and SOAP-WSDL Applied in the Image Management System[J]. Applied Mechanics and Materials, 644-650:4.
[12] DissegnaM ,Manfrin R , Rotoloni M , et al.(2015) RAL: a RESTful M2M communications framework for IoT[C]// 2015 International Wireless Communications and Mobile Computing Conference (IWCMC). IEEE.

IOP Conf. Series: Materials Science and Engineering **787** (2020) 012005 doi:10.1088/1757-899X/787/1/012005

Research on Dynamic Simulation System of Dangerous Goods Road Transportation Accident Process

Hongpeng CHI*, Bing GONG

BGRIMM Technology Group, Beijing, 100160, China

*Corresponding author's e-mail: hongpengchi@163.com

Abstract. Firstly, the overall scheme of the dynamic simulation system of the dangerous goods road transportation accident process is determined. According to the technical route of the system development, the functional framework of the software and the main input parameters (vehicle mass, speed, etc.) are completed, the three-dimensional model of the vehicle and scene is established, and the function of the simulation system is developed. The typical accidents such as rear end collision, fire explosion and leakage explosion in tunnel are simulated. The real-time display of accident process parameters and the visual display of accident hazard radius are realized.

1.Introduction

In the process of road transportation of dangerous goods, once there is a traffic accident, it is very easy to have accidents such as fire and explosion, which will cause catastrophic damage to the surrounding facilities and people, and then a series of secondary accidents. How to deal with road traffic accidents quickly and scientifically and find out the causes of accidents is a very important topic. With the continuous progress of computer graphics technology, virtual reality technology has been more and more widely used. Using computer simulation to reproduce the accident process and analyze the road traffic accident can provide important reference for revealing the mechanism of accident chain. In this simulation system, by creating a three-dimensional vehicle traffic accident demonstration platform, the virtual 3D scene can be reproduced, and the particle effect system is introduced to vividly present the real-time dynamic process of flame and explosion combustion. It is very helpful to explore the occurrence and development mechanism of the accident and determine the reasonable consequence evaluation method. It can provide reference for enterprises and governments in the prevention, prediction and evaluation of road transport accidents of dangerous goods and the formulation of emergency plans.

2.Simulation system scheme

2.1 System architecture

Based on the demand analysis of the simulation of dangerous goods road transportation accident process, the functional framework of the system is determined, as shown in Figure 1.

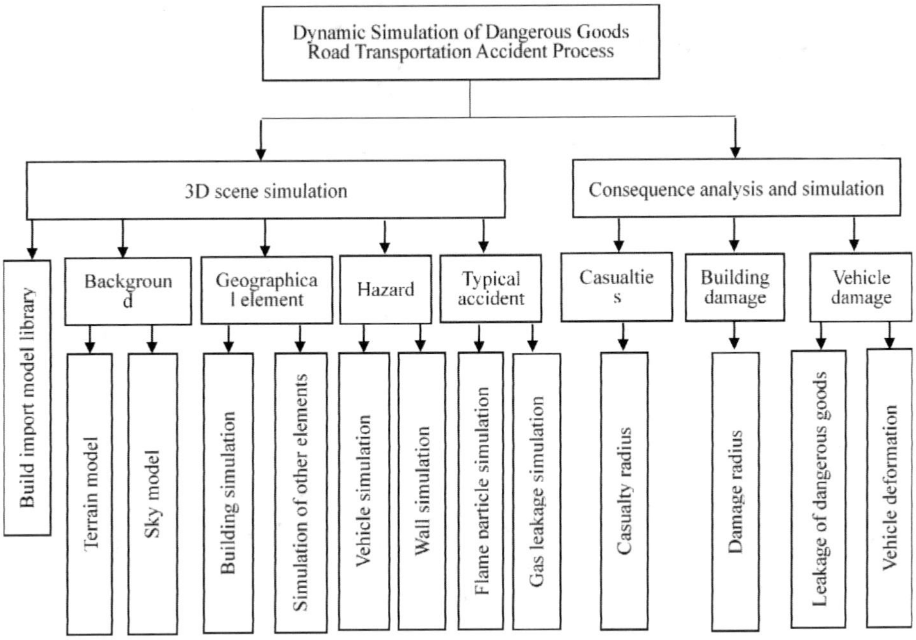

Figure 1. System function framework

2.2 System development process

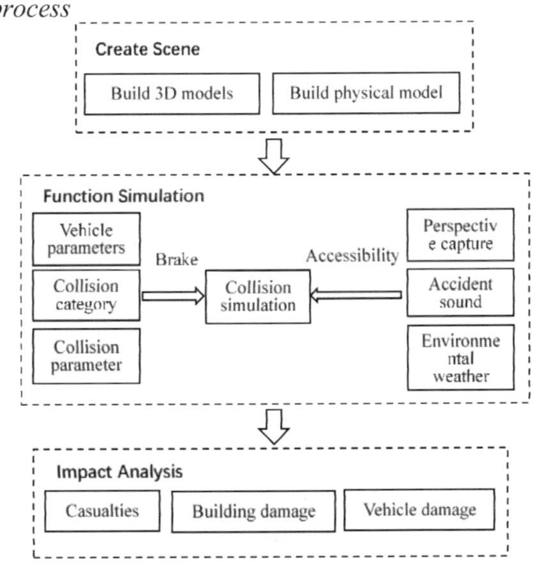

Figure 2. Flow chart of system function design

Vehicle collision is the main form of road transportation accidents, in fact, two vehicle collision is the most common traffic accident. Therefore, this system is mainly developed for the three-dimensional simulation system of the collision, leakage, combustion and explosion between the dangerous goods transport tanker and the truck. The flow of system function design is shown in Figure 2.

The establishment of scene mainly includes the establishment of transport vehicle model, road traffic environment model, collision deformation model, combustion and explosion flame particle model, leakage effect, etc. In the generation of the physical convex body collision mesh, there is a certain limit to the number of faces of the model data, so it is necessary to optimize the 3D car model used to provide the object data. The suburban and tunnel environment established in this system is used to simulate the rear end leakage, combustion and explosion accidents of dangerous goods transport tanker. With the help of the natural rendering module, the road environment is optimized to achieve efficient rendering operation. In the main elements of the physical operation engine, the scene provides a place for the physical simulation of the object. Each object participating in the physical movement is defined as a role, and the shape is used to describe and express the external form of a role. For the buildings and traffic facilities in the environment, the independent physical role is reestablished and designed as a static rigid body model.

3. Simulation model establishment

3.1. Vehicle model

Using the 3D modeling software, according to the actual parameters of the vehicle, the tractor and tank semi-trailer (head and tail) of the dangerous goods transport vehicle are respectively 3D modeled, as shown in Figure 3. On this basis, the overall appearance of the vehicle is optimized. In order to restore the scene effect more realistically, transparent materials and metal materials are designed for various parts of the vehicle, such as glass, metal shell, oil tank, chassis, tires, etc.

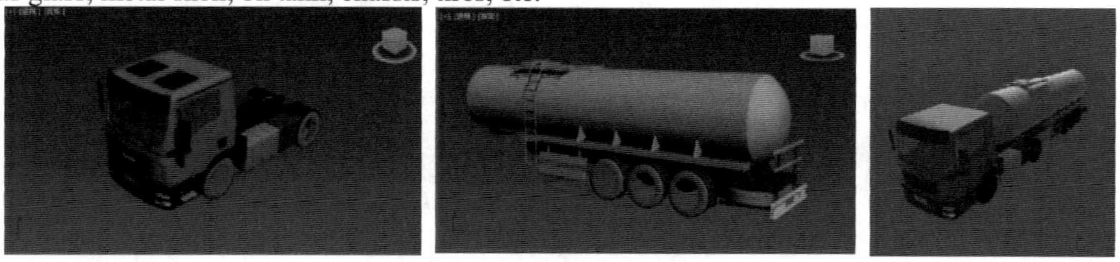

Figure 3. Vehicle model

3.2. Terrain model

The scene of dangerous goods transport vehicles mainly includes country road and city road. At present, in order to better show the process of accidents, a field road model with less shelter is established, as shown in Figure 4.

3.3. Flame and explosion particle model

Because the goods transported by dangerous goods transport vehicles are usually inflammable and explosive materials, such as gasoline, diesel, methanol, natural gas, etc., the system mainly simulates and demonstrates the fire and explosion accidents caused by accidents. The simulation of flame needs to be realized by establishing particle simulation system. Combined with the pictures and videos of real accidents, the system has established a more realistic effect of oil fire, as shown in Figure 5.

Explosion is one of the most serious consequences in the accident of dangerous goods transportation vehicles. It has strong destructive power, great harm and wide range of casualties. Therefore, the simulation of explosion process is one of the key contents of the system. When explosion, the most intuitive effect is the huge flame and smoke caused by explosion. In order to present this effect better, we need to create an example model of explosion separately. This system establishes the explosion effect model through the particle simulation method and optimizes it, as shown in Figure 6.

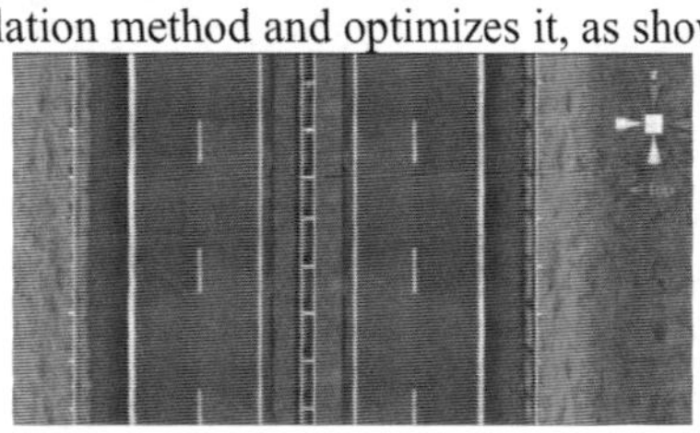

Figure 4. Road model Figure 5. Flame particle model Figure 6. Explosion particle model

4. Interactive control

For a practical software system, its interface is a very important part. In this software, the status monitoring module, visual angle control board, simulation process control button and parameter setting interface are set for the user, as shown in Figure 7.

IOP Conf. Series: Materials Science and Engineering **787** (2020) 012005 doi:10.1088/1757-899X/787/1/012005

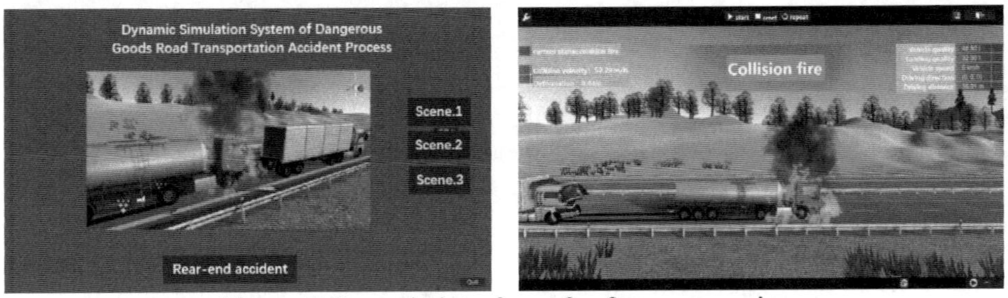

Figure 7. Scenario interface of software operation

4.1. Status monitoring function

In order to present the simulation effect in real time, the state information monitoring board is added in the simulation process of specific scene. It is used to reflect vehicle operation information and collision information, mainly including vehicle status, loading quality, driving speed, driving direction, driving distance, braking distance, collision speed, collision deformation and injury radius. In addition, in order not to affect the embodiment of the simulation process, the state parameter monitoring panel is semitransparent on the left and right sides of the scene.

4.2. Observation function control

The observation function mainly includes observation angle control, vehicle model display and scene capture. In order to facilitate the observation of the accident process, visually observe the rendered picture, experience the collision, observe the car shape after the collision, etc. By adding different cameras to the system and calling through the channel conversion module, it is convenient for users to switch multiple perspectives at any time, including the main perspective, follow perspective, global perspective, rear perspective and front perspective, which can realize all-round observation and interactive experience. In addition, in order to facilitate the user to observe the simulation process, a hidden pop-up button is set for the perspective control panel.

4.3. Parameter setting function

For parameter setting, the user can complete it through the parameter setting interface. Click the wrench button at the upper left corner of the interface to enter the parameter setting interface, as shown in Figure 8. In order to facilitate the user to set the parameters, three panels are designed: the head parameter setting panel, the tail setting panel and the impacted vehicle setting panel. The panel can be switched through the left and right arrows below.

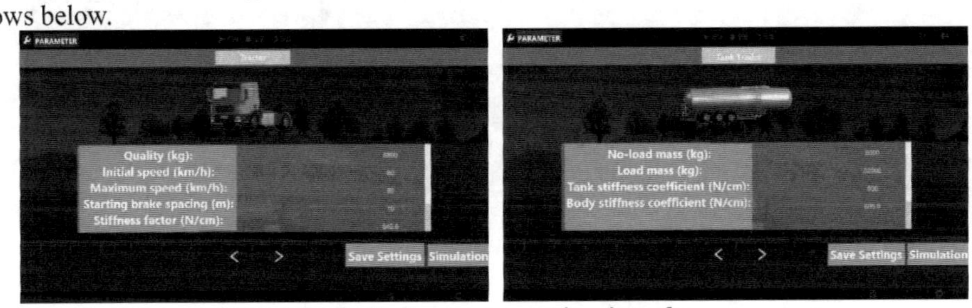

Figure 8. Parameter setting interface

After setting the parameters, users can click the Apply button to apply the set parameters to the scene. In addition, you can save the current parameter information through the Save button. The next time you open the software, the parameters will be automatically updated to the saved configuration. At the same time, the parameter setting interface can view the detailed technical parameters of the vehicle.

4.4. Simulation control

In order to facilitate the control of the simulation process, the control buttons are set, including start / pause button, reset button and repeat button.

5. Typical accident simulation

5.1. Main scene simulation

5.1.1. Simulation of collision process. As most of the material of the vehicle is metal, once there is a collision accident, the vehicle will have different degrees of deformation. The size of the shape variable is determined by the relative running speed of the vehicle before collision, the friction coefficient of the ground, the hardness of the impacted part of the vehicle and the mass of the vehicle. Therefore, it is necessary to build a collision effect that can be simulated dynamically according to the parameters. Through the establishment of code script, this effect is achieved, the effect is as follows.

5.1.2. Fire process simulation. When the vehicle collides, it is necessary to judge the energy conversion process according to the collision situation. Once a high temperature is generated at the oil, it is easy to produce a fire. Therefore, in view of this situation, the effect of fire after vehicle collision is completed, and the time of fire, the size of fire can be set and correlated through parameters.

5.1.3. Hazard impact radius. After the explosion of dangerous goods, the influence radius of blast wave is displayed intuitively.

5.2. Simulation effect.

The system can visually display the process of combustion and explosion accidents caused by vehicle collision leakage, and can output real-time data such as braking distance, collision speed, collision deformation, vehicle mass, loading mass and injury radius.

The simulation results of rear end, combustion and explosion accidents in suburban environment are shown in Figure 9.

Figure 9. Rear end combustion and explosion effect of tank car

The rear end, leakage and explosion effects of tank cars in the tunnel are shown in Figure 10.

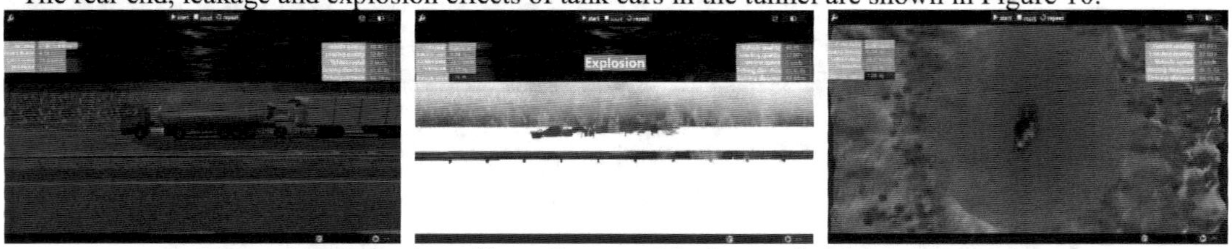

Figure 10. Explosion effect of rear end leakage of tank car in tunnel

6. Conclusion

The research of this system is based on the 3D development platform, and the 3D modeling method and C# programming technology are deeply studied. Based on the analysis and calculation of dynamics and thermodynamics in the accident of dangerous goods transport vehicle, a three-dimensional demonstration platform of vehicle accident is established by integrating the law of energy conversion. The platform can be used for 3D reconstruction of road traffic accidents. Through this platform, we can fully and truly feel the movement of vehicles in the accident, so as to intuitively reveal the behavior and potential danger of vehicles in the traffic accident. By analyzing these behaviors and risks, we can make contributions to the prevention of traffic accidents.

IOP Conf. Series: Materials Science and Engineering **787** (2020) 012005 doi:10.1088/1757-899X/787/1/012005

Acknowledgments
The authors are grateful to the financial supported by National Key Research and Development Program of China（Project Number is 2017YFC0804801）.

References

[1] Fuquan P, Shushan C, Lixia Z, (2018) Car Crash Simulation System Based on Physics Engine: Science Technology and Engineering, 18(14) : 269-274
[2] Dario V, Begani F, (2013) Energy loss in vehicle collisions from permanent deformation: An extension of the triangle method, Vehicle System Dynamics, 51(6): 857-876 .
[3] Teoh E R, Lund A K, (2011) IIHS side crash test ratings and occupant death risk in real-world crashes, Traffic Injury Prevention, 12 (5) :500-507
[4] Tongqiang D, (2005) Research on the theory model and method of accident reconstruction , Changchun: Jilin University .
[5] Ge C, Yulong P, Zhongyi Z, (2014) Reconstruction analysis model for automobile-motorcycle collision accidents, China Safety Science Journal, 24(9) : 46-52
[6] Fails A, Minton R, (2001) Comparison of Eurocap assessments with injury causation in accident, In: IRCOBI 2000. Montpellier.

IOP Conf. Series: Materials Science and Engineering **787** (2020) 012006 doi:10.1088/1757-899X/787/1/012006

Image Haze Removal Using Dark Channel Prior based on Guided Image

Long Tang[1*]

[1]College of Automation, Chongqing University, Chongqing, 400044

[*]Corresponding author's e-mail: 250883176@qq.com

Abstract. In foggy weather, the interference of fog has a serious impact on the accurate monitoring of vehicle targets. To address this problem, this paper proposed a method which is image haze removal based on the guided image. In the process of this method, the clear images are as the guide images, which is designed to effectively redress the bright region which is not meet the dark color prior by the prior information of the guide image, to avoided the error estimates of this kind of area of transmissivity. Experimental results demonstrate that our method achieves significant improvement in both image quality and image detail when compared with classical image haze removal method.

1. Introduction

Chongqing is a famous foggy city. The monitoring system of highways cannot accurately monitor running vehicles within low contrast due to atmospheric scattering in the foggy weather. Therefore, simple and effective image haze removal method is of great significance to improve the accuracy of the highway monitoring system.

At present, there are two main types of image haze removal methods. One of methods is based on image filter [1-3], for example, guided image filter is a popular algorithm. Filtering the image directly without obtaining atmospheric condition information can't achieve satisfactory results, because image contrast is closely related to atmospheric condition information. The other one is based on the physical model of atmospheric scattering [4], which is a haze removal model using the law of atmospheric scattering. This method is able to take advantage of prior knowledge of weather conditions to process images and enhance low-contrast images accurately. However, the abovementioned method has some obvious disadvantages. Images are taken by monitoring system usually do not have any information on depth of field and atmospheric conditions causing uncertainty in the image haze effect. To this end, scholars have studied the method based on statistical haze removal methods [5, 6], and achieved insightful findings which require adequate and accurate color information and variability. When the haze is very dense, the color is very thin and the difference is not obvious enough, therefore the estimated transmissivity of this method is not reliable [4]. In order to overcome these limitations, reference [7] has proposed a simple and effective dark channel prior image haze removal method based on the statistical law of the haze-free image database, which achieved a good effect on these images, but this algorithm is based on the assumption of the dark channel. When the bright region of this hypothesis is not satisfied, the estimated transmissivity of the algorithm is too small, and the recovery result does not achieve satisfactory results. Therefore, in view of the above problems, this paper proposes an image haze removal method that makes full use of prior information based on the information of the guided image and the dark channel prior haze removal model.

2. The model of haze removal using dark channel prior [7]

In the imaging model, the formula of an atomization image is described as following:

$$I(x)=J(x)t(x)+A(1-t(x)), \quad (1)$$

where, I is the observed image, J is the scene radiance, A is the global atmospheric light, and t is the medium transmission describing the portion of the light that is not scattered and reaches the camera. The main goal of the haze removal method is to recover J from I.

The dark channel prior is the conclusion obtained by statistical analysis of a large number of haze-free images. In many local areas of an image, some pixels have at least one color channel with a small value. Therefore, the minimum value of the scene radiance in this area should be small. For images, the following formula can be represented:

$$J^{dark}(x)=\min_{c\in\{r,g,b\}}\left(\min_{y\in\Omega(x)}\left(J^c(y)\right)\right), \quad (2)$$

where, J^c is a color channel of J, and $\Omega(x)$ is a local patch centered at x. According to statistical knowledge, the intensity of J^{dark} is small and tends to be zero, if J is a haze-free outdoor image, and so, J^{dark} is defined as the dark channel of J, subsequently the above statistical observation or knowledge is called the dark channel prior.

From the above description, it is assumed that the global atmospheric light A is given, and assumed that the transmissivity in the local region is constant. Taking the minimize operation in the local channel on the haze image, so that equation (1) can be obtained as follows:

$$\min_{y\in\Omega(x)}\left(\frac{I^c(y)}{A^c}\right)=\tilde{t}(x)\min_{y\in\Omega(x)}\left(\frac{J^c(y)}{A^c}\right)+\left(1-\tilde{t}(x)\right), \quad (3)$$

Then, we take the minimize operation with the three color channels on the above equation and obtain:

$$\min_c\left(\min_{y\in\Omega(x)}\left(\frac{I^c(y)}{A^c}\right)\right)=\tilde{t}(x)\min_c\left(\min_{y\in\Omega(x)}\left(\frac{J^c(y)}{A^c}\right)\right)+\left(1-\tilde{t}(x)\right), \quad (4)$$

According to the statistics knowledge of the dark channel prior, in the haze-free image, the dark channel prior J^{dark} of the image is closed to zero, which can be expressed as following:

$$J^{dark}(x)=\lim\left(\min_{c\in\{r,g,b\}}\left(\min_{y\in\Omega(x)}\left(J^c(y)\right)\right)\right)=0, \quad (5)$$

Since the atmospheric light factor is constant, and $A>0$. Therefore, substituting into formula (5) to the following formula:

$$\lim\left(\min_{c\in\{r,g,b\}}\left(\min_{y\in\Omega(x)}\left(\frac{J^c(y)}{A^c}\right)\right)\right)=0, \quad (6)$$

Substituting formula (6) into formula (4) and then simplifying, and the derived equation can be expressed as:

$$\tilde{t}(x)=1-\min_{c\in\{r,g,b\}}\left(\min_{y\in\Omega(x)}\left(\frac{J^c(y)}{A^c}\right)\right), \quad (7)$$

In order to avoid the decreasing in the image quality caused by the foggy image, a constant $\omega(0<\omega<1)$ is added to the formula (7) listed in the literature [7], subsequently the formula (7) is represented as the following form:

$$\tilde{t}(x)=1-\omega\min_{c\in\{r,g,b\}}\left(\min_{y\in\Omega(x)}\left(\frac{J^c(y)}{A^c}\right)\right), \quad (8)$$

In the reference [7], the soft matting algorithm [8] is applied in the transmittance function *t(x)* in order to improve the results. Therefore, the haze removal model can be obtained by the above descriptions as shown in the following formula:

$$J(x)=\frac{I(x)-A}{\max\left(t(x),t_0\right)}+A \quad , \tag{9}$$

In addition, the t_0 is usually set to be 0.1. First, select the pixel with the largest gray value of 0.1% in J^{dark}, and then compare the maximum value of these pixels in the original image I as the value of A listed in the formula(9).

Using the above model to process the foggy image in the highway monitor, the results are shown in Figure 1.

(a)

(b)

Figure 1. (a) and (b) are the initial image and the experimental result respectively.

It can be seen from the results in Figure. 1 that the above algorithm can yield better results, and the sharpness of the picture is quite high, but the vehicle license plate number information is blurred, and the monitoring of the vehicle does not achieve satisfactory results. Therefore, it is necessary to improve the above haze removal algorithm in order that the processing result can meet the monitor expectation.

3. Image Haze Removal Using Dark Channel Prior based on Guided Image

It can be seen from the results of Figure. 1 that the haze removal using dark channel prior results in some details in the image to be lost, and the information in some areas is significantly missing which is due to the large area of bright areas, because the highway is essentially a large white area. In the foggy environment, the pixel values of these areas are also much larger than those of other areas. It is difficult to find dark prior points with a pixel point of 0 in these areas, which makes the dark prior assumption unreasonable in these areas. In order to overcome this shortcoming, this paper uses the information of the guided image as the dark prior hypothesis, that is, the guided image is the dark channel image of the haze-free image, and replaces this guide map with *t(x)* in the formula (9).

3.1 The dark channel image of guided image

We use the dark channel image of the haze-free image as the guided image, which is the gray image composed of the minimum values of the three channels of RGB of the image, and then the image obtained by filtering the minimum value. Figure.2 and 3 show dark channel image results and histogram comparisons. The results are presented as follows:

(a) (b)

Figure 2. (a) and (b) are dark channel images of foggy and haze-free images, respectively.

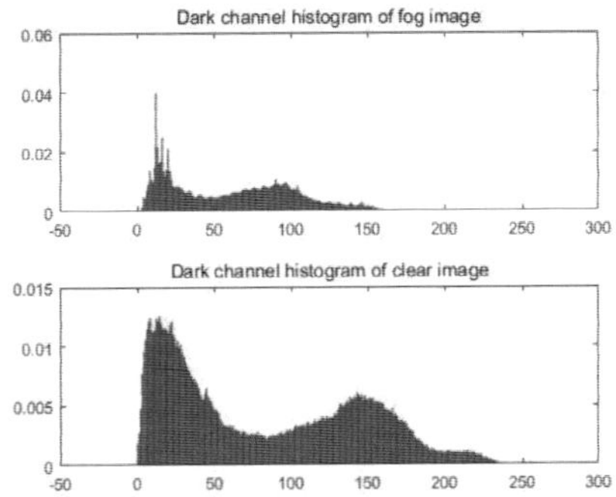

Figure 3. Histogram comparison of two dark channel images

It can be seen from the dark channel diagram of Figure. 2 that the dark channel image of the fog image shows a certain gray color, while the fogless image is a large amount of black (the pixel is close to 0). And it can be seen from the histogram of Figure. 3 that most of the pixels of the dark channel image of the haze-free image are concentrated between the interval values of [0, 50], therefore, it is justified that using the dark channel image of the haze-free image as the guided image.

3.2 Guided image filtered image results

According to the description of the above chapters, it is reasonable to assume that the dark channel image of the haze-free image is used as the dark prior Therefore, we replace t(x) in the haze removal model formula (9) with a dark channel image of the haze-free image. The experimental results are shown in Figure.4 below.

(a)

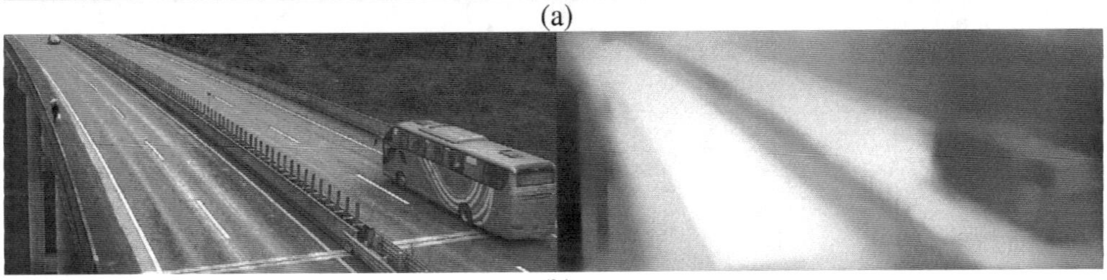

(b)

Figure 4. (a) and (b) are haze removal images and transmissivity images

It can be seen from the results in Figure. 4 that the improved algorithm can obtain better results, especially the details of the vehicle information are well preserved and clearer, which is beneficial to the subsequent vehicle monitoring work. Therefore, using the dark channel image of the guided image as a dark prior hypothesis can help to protect the image detail information.

4.Analysis of results

All the experimental images in this paper are collected from the Chongqing highway monitor. The size is uniformly cut to 800*337, and the guided images used are all haze-free images of the corresponding scenes (because the images are collected from monitor, which exist in both foggy and haze-free environments). The model is solved using Matlab 2012b, the processor is Intel i7-9820X, memory 16G, DDR4.

In order to further verify the effectiveness of the presented method, the paper compares the experimental results of the two methods with several indexes of image information. It mainly includes: entropy (EN), peak signal-to-noise ratio (PSNR), edge gradient ratio (QF) and average gradient (AVG). Subsequently, we use these indexes to analyze the experimental results of the two methods, in which we separately processed 20 fog images, and finally obtained the average value of each index for analysis and comparison. The specific results are shown in Tables 1, 2, 3 and 4.

(1)Entropy (EN): Indicates the amount of average information contained in an image. The larger the value, the more information is contained. The results are shown in Table 1 below.

Table 1. The EN of the experimental results of the two methods

Classical method	EN=7.1214
Our method	EN=**7.5127**

(2)Peak signal-to-noise ratio (PSNR): The higher the PSNR, the better the processing effect and quality.

Table 2. The PSNR of the experimental results of the two methods

Classical method	PSNR=57.8991
Our method	PSNR=**61.9881**

(3)Edge gradient ratio (QF): Evaluate edge or gradient quality, the larger the edge, the more obvious.

Table 3. The QF of the experimental results of the two methods

Classical method	QF=4.9603
Our method	QF=**8.8810**

(4)Average Gradient (AVG): Also known as sharpness, it reflects the small detail contrast and texture variation features in the image, and also reflects the sharpness of the image. The value is bigger more than better.

Table 4. The AVG of the experimental results of the two methods

Classical method	AVG=8.6390
Our method	AVG=**11.4728**

Integrating the above results, it can be seen from Figure. 5 that the method is superior to the traditional method in several index values.

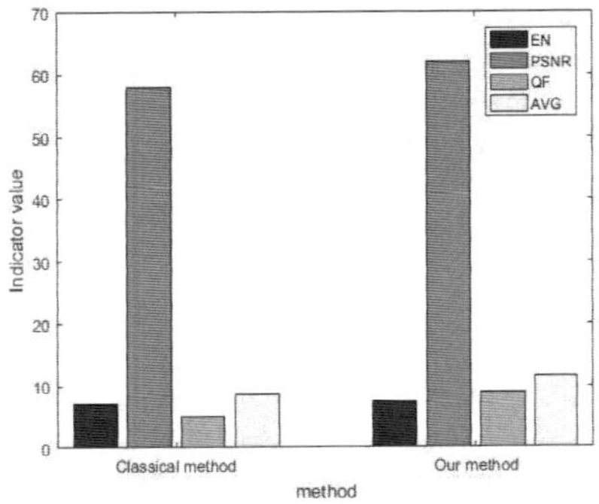

Figure 5. Column chart display of several indicators

It can be seen from the above results that the preservation of the gradient details of the processed image is more perfect, therefore the integrity of the detailed information of the vehicle can be, which facilitate in monitering the vehicle of the highway, and improving the recognition of the target information of the vehicle accuracy.

5.Conclusion

Aiming at the problem of vehicle monitor on highways in foggy weather, this paper presented a method, named as image haze removal, using dark channel prior based on guided image. The haze-free image is used as the guided image, that is, the dark channel image of the haze-free image is used as the dark prior information, which replaces the prior information obtained by the mapping algorithm in the conventional method, therefore the haze removal of the fog image can be achieved The experimental results demonstrated that the method effectively improves the retention and integrity of the detailed information in the image and improves the accuracy of vehicle monitor. Since the method is implemented under the condition that the haze-free image is needed as a priori information, the future work should focus on the improvement of the model for this situation, and improve the adaptability and unsupervised of the method.

References

[1] Kaiming He, Jian Sun , Xiaoou Tang . Guided Image Filtering[J]. IEEE Transactions on Pattern Analysis and Machine Intelligence, 2013, 35(6):1397-1409.
[2] Li Zhengguo, Zheng Jinghong. Single Image De-Hazing Using Globally Guided Image Filtering[J]. IEEE Transactions on Image Processing, 2018, PP(99):1-1.
[3] Liu Jiaying , Zhang Yongqing , Guo Zongming , et al. Guided image filtering using signal subspace projection[J]. IET Image Processing, 2013, 7(3):270-279.
[4] Jiang Jian-guo , Hou Tian-feng, Qi Mei-bin.Improved algorithm on image haze removal using dark channel prior[J].Journal of circuits and system, 2011, 16(2):7-12.
[5] Fattal Raanan. Single image dehazing[J]. Acm Transactions on Graphics, 2008, 27(3):1.
[6] Jiawan Zhang, Liang Li,Guoqiang Yang,Yi Zhang,Jizhou Sun. Local albedo-insensitive single image dehazing[J]. Visual Computer, 2010, 26(6-8):761-768.
[7] Kaiming He, Jian Sun,Xiaoou Tang. Single image haze removal using dark channel prior[C]// IEEE Conference on Computer Vision & Pattern Recognition. 2009.
[8] Anat Levin,Dani Lischinski ,Yair Weiss . A Closed-Form Solution to Natural Image Matting[J]. IEEE Transactions on Pattern Analysis and Machine Intelligence, 2008, 30(2):228-242.

IOP Conf. Series: Materials Science and Engineering **787** (2020) 012007 doi:10.1088/1757-899X/787/1/012007

Preliminary Research on VR Technology Application to Inspect the Physical Evidence of Vehicle Body Trace at the Road Traffic Accident Scene

XIE Hua-wei*

Fujian Police College, Fuzhou, Fujian, 350007, China

*Corresponding author's e-mail: momoxhw@163.com

Abstract. This thesis discusses a VR-based system designed for the policing at the grass- root level at the road traffic accident scenes. Basing on theoretical studies and technical comparisons, a VR-assisted simulation system for investigating the road traffic accident scenes is created on the aspects of system architecture performance, databases, back-stage management modules. The VR-based system plays a favorable role in assisting the scene investigators and instructors. This system can improve the efficiency of scene investigators and increase the effects of related teaching and simulative training remarkably.

1. VR Technology Profile

VR (Virtual Reality) means virtual simulation. It is a computer simulation system that can create and experience the virtual world. Through the utilization of the computer graphics system and various display and control interface devices, it provides immersive feelings in the interactive three-dimensional environment generated on the computer. [1]

Based on core technologies such as dynamic environment modeling, stereoscopic display, system development tool application, and real-time three-dimensional image generation, VR technology aims to solve the accuracy of indicating the virtual environment, the interactive naturalness between man and the virtual environment, and the real-time display intelligence technology, etc. With conception, immersion, and real-time interaction as its three characteristics, it enables users to sense the virtual environment personally so that they can explore and cognize objective things. [2]

2. The Combination between VR and the Physical Evidence Inspection of Vehicle Body Trace at the Road Traffic Accident Scene

Road traffic accident refers to the events of casualty or property loss caused by vehicle faults or accidents on the road. [3]The physical evidence inspection of vehicle body trace at the road traffic accident scene refers to an investigation activity that checks, understands, and examines the traces left on the vehicle body or those related to the accident (such as collision traces, scratch traces, and roller compaction traces) as well as discovers and fixes the physical evidence of traces related to an accident after its occurrence.

Due to the collisions and scratches of vehicles in every road traffic accident, it is essential to form corresponding vehicle body traces under the function of interactive forces. People can inspect vehicle body traces. Combined with ground traces, tire traces, and human body traces, the contact parts, direction, and the stress of vehicles can be preliminarily determined, the accident vehicle can be affirmed, and the accident process can be restored. All of them will provide strong evidence for determining the accountability of the traffic accident.[4]

By integrating VR technology into the physical evidence inspection of the vehicle body trace at the scene of the road traffic accident, some road traffic accidents that inspectors fail to arrive at the scene in time can be inspected timely in a long distance. They only need to use peripheral devices (such as VR head-mounted display) to witness the restored traffic accident scene through the handling of certain technical means and conduct their inspection work further. The system can also be set to inspect local vehicle body traces so as to determine the object as well as the magnitude and direction of the force leading to these traces. Besides, it can be used to estimate where these traces stem from. Consequently, it is conducive to initially analyzing the cause of the accident, saving human and material resources, and enhancing the efficiency of case handling.

3. System Design

3.1. Overall Scheme Design

The virtual simulation system for inspection at the scene of the road traffic accident based on VR technology is practical, efficient, intelligent, extensible, and highly reductive. It consists of the system, hardware, the server, and VR devices, etc.

The system can be used to take the panoramic images at the scene of the road traffic accident, collect image materials regarding the overview and details of the accident scene, or record videos that are involved in the scene of the road traffic accident through a special device. Then by transforming from 2D to 3D in terms of technical means, it can build a model to restore the accident scene. This enables users (e.g. grass-roots on-site inspection policemen) to realize remote inspection by virtue of VR peripheral devices through the Internet and the system platform. Each inspection step is withdrawable. The system records all inspection steps and uploads the data to the system server in real-time for independent storage.

According to the tips of the system, inspectors scan the whole and details (such as collision parts) at the scene of the road traffic accident with a special handheld device. Then the scanned data is uploaded to the server in real-time through the network. The server automatically analyzes and processes the data, conducts 3D modeling to restore the scene, and sets hotspots for key parts.

With the web version as its main form, the system can match the PC terminal. Its special device can scan and upload the data, and the server builds a model and restores the scene based on the data.

3.2. Programming Language

The programming language C# used in making 3D panorama is excerpted as follows:

```
if (tempPanrama == null)
        {
            tempPanrama = new RenderTexture(cubemapSize, cubemapSize, 24);
            tempPanrama.dimension = UnityEngine.Rendering.TextureDimension.Cube;
            tempPanrama.useMipMap = false;
            tempPanrama.autoGenerateMips = false;
            tempPanrama.antiAliasing = 1;
            tempPanrama.wrapMode = TextureWrapMode.Clamp;
            tempPanrama.filterMode = FilterMode.Bilinear;
            tempPanrama.Create();
        }
        if (panrama == null)
        {
            panrama = new RenderTexture(length, width, 24);
            panrama.antiAliasing = 1;
            panrama.wrapMode = TextureWrapMode.Clamp;
            panrama.filterMode = FilterMode.Trilinear;
```

```
            panrama.hideFlags = HideFlags.HideAndDontSave;
            panrama.Create();
        }
        if (texture2D == null)
        {
            texture2D = new Texture2D(length, width, TextureFormat.RGB24, false);
            texture2D.hideFlags = HideFlags.HideAndDontSave;
            texture2D.wrapMode = TextureWrapMode.Clamp;
            texture2D.filterMode = FilterMode.Trilinear;
            texture2D.hideFlags = HideFlags.HideAndDontSave;
            texture2D.anisoLevel = 0;
        }
        tempCamera.transform.localPosition = Vector3.zero;
        tempCamera.transform.localRotation = new Quaternion(0f, 0f, 0f, 0f);
        tempCamera.Render();
        tempCamera.RenderToCubemap(tempPanrama);

        Graphics.Blit(tempPanrama, panrama, m);//The bitmap is transferred here have to go through
this particular material——shader.
To transfer bitmaps here, it is essential to go through this specific material- shader.
        RenderTexture.active = panrama;
        texture2D.ReadPixels(new Rect(0, 0, length, width), 0, 0, false);
        texture2D.Apply();
        RenderTexture.active = null;
        byte[] bytes = texture2D.EncodeToJPG();
        string fileName = path + "/" + System.DateTime.Now.ToString("yyyy-MM-dd-HH-mm-ss")
+ ".jpg";
        System.IO.File.WriteAllBytes(fileName, bytes);
```

3.3. Effect Image

Image 1. Trace extraction of Simulated Road Traffic Accident Based on VR Technology

Image 2. Vehicle Body Trace Measurement of Simulated Road Traffic Accident Based on VR Technology

4. Vision

In the following on-site inspection of the public security unit, the notice will be automatically issued to relevant units through the overall analysis of police information under the aid of AI. After the police arrive at the scene, they will select the type of road traffic accident and determine its severity as per the actual situation as well as place the on-site inspection device. According to the given information, the device formulates the scanning scheme and plans the scanning route. The scanned data is transmitted to “the virtual simulation system for inspection at the scene of the road traffic accident” in real-time through the 5G network. The system server automatically carries out 3D modeling to restore the outline of the road traffic accident according to the collected data information as well as automatically identifies key parts and sets hotspots. On-site inspectors can also take photographs on key parts, which will be voluntarily uploaded to the server, with the vector graph as the type of these photographs.

On-site inspectors can log in to this system with the special account password or the police “Key” on their terminal device and wear the VR peripheral device to inspect the scene of the formed virtual road traffic accident. While inspecting key parts, if they cast their eyes on the set hotspots for seconds, they are able to see the 3D model set up by the system according to the accessed information of the special device.

In the process of inspection, inspectors can apply inspection tools (such as scale, protractor, and scale, etc.) that the system offers as the supplement. At the end of the inspection, they can write the inspection report and the traffic accident confirmation online. The system records and backs up the information from the scanning personnel and inspectors at the scene of the road traffic accident as future reference. Meanwhile, this information will be uploaded to the road traffic accident scene inspection platform and the OA system. When inspectors wrap up all inspection work, they can submit the data and information to specific objects (leaders). As system administrators, they can check and consult the relevant data on the inspection of any road traffic accident within their scope of authority.

5. Conclusion

The integration of AI and VR can effectively boost the work efficiency of on-site inspectors from the public security unit. However, due to the existing problems such as the lack of industrial ecology, the failure of transmitting tremendous scanning data in real-time, immature image recognition technology, and high demand for hardware, it is still not easy to fulfill these functions. Nevertheless, with the constant development of science and technology in this era, it is just over the horizon to realize them.

Acknowledgments
This work was supported by the Education Department in Fujian (FBJG20180077) .

References

[1] Elbamby M S, Perfecto C, Bennis M, et al. Toward Low-Latency and Ultra-Reliable Virtual Reality[J]. IEEE Network, 2018, 32(2):78-84.

[2] Chen M, Saad W, Yin C. Virtual Reality Over Wireless Networks: Quality-of-Service Model and Learning-Based Resource Management[J]. IEEE Transactions on Communications, 2018, 66(11):5621-5635.

[3] Zhang T, Mccarthy Z, Jow O, et al. Deep Imitation Learning for Complex Manipulation Tasks from Virtual Reality Teleoperation[J]. 2017.

[4] Domenichini L, Branzi V, Meocci M. Virtual testing of speed reduction schemes on urban collector roads[J]. Accident; analysis and prevention, 2017, 110:38.

IOP Conf. Series: Materials Science and Engineering **787** (2020) 012008 doi:10.1088/1757-899X/787/1/012008

Longitudinal and Lateral Velocity Estimation based on Covariance Inverse Robust Fusion

Junhao Huang*, **Wei Wei, Junshuai Hu**

Air Force Logistics College, Xuzhou, Jiangsu Province, 221000, China

*Corresponding author's e-mail: 396062782@qq.com

Abstract. Accurate estimation of vehicle longitudinal and lateral velocity is meaningful for internal system. Aiming at improving the accuracy of velocity estimation, the Covariance inverse (CI) based fusion method is applied. A vehicle dynamics model is established and the CI method is analysed. Finally, a joint simulation of CarSim and MATLAB/Simulink is conducted and the results show the advantages of proposed method.

1. Introduction

The application of automobile electronic control system has a good effect on the active safety of automobile, in which vehicle sensors play an important role. The on-board sensor can directly measure part of the driving state information, but because of the external environment interference or the error of the sensor itself in the driving process, it cannot accurately measure the important information such as the sideslip angle of the centre of mass, the longitudinal and lateral speed, the longitudinal and lateral force of the tire. Based on the information fusion technology, it is becoming the focus of people's attention to estimate the key state information of automobile, accurately acquire the motion state information which is difficult to measure directly, and realize the product application of active safety products in mass production vehicles [1].

The characteristics of multi-sensor information fusion are: through performance complementation, the time coverage, space coverage and frequency coverage of sensor detection are expanded; through performance redundancy, the confidence of target detection decision-making is improved, the fuzziness is reduced, the false alarm rate is reduced, the reliability and fault tolerance of the state monitoring system are enhanced, and the anti-interference ability of the system is improved [2].

2. Vehicle dynamics model

The first task of establishing vehicle dynamics model is to establish a proper vehicle body coordinate system. In this paper, the car body coordinate system is specified as follows, as shown in Figure 1, where absolute coordinate system *X-O-Y* represents geodetic coordinate system, and *X-O-Y* represents car body coordinate system (consolidated on car body). The origin of the coordinate is consistent with the centre of mass of the vehicle. The x-axis represents the longitudinal coordinate axis of the vehicle, and the driving direction of the vehicle is the positive direction of the coordinate axis; the y-axis represents the lateral coordinate axis of the vehicle, which is perpendicular to the longitudinal driving direction of the vehicle, and it is specified that the left-hand direction is positive when the vehicle is driving.

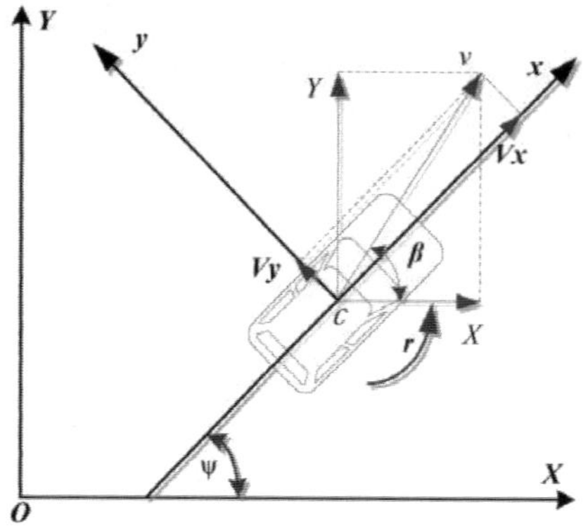

Figure 1. Vehicle dynamics model

2.1. Key driving state parameters

In this paper, the key driving state parameters for ESP control are estimated. First of all, three degrees of freedom, longitudinal, lateral and yaw, should be considered in the active safety control esp. Secondly, in order to improve the accuracy and simplify the model, the wheel differential control is considered, so the rotational degrees of freedom of four wheels need to be considered. Based on the above factors, this paper establishes a seven degree of freedom vehicle model as shown in Figure 1, and makes the following assumptions: (1) the origin of vehicle coordinate system coincides with the centroid of vehicle model; (2) the degrees of freedom of vehicle pitch, roll and vertical direction are ignored; (3) the suspension is simplified as a rigid body, and the transmission system is a linear system. Under the control of steering wheel, the two front wheel angles are the same; (4) The influence of longitudinal friction resistance on state estimation is ignored.

Table 1. Vehicle parameter symbol meaning

Parameter symbols	Parameter names	unit
v_x	longitudinal velocity	*m/s*
v_x	lateral velocity	*m/s*
r	yaw rate	*rad/s*
β	sideslip angle of centroid	*rad*
ω_{ij}	wheel angular speed	*rad/s*
T_{bij}	wheel braking torque	*N/m*
δ	front wheel corner	*rad*

2.2. Establishment of state estimation model

According to the vehicle model, the state vector is defined as

$$\boldsymbol{X}_k = [v_x, v_y, r, \omega_{fl}, \omega_{fr}, \omega_{rl}, \omega_{rr}] \tag{1}$$

$$\boldsymbol{U}_k = [\delta, T_{bfl}, T_{bfr}, T_{brl}, T_{brr}] \tag{2}$$

The following relationship exists between the vehicle's sideslip angle and the longitudinal and lateral speeds

$$\beta = \arctan\frac{v_y}{v_x} \tag{3}$$

Combined with the idea of state parameter estimation and tire model, the above vehicle dynamics model is sorted into state equation and measurement equation.

The standard form of the equation of state is as follows:

$$\boldsymbol{X}_k = f(\boldsymbol{X}_k, \boldsymbol{U}_k, k) + \boldsymbol{W}_k \tag{4}$$

The standard form of measurement equation is as follows:

$$\begin{aligned}\boldsymbol{Z}_k &= [a_x, a_y, r, \omega_{fl}, \omega_{fr}, \omega_{rl}, \omega_{rr}]^T \\ &= h[\boldsymbol{X}_k, \boldsymbol{F}_k, \boldsymbol{U}_k] + \boldsymbol{V}_k\end{aligned} \tag{5}$$

where $\boldsymbol{W}_k$ represents the process noise of the system state equation and $\boldsymbol{V}_k$ represents the measurement noise of the system measurement equation. a_x, a_y represents the longitudinal and lateral acceleration.

3. Covariance Cross Robust Fusion

The CI algorithm proposed by Julier [3] and Uhlmann [4] can realize the fusion estimation without knowing the cross covariance. The algorithm can improve the accuracy and robustness of the fusion estimation.

CI fusion Kalman filter is defined as

$$\hat{x}_{CI} = \boldsymbol{\Omega}_1 \hat{x}_1 + \boldsymbol{\Omega}_2 \hat{x}_2 \tag{6}$$

$$\boldsymbol{P}_{CI}^{-1} = \omega \boldsymbol{P}_1^{-1} + (1-\omega)\boldsymbol{P}_2^{-1} \tag{7}$$

where $\boldsymbol{\Omega}_1 = \omega \boldsymbol{P}_{CI} \boldsymbol{P}_1^{-1}$, $\boldsymbol{\Omega}_2 = (1-\omega)\boldsymbol{P}_{CI}\boldsymbol{P}_2^{-1}$, $\omega \in [0,1]$

The minimum performance index is

$$\min_{\omega} tr\boldsymbol{P}_{CI} = \min_{\omega\in[0,1]} tr\{[\omega \boldsymbol{P}_1^{-1} + (1-\omega)\boldsymbol{P}_2^{-1}]^{-1}\} \tag{8}$$

The optimal weight coefficient matrix can be obtained by 0.618 method.

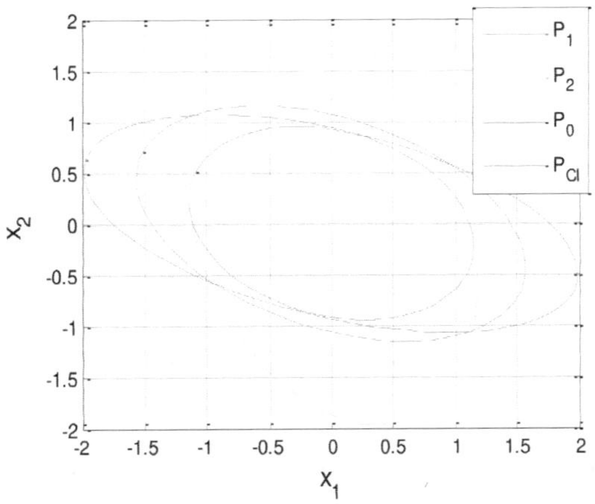

Figure 2. Comparative analysis of covariance ellipse

Presume that $\hat{x}_1$, $\hat{x}_2$ are unbiased estimation of a random vector and the corresponding nonnegative definite covariance matrices are $\boldsymbol{P}_1$ and $\boldsymbol{P}_2$. If $\boldsymbol{P}_1 \geq \boldsymbol{P}_2$, the ellipsoid coverage of spatial variance of $\boldsymbol{P}_1$ is greater than $\boldsymbol{P}_2$. Analysis of CI fusion precision by covariance ellipse. Point $v = \{v \mid v^T \boldsymbol{P}^{-1} v = c\}$ form the covariance ellipse variance of matrix P , c is a constant.

The above algorithm can be extended to the case of multi-sensor as

$$\hat{x}_{CI} = \boldsymbol{\Omega}_1 \hat{x}_1 + \boldsymbol{\Omega}_2 \hat{x}_2 + \ldots + \boldsymbol{\Omega}_n \hat{x}_n \tag{9}$$

$$\boldsymbol{P}_{CI}^{-1} = \omega_1 \boldsymbol{P}_1^{-1} + \omega_2 \boldsymbol{P}_2^{-1} + \ldots + \omega_n \boldsymbol{P}_n^{-1} \tag{10}$$

Minimum performance index is

$$\min_{\omega} tr\boldsymbol{P}_{CI} = \min_{\omega\in[0,1]} tr\{[\sum_{i=1}^{n} \omega_i \boldsymbol{P}_n^{-1}]^{-1}\} \tag{11}$$

where

IOP Conf. Series: Materials Science and Engineering **787** (2020) 012008 doi:10.1088/1757-899X/787/1/012008

$$\boldsymbol{\Omega}_i = \omega_i \boldsymbol{P}_{CI} \boldsymbol{P}_i^{-1} \tag{12}$$

$$\omega_1 + \omega_2 + \ldots + \omega_n = 1,\ \omega_i \in [0,1] \tag{13}$$

Using CI algorithm, the state estimation result and covariance matrix can be obtained.

$$\boldsymbol{X}_i^{loc}(k) = \sum_{j\in N_i \cup \{i\}} \boldsymbol{\Omega}_j \hat{\boldsymbol{X}}_j(k) \tag{14}$$

$$\boldsymbol{P}_i^{loc}(k) = (\sum_{j\in N_i \cup \{i\}} \omega_j \boldsymbol{P}_j(k)^{-1})^{-1} \tag{15}$$

where

$$\boldsymbol{\Omega}_j = \omega_j \boldsymbol{P}_i^{loc}(k) \boldsymbol{P}_j(k)^{-1} \tag{16}$$

$$\sum \omega_j = 1,\ \omega_j \in [0,1] \tag{17}$$

4. Experimental results

In order to verify the effectiveness of the above estimation algorithm, the joint simulation of CarSim and MATLAB/Simulink is carried out. The input of the system is the output of the CarSim simulation module, namely: longitudinal and lateral acceleration, yaw rate, steering wheel angle, etc.; the simulation module is established by using Matlab/Simulink. Finally, the simulation is carried out under typical working conditions, and the estimated value of this algorithm is compared with the simulation results of CarSim under the same working conditions. The vehicle working condition is set as double shift line, where there is a period of violent acceleration and deceleration in the process of vehicle driving.

The longitudinal and lateral velocity estimation ralative error are shown in Figure 3 (a) and (b).

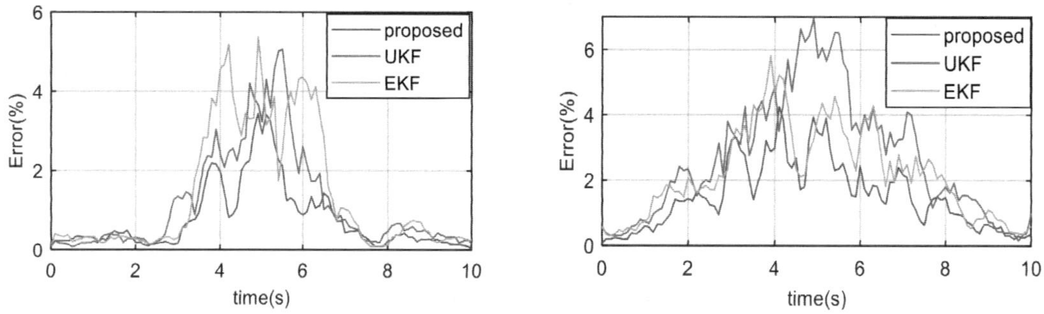

(a) longitudinal velocity estimation (b) lateral velocity estimation

Figure 3. Relative estimation error

According to Figure 3, the relative error of velocity estimation with proposed method is smaller than the UKF and EKF[5], which means a better estimation accuracy and performance.

Table 2 shows the MSE (mean square error) of the three methods. Consistent with the previous results, the MSE of the proposed algorithm in estimating the longitudinal and lateral velocity are 3.1050 and 3.3715, showing a better performance comparing with the UKF and EKF method.

Table 2. MSE of longitudinal and lateral velocity estimation

		UKF	EKF	Proposed
MSE	longitudinal velocity	5.7542	5.9127	3.1050
	lateral velocity	5.9512	6.1423	3.3715

5. Conclusion

In this paper, the CI based fusion method is applied in vehicle longitudinal and lateral velocity estimation. A vehicle dynamics model is established and the CI method is analysed. Finally, a the joint simulation of CarSim and MATLAB/Simulink is conducted and the results show the advantages of proposed method. Due to the limitation of the conditions, the road conditions selected in this paper are typical. The following research work should select more complex conditions, such as fishhook condition, single line transfer condition, etc., to carry out simulation verification and real vehicle verification for the performance of each algorithm, so as to make the conclusion more representative.

References

[1] Chen C K , Le A T. (2016) Vehicle Side-Slip Angle and Lateral Force Estimator Based on Extended Kalman Filtering Algorithm[M]. AETA 2015: Recent Advances in Electrical Engineering andRelated Sciences. Springer International Publishing.
[2] Balazs J A, Velásquez J D. (2016) Opinion Mining and Information Fusion: A survey[J]. Information Fusion, 27(C):95-110.
[3] Julier S J, Uhlmann J K. (1997) Non-divergent estimation algorithm in the presence of unknown correlations. Proc of 1997 IEEE American Control Conf. Albuquerque:2369-2373.
[4] Uhlmann J K. (1996) General data fusion for estimates with unknown cross covariance. Proc of the SPIE Aerosence Conf. SPIE, 2755: 536-547.
[5] Lebeda. (2015) Application of the particle filters for identification of the non-Gaussian Systems[C].Proceedings of the 16th International Carpathian Control Conference: 282-285.

IOP Conf. Series: Materials Science and Engineering **787** (2020) 012009 doi:10.1088/1757-899X/787/1/012009

Coupling relationship between transportation and tourism systems—a case study in the Huizhou City, China

Jingni Zeng[1], Qiangqiang Rong[1*], Wencong Yue[1], Xuejun Dai[2] and Meirong Su[1]

[1]Research Center for Eco-environmental Engineering, Dongguan University of Technology, Dongguan, Guangdong, 523808, China

[2]School of Geography and Tourism, Huizhou University, Huizhou, Guangdong, 516007, China

*Corresponding author's e-mail: rongqq@dgut.edu.cn

Abstract. Coupling relationship between transportation and tourism systems is significant to the sustainable development of regional economy, society and environment. Based on the coupling coordination degree model and scenario analysis method, the coupling relationship between transportation and tourism systems in the Huizhou City was studied in this research. The results showed that during 2014 and 2015, the coupling coordination degree between the two systems in the Huizhou City was at a moderate imbalance level. In terms of the districts and counties in the city, Huicheng District was at a moderate imbalance level, while Huiyang District, Longmen, Boluo and Huidong Counties were at a slight imbalance level. Results from the scenario analysis indicated that in order to improve the coordination between the transportation and tourism systems, decision makers in Huicheng and Huiyang Districts are supposed to increase effort to develop tourism while those in Longmen, Boluo and Huidong Counties are expected to pay more attention to transportation development. This research will help provide a decision-making basis for the urban planning in the Huizhou City. It will also provide a theoretical and methodological support for the other areas with similar concerns.

1. Introduction

With the rapid development of social economy and significant improvement of people's living standards, the output value of travel industry in China has continuously increased in recent decades. Concurrently, transportation infrastructure has also been constantly improved. However, the development of transportation and tourism systems in many cities in China are incongruous, which have seriously restricted the sustainable development of social economy. Thus, it is necessary to scientifically describe the transportation status and its interaction with tourism development. Analysis of coupling relationship between transportation and tourism systems will help provide a decision-making basis for urban planning, which is of great importance to the regional management.

In recent years, there have been many quantitative research on the relationship between transportation and tourism systems. A considerable number of studies were conducted on the development of intelligent models and systems, which have been widely applied to many research areas in multiple spatial scales[1-2]. For example, an Intelligent Transport and Tourism System was proposed in Hadoop environment to provide information such as lodging facilities available and items to be purchased in a city in an efficient manner[1]. Also, the impact of a certain traffic form on tourism were investigated in some research[3-4]. A high speed rail way can reduce the transportation costs, which will further reinforce the spatial competition between tourism destinations[3]. In addition, transportation service for tourism in some special areas, such as remote islands, congested tourism traffic corridors and new tourism towns, were analyzed in the previous research[5-7]. However, essential connection between the two systems and the subjective tendency of decision makers on the evaluation results of coupling relationship need to be further investigated, which are very important to the developing plan of tourism transportation.

In this research, coupling and coordination relationship between transportation and tourism systems will be explored based on the coupling and coordination degree model. Taking the Huizhou City as the study area, the temporal and spatial changes of the relationship between the two systems in the city were analyzed. Also, the subjective tendency of decision makers was considered through scenario analysis. Thus, this research will provide some alternatives for the coupling optimization of tourism and transportation industries in the Huizhou City. Also, it can provide a theoretical support for the regional management of other areas with similar concerns. It is of great scientific and practical significance.

2. Methodology

2.1 Description of the coupling and coordination degree model

The coupling and coordination degree model was used in this research to evaluate the coupling coordination degree between transportation and tourism systems due to its advantages in interpreting the relationship between two or more systems[8]. The model framework is given as follows[9]:

$$x_{ij}' = (x_{ij} - x_{min})(x_{max} - x_{min})^{-1} + 0.01 \tag{1}$$

$$p_{ij} = x_{ij}'\left(\sum_{i=1}^{m} x_{ij}'\right)^{-1} \tag{2}$$

$$h_j = -(\ln m)^{-1}\sum_{i=1}^{m} p_{ij}\ln p_{ij} \tag{3}$$

$$w_j = (1-h_j)\left(\sum_{j=1}^{n}(1-h_j)\right)^{-1} \tag{4}$$

$$U = \sum_{j=1}^{n} w_j x_{ij}' \tag{5}$$

$$D = (C \times T)^{1/2} \tag{6}$$

$$C = ((U_1 \times U_2)(U_1 + U_2)^{-2})^{1/2} \tag{7}$$

$$T = \alpha U_1 + \beta U_2 \tag{8}$$

where, x_{ij} and x_{ij}' represent the original value and the standardized value of index j in year i, respectively; x_{max} and x_{min} indicate the maximum and minimum value of index j among all years, respectively; p_{ij} denotes the proportion of the index j in year i; h_j denotes the information entropy of the index j; w_j denotes the weight of the index j by information entropy weight (IEW)[10]; n is the number of indexes, and m is the number of years; Suppose x_{ij}' represents the standardized index of transportation or tourism system. U is the comprehensive evaluation value of transportation or tourism system; D represents the coupling coordination degree, and $D \in [0,1]$; C represents the coupling degree; U_1 and U_2 refer to the comprehensive evaluation value of transportation and tourism systems, respectively; T reflects the overall level of transportation and tourism systems; α and β represent the contribution of transportation and tourism systems, respectively.

2.2 Establishment of the comprehensive evaluation index systems

Two sets of evaluation index systems were formed for transportation and tourism systems, respectively. Specifically, the transportation system contains 10 indicators including passenger traffic volume, highway passenger traffic volume, turnover of passenger traffic, civil vehicles ownership, private vehicles ownership, number of passenger vehicles, length of highways on operation, cargo volume, highway freight volume and length of urban paved pavement. As for tourism system, there are 6 indicators in the evaluation index system, including total tourism revenue, revenue from domestic tourism, foreign exchange income from international tourism, number of tourist hotels, number of inbound overnight visitors and number of domestic overnight visitors. Concurrently, the coupling relationship between the transportation and tourism systems at both city and county levels will be discussed in this research. Due to the data availability and the inadequate of the existing data statistical systems, some indicators will be deleted in the index systems established for county level evaluation. Specifically, only passenger traffic volume, turnover of passenger traffic, length of highways, highway density and cargo volume will be considered in the transportation system, while total number of visitors, total tourism revenue, number of visitors in A-grade scenic spots and operating income in A-grade scenic spots will be considered in tourism system.

2.3 Discriminating standard of the coupling coordination degree between transportation and tourism systems

In order to facilitate the analysis of the coordinated development stage between the two systems in each region, the values of coupling coordination degree are divided into the following four levels[11]: high imbalance ($0 < D \leq 0.2$), moderate imbalance ($0.2 < D \leq 0.3$), slight imbalance ($0.3 < D \leq 0.4$), approaching imbalance ($0.4 < D \leq 0.5$), reluctant coordination ($0.5 < D \leq 0.6$), primary coordination ($0.6 < D \leq 0.7$), intermediate coordination ($0.7 < D \leq 0.8$) and high coordination ($0.8 < D \leq 1$)

3. Case study

With an area of 11,599 km^2, Huizhou City is located in the east coast of Guangdong-Hong Kong-Macao Greater Bay Area. It is a new second-tier city with 5 counties, including Huicheng, Huiyang Districts, and Huidong, Boluo and Longmen Counties. The population of the city is approximately 4.83 million. By the end of 2018, the total length of highways in the Huizhou City was 14001 km. The annual cargo throughput of coastal ports was 74.53 million tons. The Pingtan Airport of the Huizhou City has 36 flights, connecting 31 cities. Until 2018, three trunk railways, one intercity railway, and one freight railway had been built in the Huizhou City. Also, there are two trunk railways under construction. The total number of tourists in 2018 was 58.9 million and the total tourism revenue was 50 billion Yuan. Although there are abundant tourist resources in Huizhou and the construction of transportation infrastructure is progressing steadily, the imbalance between regional transportation and tourism development still exists. Thus, it is very important to study the interrelation between the two systems. The relevant data of the indicators in this research were mainly obtained from the statistical yearbooks provided by the Bureau of Statistics, as well as the Culture and Tourism Bureau of the Huizhou City and each county. Years 2014 and 2015 were considered in this research.

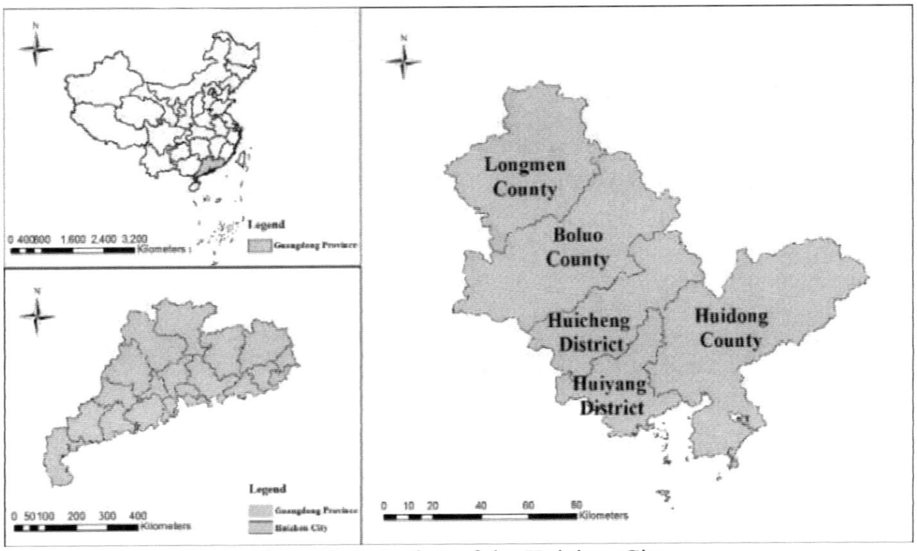

Figure 1. Location of the Huizhou City.

4. Results and discussion

In order to analyze the subjective tendency of decision makers on the evaluation results of coupling relationship, scenario analysis method was used in this research. Specifically, three scenarios were set as follows. Scenario 1: the contribution of tourism to the coordinated coupling of the two systems is greater than that of transportation (assume $\alpha = 0.4, \beta = 0.6$); Scenario 2: the contribution of transportation and tourism systems were supposed to be of the same importance ($\alpha = \beta = 0.5$); and Scenario 3: the contribution of transportation to the coordinated coupling of the two systems is greater than that of tourism (assume $\alpha = 0.6, \beta = 0.4$).

4.1 The overall coupling coordination degree between the transportation and tourism systems in the Huizhou City

The comprehensive evaluation values of the transportation and tourism systems (i.e., U_1 and U_2) in the Huizhou City during 2014 and 2015 were both less than 0.2, indicating that there was still much room for the improvement in the development of the two systems (Table 1). The comprehensive evaluation values of the transportation system were slightly lower than those of tourism system. Therefore, as for the regional coupling coordination status of the Huizhou City during this period, the development of the transportation system was overall slightly lagged behind. In addition, the results of the scenario analysis showed that the coupling coordination degrees of the two systems in the Huizhou City reduced with the decreasing β values (i.e., contribution of tourism system). When the decision maker tends to pay more attention to the tourism system, the resulting coupling coordination degree will be higher.

Table 1. Coupling coordination degrees between transportation and tourism systems in the Huizhou City from 2014 to 2015.

Years	U_1	U_2	C	Scenarios	D	Levels
2014	0.0791	0.1272	0.4862	1	0.2291	Moderate imbalance
				2	0.2239	
				3	0.2186	
2015	0.0790	0.1272	0.4861	1	0.2291	Moderate imbalance
				2	0.2239	
				3	0.2186	

4.2 Coupling and coordination degree of the transportation and tourism systems at the county level

During 2014 and 2015, the coupling coordination degrees between the transportation and tourism systems in the counties were at a relatively slight imbalance level except for Huicheng District. The comprehensive evaluation values of the tourism system in Huicheng District were much lower than those in the other four counties (Tables 2 and 3). The comprehensive evaluation values of the transportation system in Huidong, Boluo and Longmen Counties were lower than those of the tourism system, indicating that the development of transportation in the three counties was slightly lagged behind. Although there were several newly built highways in 2015, which have had a positive impact on the regional connectivity, the development speed of the road-network was slow. Therefore the role of transportation in tourism was still limited. Concurrently, the results of the scenario analysis showed that the coupling coordination degrees between transportation and tourism systems in Huicheng and Huiyang Districts would reduce with the decreasing β values (figure 2). Thus, when the decision makers attach more importance to tourism system, the resulting coupling coordination degree between the two systems will be higher. Conversely, the coupling coordination degrees between the two systems in the Huidong, Boluo, and Longmen Counties would reduce with the decreasing α values (i.e., contribution of the transportation system). When the decision makers pay more attention to transportation system, the coupling coordination degree between the two systems in these three counties will be higher.

Table 2. Comprehensive evaluation value and coupling degrees between transportation and tourism systems at the county level.

Districts	2014			2015		
	U_1'	U_2'	C	U_1'	U_2'	C
Huicheng District	0.2133	0.1187	0.4793	0.2176	0.1243	0.4811
Huiyang District	0.1971	0.2279	0.4987	0.2186	0.2115	0.5000
Huidong County	0.1982	0.2563	0.4959	0.2179	0.2748	0.4967
Boluo County	0.1943	0.2422	0.4970	0.2144	0.2771	0.4959
Longmen County	0.1944	0.2616	0.4945	0.2128	0.2759	0.4958

Table 3. Coupling coordination degrees and levels between transportation and tourism systems at the county level.

Districts	D (Levels) in 2014			D (Levels) in 2015		
	Scenario 1	Scenario 2	Scenario 3	Scenario 1	Scenario 2	Scenario 3
Huicheng District	0.2739 (Moderate imbalance)	0.2820 (Moderate imbalance)	0.2900 (Moderate imbalance)	0.2788 (Moderate imbalance)	0.2868 (Moderate imbalance)	0.2945 (Moderate imbalance)
Huiyang District	0.3279 (Slight imbalance)	0.3255 (Slight imbalance)	0.3232 (Slight imbalance)	0.3273 (Slight imbalance)	0.3279 (Slight imbalance)	0.3284 (Slight imbalance)
Huidong County	0.3400 (Slight imbalance)	0.3357 (Slight imbalance)	0.3314 (Slight imbalance)	0.3538 (Slight imbalance)	0.3498 (Slight imbalance)	0.3457 (Slight imbalance)
Boluo County	0.3329 (Slight imbalance)	0.3293 (Slight imbalance)	0.3257 (Slight imbalance)	0.3536 (Slight imbalance)	0.3491 (Slight imbalance)	0.3446 (Slight imbalance)
Longmen County	0.3407 (Slight imbalance)	0.3358 (Slight imbalance)	0.3308 (Slight imbalance)	0.3525 (Slight imbalance)	0.3481 (Slight imbalance)	0.3435 (Slight imbalance)

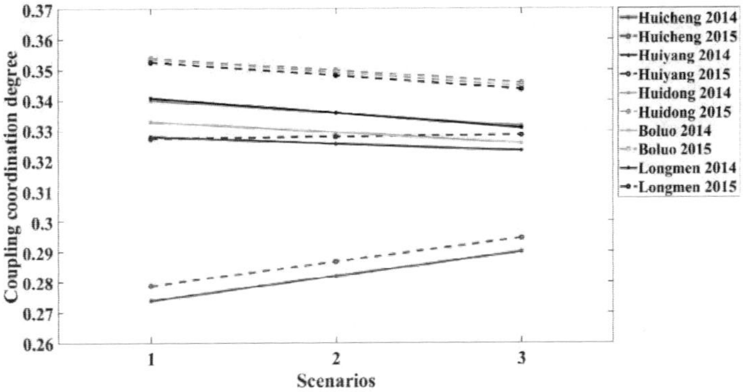

Figure 2. Coupling coordination degrees between transportation and tourism systems of the 5 counties in the Huizhou City.

5. Conclusions

In the context of national tourist transportation development, it is important to increase the level of coupling relationship between transportation and tourism systems. Taking the Huizhou City as the study area, the coupling coordination degree between the two systems during 2014 and 2015 were obtained based on the coupling coordination degree model. The results showed that the coupling coordination degrees between the two systems in the Huizhou City were at a moderate imbalance level. As for the coupling coordination degree of the two systems at the county level, the degrees in Huicheng District were at a moderate imbalance level, while those of Huiyang District, Longmen, Boluo and Huidong counties were at a slight imbalance level. The results of the scenario analysis showed that it is favorable to the coordination between the two systems in Huicheng and Huiyang Districts when the decision makers attach more importance to the tourism development. On the contrary, decision makers in Longmen, Boluo and Huidong counties are supposed to pay more attention to the development of transportation in order to enhance the coupling relationship between the two systems. This research will provide a decision-making basis for the regional management of tourist transportation development.

Acknowledgements

This work was financially supported by the National Key Research Program of China (Grant No. 2016YFC0502806), the National Natural Science Foundation of China (51809045 and 41801203), the Natural Science Foundation for Distinguished Young Scholars of Guangdong Province (No.2017A030306032), Guangdong Innovation Team Project for Colleges and Universities (No. 2016KCXTD023), GDUPS (2017), the Scientific Research Foundation for High-level Talents and Innovation Team in Dongguan University of Technology (No. KCYKYQD2016001).

References

[1] Ragunathan, T., Battula, S. K., Vedika, J., et al. (2015) ITTS: Intelligent transport and tourism system. J. Procedia Computer Science, 50: 191-196.

[2] Nguyen, V. T., Tetsuo, S. (2017) The effect of transportation on tourism promotion: Literature review on application of the Computable General Equilibrium (CGE) Model. J. Transportation Research Procedia, 25: 3096-3115.

[3] Masson, S., Petiot, R. (2009) Can the high speed rail reinforce tourism attractiveness? The case of the high speed rail between Perpignan (France) and Barcelona (Spain). J. Technovation, 29: 611-617.

[4] Turton, B. J., Mutambirwa, C. C. (1996) Air transport services and the expansion of international tourism in Zimbabwe. J. Tourism Management, 17(6): 453-462.

[5] Mafrudoh, L., Gultom, S., Pratiwi, S. W., et al. (2018) The accessibility of transportation to support tourism in the municipality of Mentawai Islands. J. Advances in Engineering Research, 2018,

147: 250-259.
[6] Zheng, H., Nava, E., Chiu, Y. C. (2011) Evaluating active traffic and demand management strategies for congested tourism traffic corridor. In: International IEEE Conference on Intelligent Transportation Systems. Washington, DC. 2015-2020.
[7] Li, G. C., Li, B., Ju, M. Y., et al. (2017) Discussion on integrated traffic planning (ITP) of new tourism town upon sustainable development and livable request. J. Transportation Research Procedia, 25: 3398-3411.
[8] Xu, D., Hou, G. L. (2019) The Spatiotemporal Coupling Characteristics of Regional Urbanization and Its Influencing Factors: Taking the Yangtze River Delta as an Example. J. Sustainability, 11: 822.
[9] Tang, Z. (2015) An integrated approach to evaluating the coupling coordination between tourism and the environment. J. Tourism Management, 46: 11-19.
[10] Shannon, C. E. (1948) A mathematical theory of communication. J. The Bell System Technical Journal, 3: 379-423.
[11] Liu, C. M., Zhang, R. Y., Wang, M., et al. (2018) Measurement and Prediction of Regional Tourism Sustainability: An Analysis of the Yangtze River Economic Zone, China. J. Sustainability, 10: 1321.

IOP Conf. Series: Materials Science and Engineering **787** (2020) 012010 doi:10.1088/1757-899X/787/1/012010

Road Resource Optimization and Restructuring Based on BP Neural Network

Hailong Li[1], Zhiwei Gao[2*], Xiaodong Tan[2], Yilong Ma[2], Xiaoyi Yang[2]

[1]Infrastructure equipment department of CRRC Qiqihar Vehicle Co., Ltd, Qiqihar, Heilongjiang, 164700, China

[2]School of Mechanical Engineering, Dalian Jiaotong University, Dalian, Liaoning, 116021, China

942417274@qq.com

Abstract. In order to solve the problem of uneven distribution of road resources during rush hour of commuter traffic in cities, a method called "tidal lane" was designed to optimize and restructure the road resources after calculation and measurement of the road occupation rate and the speed of the vehicle. The algorithm based on BP neural network is designed to judge whether the road is changing lanes or reversing with the input of road occupation rate and vehicle speed. The signal is controlled according to the results obtained by the algorithm to achieve the optimization and restructuring of road resources. After the simulation, the trained BP neural network is obtained, and then the test results are obtained. Finally, the correct rate of the network is calculated to reach 97%, which meets the design requirements and proves the reliability and correctness of the method.

1. Introduction

According to the survey traffic flow data, there will be imbalance in the use of road resources in two directions of a section of road in a certain period of road traffic. The traffic flow in one direction is less, the existing resources of the road are surplus, while the traffic flow in the other direction is in a state of slow congestion. In view of this phenomenon, the concept of "tidal lane" is put forward[1], and the intelligent road resource optimization and restructuring are realized by using the control method based on BP neural network algorithm through the signal lamp control. According to the investigation and analysis, the design is reasonable, and the simulation test of the algorithm verifies the feasibility of the design, and hopes to achieve the universality of the design in the intelligent transportation in the research.

2. Road resource optimization and restructuring design

2.1 Scheme design

"Tidal lane" is to divide road resources into one lane in the direction of waste or incomplete utilization, and provide it to the other direction of slow congestion in traffic, so as to relieve the driving pressure in the direction of traffic congestion and improve the utilization rate of existing road resources. Taking the lane change in direction B as an example. Generally, there are three lanes in A and B directions, now it is necessary to reorganize the direction of the lane in the B direction, and control the traffic lights of the lanes in the B direction near the double yellow line. The original indicator light is traversable, that is, it is a green light. After the restructuring, the indicator light will turn red, and it will be prohibited to continue driving along this lane. On the contrary, after lane change in direction A, it indicates the status of traffic lights and the direction of road travel from two directions, as shown in Figure 1.

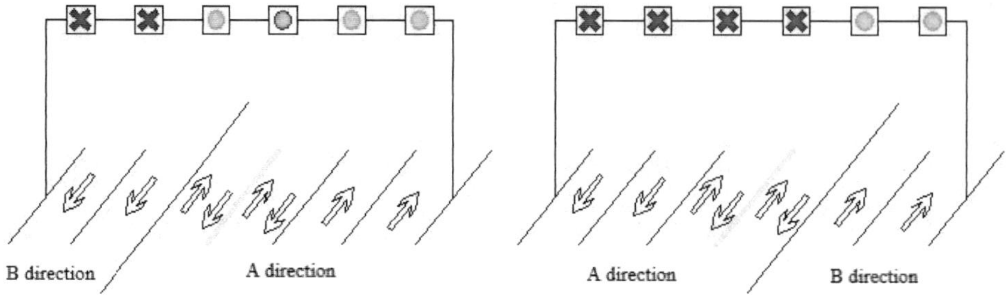

Figure 1. Schematic diagram of road plan design

2.2 Change lane restructuring condition

(1) When there is no restructuring on the road and the number of original lanes is maintained, the conditions for the restructuring of the lane change are: When $min[C_A(k), C_B(k)] \le a$, $max[C_A(k), C_B(k)] \ge b\ (a < b)$ and $|\Delta C'_{AB}(k)| = \frac{|C_A(k)-C_B(k)|}{min[C_A(k),C_B(k)]} > x$, Where $C_A(k)$ is the occupying coefficient of the kth period in the A direction, $C_B(k)$ is the occupying coefficient of the kth period in the B direction, and $|\Delta C'_{AB}(k)|$ is the relative difference of the difference of the occupying coefficients in the two directions A and B in the kth period Absolute value, in order to prevent the road traveling direction from changing too frequently, it is necessary to satisfy the above conditions in two consecutive detection cycles, and the value of $max[C_A(k), C_B(k)]$ is the same direction in two consecutive cycles. The coefficient changes the existing driving state.

(2) When the restructuring has taken place on the road and the number of original lanes has changed, the conditions for the road restructuring of the road section are: (a) $\Delta D'_{AB}(k) = \left(D_{min[N'_A,N'_B]}(k) - D_{max[N'_A,N'_B]}(k)\right)/D_{max[N'_A,N'_B]}(k) \ge y$; (b) $D_{min[N'_A,N'_B]}(k) > c$, and $\Delta D'_{AB}(k) > z$; (c) $D_{max[N'_A,N'_B]}(k) < d$; Where $D_{min[N'_A,N'_B]}(k)$ is the road occupation rate of the direction in which the lane is reduced after the restructuring of the road, $D_{max[N'_A,N'_B]}(k)$ is the road occupation rate of the direction in which the lane is increased after the road restructuring, and $\Delta D'_{AB}(k)$ is the A and B in the kth period after the road restructuring The relative difference between the difference of the channel coefficients in both directions.

The value of a, b, c, d, x, y, z is given according to the actual traffic congestion situation, here a=9, b=8, c=14, d=6, x=0.5, y=0.6, z=0.5.

3. Algorithm Design Based on BP Neural Network

(1) Preparation of training samples

In this paper, through the 400 sets of data collected, the road occupation rate and the vehicle speed of the lanes in the A and B directions are calculated and the lane change value is obtained. Among them, 280 sets of data are used as data training samples, 60 sets of data are used as debugging samples, 60 sets of data are used as test samples, and there are also new 100 sets of data to retest existing networks.

(2) Input layer and output layer design

Once you have selected the input and output variables, you need to describe them. Input and output variables are described differently depending on the needs of the network design. The input and output sample data used in this paper is mostly digital. Such data as the input neuron has a more direct and more intuitive effect on the target output node. When comparing, the input data has more obvious influence on the output data. In this paper, the input neuron is 4, and the BP neural network model structure is shown in Figure 2.

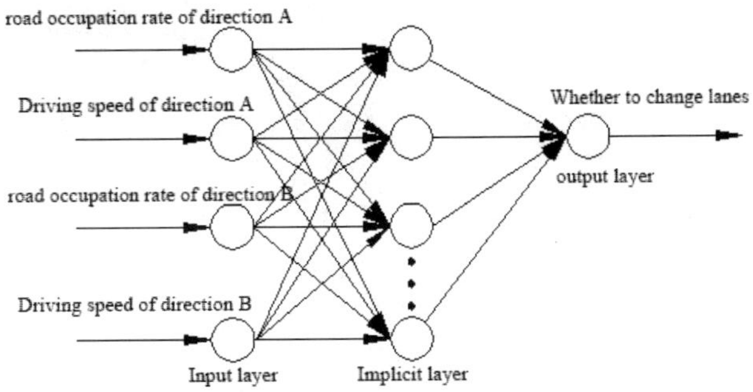

Figure 2. BP neural network model structure diagram

(3) Hidden layer design

The design of the hidden layer directly affects the characteristics of the neural network and the results of the output function[3]. There is no specific choice analytic formula to determine the number of learning neurons in the selection design of the hidden layer. The neuron design of the hidden layer basically depends on multiple simulation results to find a relatively appropriate value. The complexity of the problem, the number of input and output units, all of which directly affect the number of neurons in the hidden layer[4]. Theoretically, there will be a most appropriate number of hidden layer units. The number of hidden layer neurons in this paper is designed to be 4.

(4) Selection of initial weight

The choice of the initial value of the nonlinear system determines whether the learning of the network satisfies the error requirement, how long the training time and whether it can converge. In order to ensure that the weight of each neuron can be adjusted where the amount of S-type activation function changes is large. Usually, in the interval $(-2.4/F, 2.4/F)$, F is the number of input layer neurons, so the initial weight selection interval in this paper is the random number in $(-0.6,0.6)$ as the initial weight.

(5) Transfer function selection

The general hidden layer uses the Sigmoid function, while the output layer uses a linear function.

(6) Selection of training methods

In addition to the standard steepest descent method, BP network has several improved training algorithms. The training algorithm is related to the problem itself and the number of training samples[5]. The choice of serial or batch training methods is also something that needs to be determined during the neural network design process. This paper chooses LM algorithm and batch training method.

(7) Selection of expected error

During the whole network training period, the selection of expected error should be moderate, neither too large nor too small, so we can choose two expected value errors for network training, and get a more appropriate expected value error through continuous calculation and adjustment. In this paper, 1e-10 is selected as the adjusted expected error.

Algorithm flow chart shown in Figure 3.

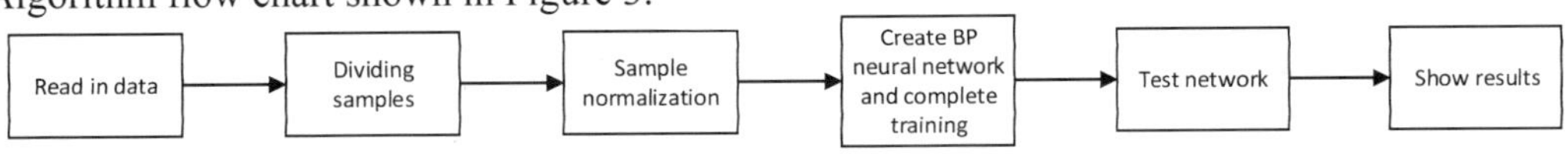

Figure 3. algorithm flow chart

4. Algorithm experiment based on BP neural network

4.1 Algorithm simulation and analysis

Set the maximum number of iterations to 1000, the initial learning rate is 0.01. Then train the network and get the structural diagram of the design as shown in Figure 4.

IOP Conf. Series: Materials Science and Engineering **787** (2020) 012010 doi:10.1088/1757-899X/787/1/012010

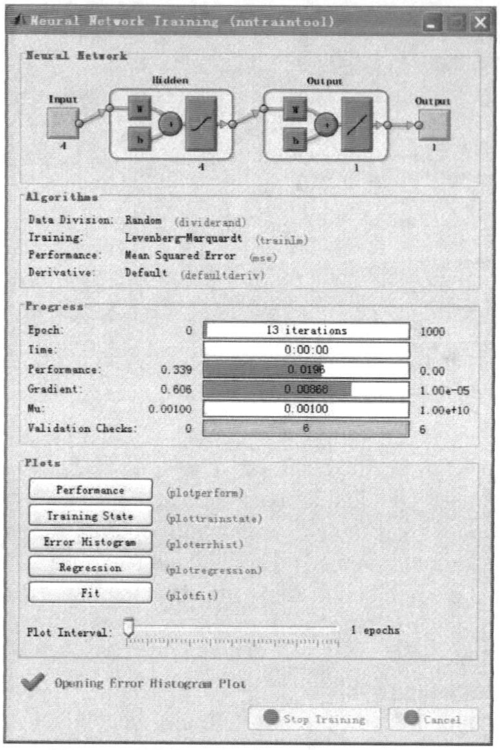

Figure 4. BP neural network training

The error reduction curve obtained after BP neural network training is shown in Figure 5. It can be concluded in the network parameter curve that this training stop occurs in the 14th iteration without obvious over-fitting. The best performance is 7 iterations. Test samples and verification sample errors have similar characteristics. Stop verification during training and increase error by 6 iterations. Some parameter drop states and network training state curves during network training are shown in Figure 6.

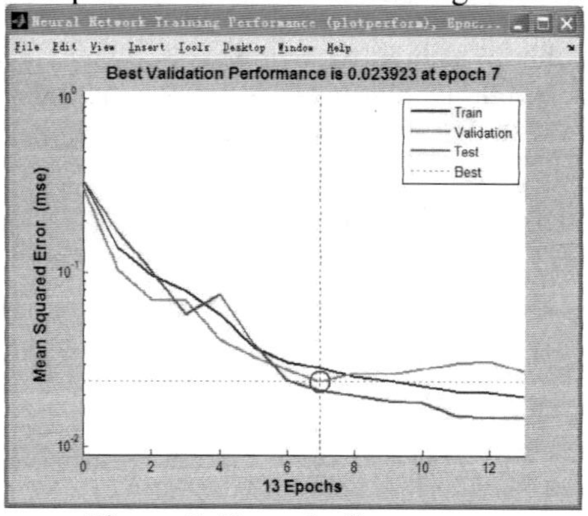

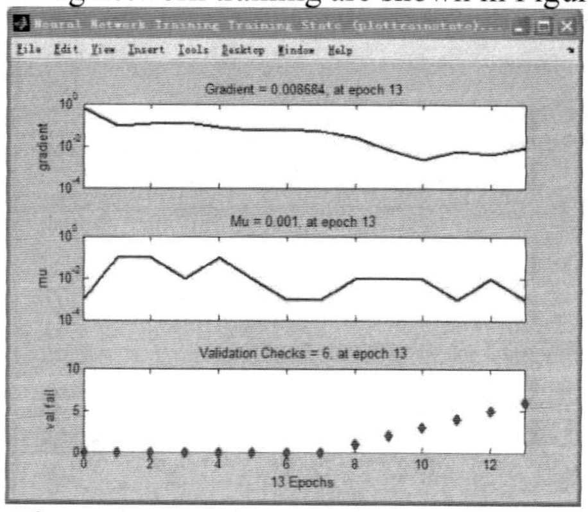

Figure 5. Error decline curve Figure 6. Neural Network Training State

Figure 7 shows the distribution of training values, test values, actual values, and predicted values of the output values.

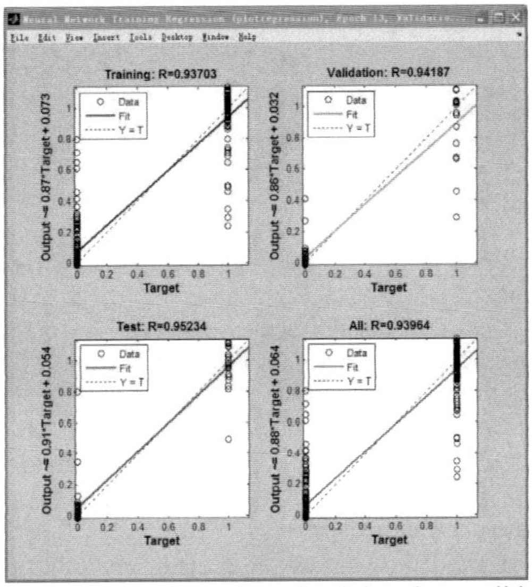

Figure 7. Regression curve of restructuring value of training data, validation data and testing data

Figure 7 is used to verify the performance of the trained network. For a perfect fit, the data should drop along the 45 degree line, and the network output is equal to the target.

4.2 Correct rate analysis

100 groups of data are input into the network for calculation, and the value of the lane change calculated through the network is compared with the target value to obtain the correct rate of the trained network. Take 0.5 as the threshold, and the output less than or equal to the threshold is judged as 0 (no lane change). An output greater than or equal to the threshold is judged as 1 (lane change). A set of data with different training values and target values is selected to form a numerical comparison table. Only 3 out of 100 groups of test data have different results, namely, the 48th group, the 51th group, and the 96th group are as shown in Table 1, so the correct rate of the trained network is 97%. This result has reached the design requirements.

Table 1. Comparison of judgment error data

Serial number	A direction		B direction		Training value	Target value	Correct rate
	Road occupation rate	Speed (km/h)	Road occupation rate	Speed (km/h)			
48	6.5	27	13	23	1	0	0%
51	13	23	6.5	27	1	0	0%
96	4.3	44	6.2	10	0	1	0%

5. Conclusion

In this study, the design method of "tidal lane" is taken as the framework, and the lane change judgment based on BP neural network algorithm is mainly studied. The training network obtained through the experiment can be applied to new data, and the accuracy meets the design requirements, which shows the feasibility and universality of the algorithm, and it is completely feasible to control the traffic lights through the algorithm, which solves the problem of automatic and intelligent road resource optimization and restructuring. This paper combines the road design of road resource optimization and restructuring with the artificial intelligence algorithm to make the road resource utilization ratio higher and the traffic trip more intelligent, which is the inevitable trend in the future.

IOP Conf. Series: Materials Science and Engineering **787** (2020) 012010 doi:10.1088/1757-899X/787/1/012010

Reference

[1] Tan X D, Yang X Y, Song P W. Optimization and Restructuring of the Road Resource with Construction Materials Based on Road Environment[J]. Advanced Materials Research, 2012, 600:152-156.
[2] Chmiel W, Dańda, Jacek, Dziech A, et al. INSIGMA: an intelligent transportation system for urban mobility enhancement[J]. Multimedia Tools and Applications, 2016, 75(17):10529-10560.
[3] Domijan D. A computational model of fMRI activity in the intraparietal sulcus that supports visual working memory[J]. Cogn Affect Behav Neurosci, 2011, 11(4):573-599.
[4] Yang W, Wang S, Hu J, et al. Securing Deep Learning Based Edge Finger-vein Biometrics with Binary Decision Diagram[J]. IEEE Transactions on Industrial Informatics, 2019:1-1.
[5] Dutta R K, Dutta K, Jeevanandham S. Prediction of Deviator Stress of Sand Reinforced with Waste Plastic Strips Using Neural Network[J]. International Journal of Geosynthetics and Ground Engineering, 2015, 1(2):11.
[6] Xia X, Li T. A fuzzy control model based on BP neural network arithmetic for optimal control of smart city facilities[J]. Personal and Ubiquitous Computing, 2019.
[7] Gu Y L, Wang X C, Xu J X. Traffic Data Fusion Research Based on Numerical Optimization BP Neural Network[J]. Applied Mechanics and Materials, 2014, 513-517:1081-1087.

IOP Conf. Series: Materials Science and Engineering **787** (2020) 012011 doi:10.1088/1757-899X/787/1/012011

Performance of Active Control and Energy Harvesting of a Novel Suspension System

Dashan Huang[1*], Jinqiu Zhang[2] and Yile Liu[2]

[1] Element 41, Unit 32272 of PLA, Deyang, Sichuan, 618408, China

[2] Department of Vehicle Engineering, Army Academy of Armored Force, Beijing 100072, China

*Corresponding author's e-mail: scoobidoobidoo@126.com

Abstract. Several processes were employed to improve the defects lie in the existing energy-regenerative suspension, such as the low efficiency of energy harvesting, the lack of unified assessment standards on the performance of energy harvesting. Firstly, a novel suspension system was established with the energy harvesting could also be implemented while controlling, which would improve the efficiency of energy harvesting; Then power flow analysis method was adopted to realize the quantitative analysis on the performance of energy harvesting, 4 evaluation criteria on performance of energy harvesting were established in considering of relative energy-regenerative ability and absolute energy-regenerative ability, which would be used as the unified assessment standards. A LQR controller was presented to control and reduce the vibration of the novel suspension system. To optimize the weight coefficients of LQR active control algorithm, an improved genetic algorithm was designed with 5 improvements on both genetic operator and genetic mechanism. The simulations on the performance of active control of the novel suspension system were preformed, the results show that the riding comfort and handling stability of vehicle are both improved with the genetic-algorithm-optimized LQR controller. The simulations on the performance of energy harvesting of the novel suspension system under different road excitations were conducted, the results show that the assessment standards could represent the energy harvesting ability of suspension system, the novel suspension system provide a better energy harvesting efficiency, the self-powered of suspension system under standard road excitation is realized.

1. Introduction

The traditional passive suspension system can't meet the demands of ride comfort and handling stability in modern vehicle, moreover, modern vehicle is incapable of providing the extensive energy required by vibration control of active suspension system. An energy harvesting device was designed in the energy-regenerative active suspension to recover the energy that were dissipated as heat in traditional suspension system, reduce the energy consumption by active suspension system relatively. Attentions were paid to the energy-regenerative active suspension system recently for it can provide good drive performance with less energy consumption. The research on energy-regenerative suspension system was conducted since 1970s. In 1978, Karnopp [1-2] showed the causes of energy dissipation in suspension system, while the requirement of input energy was discussed. In 1999, Nakano [3-4] proved that the required energy for active control could be provided by the recovered energy of suspension system entirely. In 2005, Bose Corporation [5] developed an energy-regenerative suspension system in using of electro-magnetic actuator equipped with linear motor, the features of vibration control and energy harvesting were tested and verified. The research of energy-regenerative suspension system in China was carried out mainly in several institutes and universities since 2000s. In 2005, Yu Fan [6-7], who worked in Shanghai Jiao Tong University, indicated that the required energy supplement for suspension system would decrease dramatically if the vibration energy recovered efficiently according to the comparison and analysis between the energy that damper dissipated and the energy that active control required. In 2009, Yu Changmiao [8-9], a researcher from Jilin University, noted that the dissipation power of damper is about 42.3% of the output power of engine, the harvesting of the dissipation power is of great significance and application value. In 2013, Zhang Jiye [10], who studied in Southwest Jiaotong University, designed an energy-regenerative suspension system based on linear-motor-type electro-magnetic actuator, the conditions of the suspension system energy balance were presented. A novel energy-regenerative active suspension (NEAS) was proposed, the coexistence of active control and energy harvesting in the entire process was realized. An LQR active control algorithm was provided and accompanied with an improved genetic algorithm in optimizing its weight coefficients. The flowing and transforming processes of power of suspension system under passive energy-harvesting mode and active-control mode were put forward based on power flow method, the corresponding quantitative analysis method, evaluation criteria and calculation formula of performance of energy harvesting were presented to provide a uniform standard for theoretical analysis.

2. Model of NEAS

The damper of passive suspension system was substituted by an electro-magnetic actuator to develop the current energy-regenerative active suspension system, the actuator of the system could produce required control force according to the relevant control signal to get better vibration control effectiveness and recover the vibration energy in suspension system. The electro-magnetic actuator is the key component of current suspension system to realize vibration control and energy harvesting, which was used as both active control device and energy harvesting device, the alternation between the two operating modes were determined by the active control algorithm. A compromise should be made between the vibration control effectiveness and energy harvesting efficiency, which leads to that the recovered energy are not as much as what was designed. Moreover, much additional energy should be provided to implement active control. A NEAS was designed with the active control device and energy harvesting device work separately, as shown in figure 1, a motor was used as actuator to implement active control, a generator was used as energy-regenerator to implement energy harvesting, a movement mode convertor, that was equipped with the gear rack mechanism, was employed to combine the actuator and the energy-regenerator as electro-magnetic actuator of NEAS, which made it possible that active control and energy harvesting could be implemented simultaneously and with high efficiency.

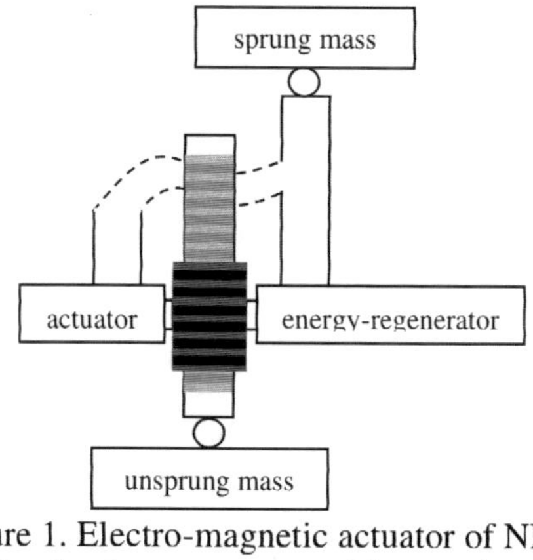

Figure 1. Electro-magnetic actuator of NEAS.

The 2-DOF (two-degree-of-freedom) quarter-car model was taken as the subject of the study for it could reveal major kinematics features of suspension system. The basic assumptions of the 2-DOF quarter-car model [11] are: (1) it is strictly symmetric on the left side and the right side of the vehicle; (2) the influence between the front side and the rear side of the vehicle are insignificant; (3) the vertical kinematics features of tyre could be simplified as spring. The 2-DOF quarter-car model of NEAS is shown in figure 2. The associated parameters include: the sprung mass m_s, the unsprung mass m_t, the suspension spring constant k_s, the tyre spring constant k_t, the mechanical damping of actuator c_{mmn}, the electro-magnetic force of actuator F_{an}, the mechanical damping of energy-regenerator c_{mgn}, the electro-magnetic damping of energy-regenerator c_{egn}. Let x_s, x_t, and x_r be the vertical displacements of the sprung mass, the unsprung mass and the road excitation respectively.

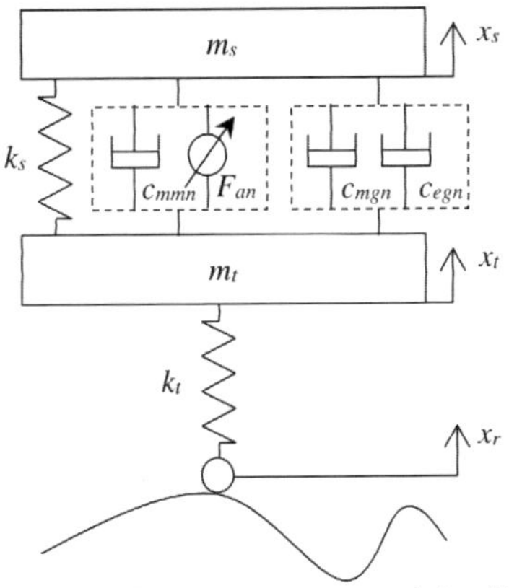

Figure 2. 2-DOF quarter-car model of NEAS.

The kinematic differential equations of NEAS can be described as in

$$m_s\ddot{x}_s + c_{mmn}(\dot{x}_s - \dot{x}_t) + c_{mgn}(\dot{x}_s - \dot{x}_t) + c_{egn}(\dot{x}_s - \dot{x}_t) + k_s(x_s - x_t) + F_{an} = 0 \tag{1}$$

$$m_t\ddot{x}_t - c_{mmn}(\dot{x}_s - \dot{x}_t) - c_{mgn}(\dot{x}_s - \dot{x}_t) - c_{egn}(\dot{x}_s - \dot{x}_t) - k_s(x_s - x_t) + k_t(x_t - x_r) - F_{an} = 0 \tag{2}$$

3. Active Control Algorithm

3.1. LQR active control algorithm

Take $\boldsymbol{X} = [x_s \quad x_t \quad \dot{x}_s \quad \dot{x}_t \quad x_r]^{\mathrm{T}}$ as state vector, then the differential equations of NEAS can be expressed in the form of the following state equation

$$\dot{\boldsymbol{X}} = \boldsymbol{AX} + \boldsymbol{BU} \tag{3}$$

where, the variables $\boldsymbol{A}$, $\boldsymbol{B}$, $\boldsymbol{U}$ are shown as follow

$$\boldsymbol{A} = \begin{bmatrix} 0 & 0 & 1 & 0 & 0 \\ 0 & 0 & 0 & 1 & 0 \\ -\frac{k_s}{m_s} & \frac{k_s}{m_s} & -\frac{c_{mmn}+c_{mgn}+c_{egn}}{m_s} & \frac{c_{mmn}+c_{mgn}+c_{egn}}{m_s} & 0 \\ \frac{k_s}{m_t} & -\frac{k_s+k_t}{m_t} & \frac{c_{mmn}+c_{mgn}+c_{egn}}{m_t} & -\frac{c_{mmn}+c_{mgn}+c_{egn}}{m_t} & \frac{k_t}{m_t} \\ 0 & 0 & 0 & 0 & -2\pi f \end{bmatrix}, \boldsymbol{B} = \begin{bmatrix} 0 \\ 0 \\ -\frac{1}{m_s} \\ \frac{1}{m_t} \\ 0 \end{bmatrix}, \boldsymbol{U} = F_{an}$$

where, f is instantaneous frequency of road excitation, the phase difference between displacement and velocity of road excitation was ignored.

Take $\boldsymbol{Y} = [\ddot{x}_s \quad x_s - x_t \quad x_t - x_r]^{\mathrm{T}}$ as output vector, the output equation of the system can be described as

$$\boldsymbol{Y} = \boldsymbol{CX} + \boldsymbol{DU} \tag{4}$$

where, the variables $\boldsymbol{C}$, $\boldsymbol{D}$ are shown as follow

$$C = \begin{bmatrix} -\frac{k_s}{m_s} & \frac{k_s}{m_s} & -\frac{c_{mmn}+c_{mgn}+c_{egn}}{m_s} & \frac{c_{mmn}+c_{mgn}+c_{egn}}{m_s} & 0 \\ 1 & -1 & 0 & 0 & 0 \\ 0 & 1 & 0 & 0 & -1 \end{bmatrix}, D = \begin{bmatrix} -\frac{1}{m_s} & 0 & 0 \end{bmatrix}^{\mathrm{T}}$$

There are mainly 3 evaluation indexes about the performance of suspension system, the sprung mass acceleration, which represents ride comfort, the dynamic load of tyre, which indicates handling stability, the dynamic deflection of suspension, which means driving safety. The LQR active control algorithm was adopted in controlling of NEAS, the objective function of LQR [6, 12] was derived from

$$J = \lim_{T\to\infty}\frac{1}{T}\int_0^T \left[q_1\ddot{x}_s^2 + q_2(x_s - x_t)^2 + q_3(x_t - x_r)^2 \right] dt \tag{5}$$

where, q_1、q_2、q_3 are the weighting coefficients of the acceleration of spring mass, the dynamic load of tyre, the dynamic deflection of suspension respectively.

Substituted equation (4) into equation (5), then the objective function becomes

$$J = \lim_{T\to\infty}\frac{1}{T}\int_0^T \left[Y^{\mathrm{T}} Q_0 Y \right] dt = \lim_{T\to\infty}\frac{1}{T}\int_0^T \left[X^{\mathrm{T}}QX + U^{\mathrm{T}}RU + 2X^{\mathrm{T}}NU \right] dt \tag{6}$$

where, $Q_0 = \begin{bmatrix} q_1 & 0 & 0 \\ 0 & q_2 & 0 \\ 0 & 0 & q_3 \end{bmatrix}$, $Q = C^{\mathrm{T}} Q_0 C$, $R = D^{\mathrm{T}} Q_0 D$, $N = C^{\mathrm{T}} Q_0 D$.

The control aim of suspension system of vehicle is that the concerned suspension states, such as sprung mass acceleration, dynamic load of tyre and dynamic deflection of suspension, are forced to return to their balance states (take it as zero state) under the LQR active control with acceptable quantity of energy consumption when they are drifting from or deviating from the zero states. The optimization control of suspension system is to calculate and utilize the optimal control force U to transfer the initial suspension state X_0 to zero state in the limitless time period $[t_0, \infty)$ while minimizing the function value of the objective function of LQR active control algorithm as is defined in equation (6). Hence, the optimization control of suspension system could be presented as in

$$U = \left\{ \min J \,\middle|\, \dot{X} = AX + BU \right\} \tag{7}$$

The equation (7) indicates a functional condition extremum problem, the optimal control force provided by the electro-magnetic actuator could be expressed as

$$U = -KX = -R^{-1}\left(B^{\mathrm{T}}P + N^{\mathrm{T}} \right) X \tag{8}$$

where, K is the optimal state feedback gain matrix, P could be worked out from the following Riccati equation

$$PA + A^{\mathrm{T}}P - (PB + N)R^{-1}\left(B^{\mathrm{T}}P + N^{\mathrm{T}} \right) + Q = 0 \tag{9}$$

3.2. Parameter optimization with genetic algorithm

The weight coefficients of LQR active control algorithm are used to be determined on the experience of the designer [13-14], which not only is rootless and irrational, but also cannot ensure that the optimal performance of active control on suspension system of vehicle would be obtained with the experience-based LQR algorithm. Genetic algorithm is a heuristic intelligence optimization algorithm based on the simulation of biological evolution, the independent variables of the function to be optimized are encoded with discrete numeric as genes, then the optimal solution or sub-optimal solution is obtained through the running of selection operator, crossover operator and mutation operator [15]. For this reason, genetic algorithm was introduced and brought into the process of LQR controller designing to optimize the weight coefficients of LQR active control algorithm taking advantage of and in using of the parallel optimization and global optimization ability of genetic algorithm. There are 5 basic issues should be discussed when using genetic algorithm to optimize the weight coefficients of LQR controller [16]:

a. chromosome encoding

Chromosomes are encoded in integer, of which the length is determined by the number of the unknown parameters in $\boldsymbol{Q}_0$, the unknown parameters are set in each gene on chromosome. Hence, the chromosome could be described as

$$\boldsymbol{Ch} = \left(q_1, q_2, q_3\right) \tag{10}$$

b. chromosome population generating

A certain number chromosome are generated randomly according to the requirement of calculating to compose the initial chromosome population, which is

$$\boldsymbol{POP} = \left(\boldsymbol{Ch}_1, \boldsymbol{Ch}_2, ..., \boldsymbol{Ch}_N\right) \tag{11}$$

where, N is the scale of the chromosome population.

c. fitness function constituting

The fitness function of genetic algorithm could be expressed by the 3 evaluation indexes on suspension system performance. However, some conformation measurements should be implemented to deal with the inconformity and discordancy of the unit and magnitude of the evaluation indexes. The passive suspension system with the identical working conditions was adopted as a comparison of the evaluation index to validate the fitness function constituted with the 3 evaluation indexes and make the fitness function effective and efficient [17-18]. Thus, the fitness function was constituted as

$$\min Fitness = \alpha \frac{\mathrm{SMA}(\boldsymbol{Ch})}{\mathrm{SMA_p}(\boldsymbol{Ch})} + \beta \frac{\mathrm{DTD}(\boldsymbol{Ch})}{\mathrm{DTD_p}(\boldsymbol{Ch})} + \gamma \frac{\mathrm{DDS}(\boldsymbol{Ch})}{\mathrm{DDS_p}(\boldsymbol{Ch})} \tag{12}$$

where, SMA, DTD, and DDS are the root mean square value of sprung mass acceleration, dynamic tyre deformation, and dynamic deformation of suspension respectively, while $\mathrm{SMA_p}$, $\mathrm{DTD_p}$, and $\mathrm{DDS_p}$ are all that of the passive suspension respectively; α, β, and γ are the weight coefficients of the fitness function. The dynamic tyre deformation is the relative position of unsprung mass and road excitation x_t - x_r, it's a multiple of the dynamic load of tyre with a constant. The dynamic tyre deformation was used to evaluate the handling stability of suspension system of vehicle instead of the dynamic load of tyre in the entire paper, for which is much easier to be obtained by sensors and will simplify the calculations on fitness function.

d. genetic operator setting

There are 3 main genetic operators in standard genetic algorithm, which are selection operator, crossover operator and mutation operator. Chromosomes are chosen under selection operator using proportion selection method, generate the chromosomes of next generation under crossover operator using single-point crossover method, amend the chromosomes of next generation under mutation operator with single-point mutation method. The crossover probability and the mutation probability are constants throughout the iteration and evolution of genes on the chromosomes in the population. The population of chromosome keeps unchanged in executing certain number of genetic operators.

e. termination criterion designing

The number of iterations is employed as the termination criterion to determine the evolution is accomplished or not.

3.3. Improvements of genetic algorithm

The identification and optimization of the weight coefficients of LQR active control algorithm is an offline processing, which means the optimization of the weight coefficients of LQR controller and the application of LQR algorithm in suspension system is relatively independent, the real-time character of genetic algorithm is inconsiderable and could be ignored. For this reason, some more complicated and efficient genetic strategies could be implemented to provide a more effective and powerful LQR algorithm. There are mainly 2 ways to improve genetic algorithm, in genetic operator and in genetic mechanism. 2 improvements were made on genetic operators, which are the selection operator with self-adaption tournament method and multi-chromosome crossover operator; 3 improvements were made on genetic mechanisms, which are initial chromosome population optimization, elite chromosome reservation, optimal chromosomes suppression.

a. selection operator with self-adaption tournament method

Selection operator is mainly focus on choosing the better chromosomes which would be passed on to the next generation of chromosome population directly or indirectly.The selection operator with self-adaption tournament method is based on the comparison of the fitness of chromosomes, of which the principle is that a certain number of chromosomes are chosen randomly to compose as a group, the chromosomes in the group compete with each other till the one with best fitness is left, which eventually is to be the chosen one of this round [19].

b. multi-chromosome crossover operator

Crossover operator is used to generate new chromosomes for the next generation of chromosome population in exchanging some parts of genes randomly in a certain way with two chromosomes in current generation of chromosome population which have been processed with selection operator.

The multi-chromosome crossover operator designed in this paper makes the best use of the genes in multiple chromosomes, which could enhance the activeness of chromosome and avoid the locality and one-sidedness of standard crossover operator. The crossover of the genes of chromosomes with multi-chromosome crossover operator is unidirectional, only the one which is set as the gene receiver alters its genes, the others which are working as the gene provider remain unchanged, the dominant chromosome for crossover in each execution as the gene provider and the gene for crossover at each time on the provider are determined by the shielding pattern [20-21].

c. initial chromosome population optimization

The initial chromosome population in standard genetic algorithm is generated randomly, of which the fitness is rather bad. As the iteration runs, the inferior genes on chromosome die out easily, quickly the chromosome population traps into local optimum. Yet chromosome with genes relatively better fitness could be generated as the standard genetic algorithm performs. Run the standard genetic algorithm repeatedly until the number of better chromosomes generated meets the require of chromosome population, then take this group of chromosomes as an initial chromosome population, which is with better genes and better fitness, could improve the optimal ability of genetic algorithm.

d. elite chromosome reservation

The chromosome with the best fitness in the current generation should be protected and reserved as an elite, it should be selected and reproduced to the next generation without alter and modify by crossover operator and mutation operator, which ensures that the best chromosome won't be decomposed by genetic operators and also the convergence of fitness is monotonically increasing, consequently the running efficiency of genetic algorithm is improved.

e. optimal chromosomes suppression

As the iteration runs, the diversity of genes on chromosome in the population is weakened and lessened at the mid-latter stage. Many chromosomes which are identical with the best one emerge and be produced, which means the optimal ability of genetic algorithm at these stages is reduced and lowered.The optimal chromosomes suppression mechanism is to alter the genes on the chromosomes which are identical with the best one with high-frequency mutation at the mid-latter stage, which could increase the diversity of genes on chromosome and improve the running efficiency of genetic algorithm at these stages notably.

IOP Conf. Series: Materials Science and Engineering **787** (2020) 012011 doi:10.1088/1757-899X/787/1/012011

4. Quantitative Analysis of Performance of Energy Harvesting

The performance of energy harvesting of suspension system relate to the suspension structure itself, operating mode and road excitation. The flowing and transforming processes of power with the corresponding computational formula in the NEAS were provided based on power flow method. The uniform standard on quantify and analyze the performance of energy harvesting of suspension system were established.

4.1. Power flow analysis

There are two operating modes existing in NEAS, the passive energy-harvesting mode and the active-control mode, the flowing and transforming processes of power are different under each operating mode.

While in passive energy-harvesting mode, as shown in figure 3, the behaviors and functions of energy-regenerator and actuator in NEAS are the same, they are both used as passive dampers to recover vibration energy. The absorption power of suspension system is mainly obtained from the output power of engine, flows towards two directions, one is dissipated by passive damping of energy-regenerator and actuator, the other is absorbed by electro-magnetic damping of energy-regenerator and actuator. Then part of the second power is dissipated by the internal-resistance of energy-regenerator and actuator, the other part is taken as output power, approximately as regenerative power of suspension system. Thus, the absorption power of NEAS while in passive energy-harvesting mode can be expressed as

$$P_{pri} = P_{pre} + P_{pdm} \tag{13}$$

where, P_{pre} is the total absorption power of electro-magnetic damping of suspension system, P_{pdm} is the total absorption power of passive damping of suspension system, which are given respectively as

$$P_{pre} = P_{preg} + P_{prem} = \left(c_{emn} + c_{egn}\right)\left(\dot{x}_s - \dot{x}_t\right)^2 \tag{14}$$

$$P_{pdm} = P_{pdmg} + P_{pdmm} = \left(c_{mmn} + c_{mgn}\right)\left(\dot{x}_s - \dot{x}_t\right)^2 \tag{15}$$

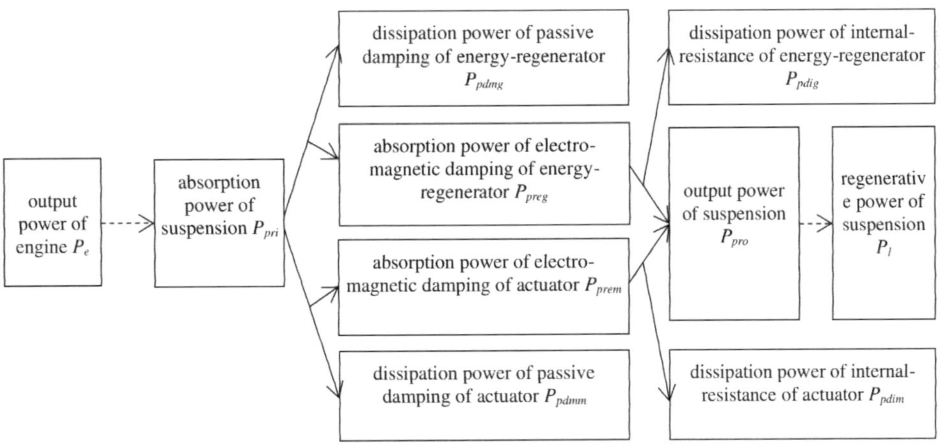

Figure 3. The power flow of NEAS in passive energy-harvesting mode.

The output power of NEAS while in passive energy-harvesting mode is shown as

$$P_{pro} = P_{pre} - P_{pdi} \tag{16}$$

where, P_{pdi} is the total dissipation power of internal-resistance of suspension system, it could be described as

$$P_{pdi} = P_{pdig} + P_{pdim} = I_{lg}^2 R_{dig} + I_{lm}^2 R_{dim} \tag{17}$$

While in active-control mode, the NEAS shows two operating states as the power flowing and transforming, which are achievable active control and attenuatable active control.

It is achievable active control when the active control force provided by the actuator is dominating the kinematic states of NEAS, which implies that the NEAS is capable of reaching the desired control state. As shown in figure 4(a), the absorption power of suspension system is mainly obtained from the output power of engine, after the dissipation of internal-resistance of actuator, it is absorbed by electro-magnetic damping of actuator to produce the control force, then the power flows towards three directions, one is dissipated by passive damping of energy-regenerator and actuator, another is absorbed by electro-magnetic damping of energy-regenerator, then part of the second power is dissipated by the internal-resistance of energy-regenerator, the other part is taken as output power of suspension system, the remained power is output by the electro-magnetic suspension system to change the kinematic states and control the vibration states.

It is attenuatable active control when the road excitation is dominating the kinematic states of NEAS, which indicates that the NEAS is incapable of reaching the desired control state, it could only reduce the deterioration tendency of vehicle vibration. As shown in figure 4(b), the power flow of attenuatable active control state is similar to passive energy-harvesting mode of NEAS, which doesn't consume the power of source, but the absorption power of electro-magnetic damping of actuator is dissipated by internal-resistance and load-resistance, which can't be recovered. Thus, the required power of NEAS while in active-control mode can be expressed as

$$P_{cn} = \begin{cases} |P_{ca}| + P_{dim} & F_{an} \cdot (\ddot{x}_s - \ddot{x}_t) > 0 \\ 0 & F_{an} \cdot (\ddot{x}_s - \ddot{x}_t) \le 0 \end{cases} \tag{18}$$

$$P_{ca} = F_{an}(\dot{x}_s - \dot{x}_t) \tag{19}$$

$$P_{dim} = I_{lm}^2 R_{dim} \tag{20}$$

The energy harvesting power of NEAS while in active-control mode is shown as

$$P_{ro} = P_{reg} - P_{dig} \tag{21}$$

$$P_{reg} = c_{egn}(\dot{x}_s - \dot{x}_t)^2 \tag{22}$$

$$P_{dig} = I_{lg}^2 R_{dig} \tag{23}$$

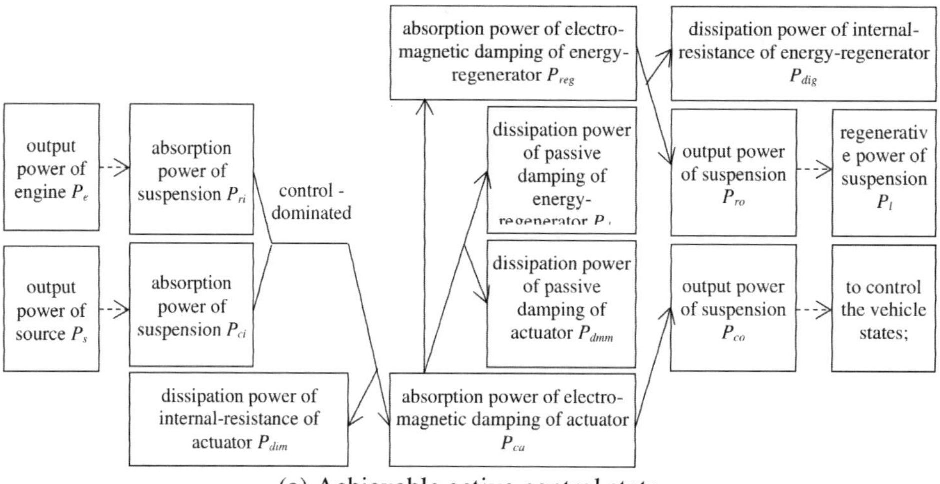

(a) Achievable active control state

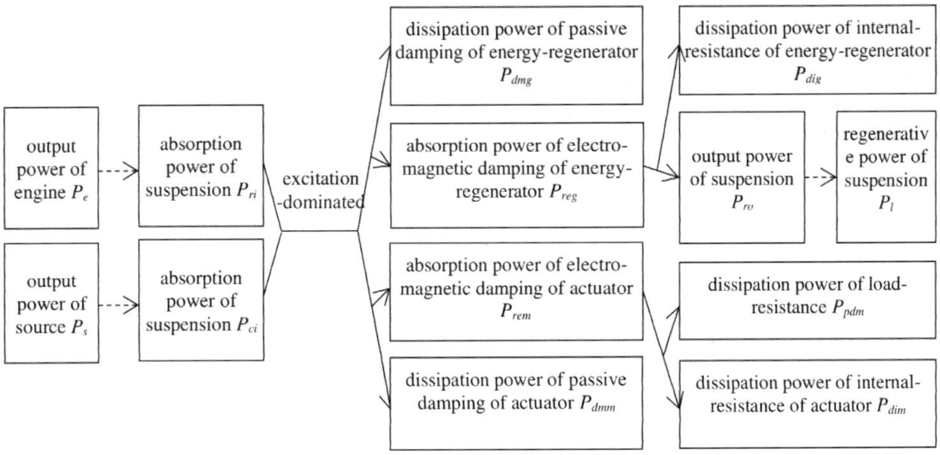

(b) Attenuatable active control state

Figure 4. The power flow of NEAS in active-control mode.

4.2. Evaluation criteria of performance of energy harvesting

Four evaluation criteria were presented to provide uniform standard for the analysis and calculation on performance of energy harvesting of suspension system based on the operating mode and power flow analysis method.

(1) Potentiality of Energy Harvesting

Define the ratio of average output power $\overline{P}_{pro}$ to average absorption power $\overline{P}_{pri}$ while the suspension system is operating in passive energy-harvesting mode as the potentiality of energy harvesting (PER), which is shown as

$$\eta_p = \frac{\overline{P}_{pro}}{\overline{P}_{pri}} \times 100\% \tag{24}$$

(2) Effectiveness of Energy Harvesting

Define the ratio of average output power of energy harvesting $\overline{P}_{ro}$ while the suspension system is operating in active-control mode to average output power $\overline{P}_{pro}$ while the suspension system is operating in passive energy-harvesting mode as the effectiveness of energy harvesting (EER), which is shown as

$$\lambda_e = \frac{\overline{P}_{ro}}{\overline{P}_{pro}} \tag{25}$$

(3) Self-power Coefficient

Define the ratio of average output power of energy harvesting $\overline{P}_{ro}$ to average required power of active control $\overline{P}_{cn}$ while the suspension system is operating in active-control mode as the self-power coefficient, which is shown as

$$\gamma = \frac{\overline{P}_{ro}}{\overline{P}_{cn}} \tag{26}$$

(4) Average Output Power of Energy Harvesting

As relative evaluation criteria, the PER, EER, self-power coefficient could evaluate the relative performance of energy harvesting of suspension system, which couldn't reflect the absolute performance of energy harvesting. For this reason, the average output power of energy harvesting $\overline{P}_{ro}$

was taken as an absolute evaluation criterion to evaluate the absolute performance of energy harvesting of suspension system.

5. Simulation and Analysis of Performance of Active Control

The variation tendencies of best fitness and average fitness of the improved genetic algorithm in optimizing the weight coefficients of LQR active control algorithm are presented in figure 5. The figure illustrates that: 1) as the iteration runs, the best fitness of chromosome of improved genetic algorithm is monotonically decreasing, which means the improved genetic algorithm is valid and effective in optimizing the weight coefficients of LQR controller; 2) the best fitness of chromosome that the improved genetic algorithm generated is 2.575, the corresponding chromosome is (1,20000,100). Thus, the weight coefficients of LQR controller could be $q_1 = 1$, $q_2 = 20000$, $q_3 = 100$.

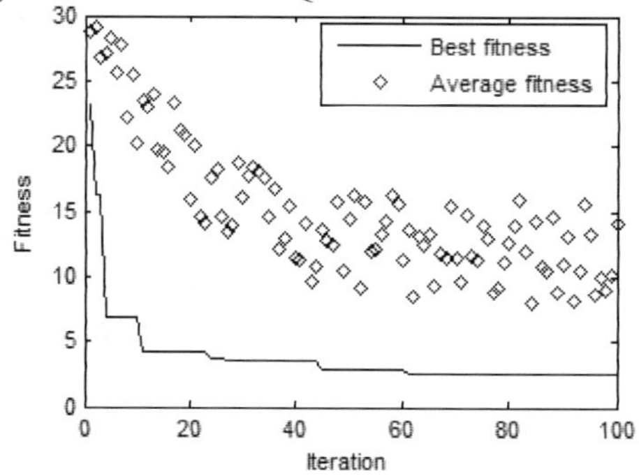

Figure 5. Variation tendencies of fitness of improved genetic algorithm.

5.1. Simulation in time domain

The running performance and the response characteristics of the suspension system with LQR active control algorithm which is optimized by improved genetic algorithm (LQR-IGA) are calculated and analyzed both in time domain and frequency domain with the comparisons of the corresponding passive suspension system (PASSIVE) and the suspension system with LQR active control algorithm of which the weight coefficients are based on the experiment of designer (LQR-EX). The weight coefficients of LQR-EX is given as $q_1 = 1$, $q_2 = 40000$, $q_3 = 500$.

The effects on vibration characteristics of suspension system of vehicle with LQR-IGA active control algorithm are analyzed. Considering the scenario that the vehicle is running on the level C road with the constant speed of 10 m/s last about 40 seconds. The comparisons on sprung mass acceleration, dynamic tyre deformation, dynamic deformation of suspension in time domain among PASSIVE, LQR-EX, LQR-IGA suspension system are shown in figure 6, figure 7 and figure 8, the corresponding states data of the response characteristics and their improvements are given in table 1. The figures and the data in Table 1 manifest that: 1) the sprung mass acceleration of suspension system with LQR-IGA control algorithm is much smaller than that of passive suspension system, that is a little smaller than the suspension system with LQR-EX control algorithm, which indicates that the riding comfort of the vehicle with LQR-IGA active control is the best of all, the riding comfort was improved by 52.66% compared with passive suspension system of vehicle; 2) the dynamic tyre deformation of suspension system with LQR-IGA control algorithm is a little smaller than that of passive suspension system, while that with LQR-EX control algorithm is a litter bigger than that of passive suspension system, which means that the handling stability of the vehicle with LQR-IGA active control is the best of all, the handling stability was improved only by 5.70% compared with passive suspension system of vehicle; 3) the dynamic deformation of suspension system with LQR-IGA control algorithm is larger than that of passive suspension system and that with LQR-EX control algorithm, but the dynamic deformation of suspensions in the 3 kinds of suspension system are all smaller than the permitted dynamic deformation (0.1 m), which indicates that the riding comfort and the handling stability of suspension system with LQR-IGA wouldn't be worsened or deteriorated according to the dynamic deformation of suspension.

IOP Conf. Series: Materials Science and Engineering **787** (2020) 012011 doi:10.1088/1757-899X/787/1/012011

Table 1. Suspension state and its improvement

Suspension state	State value and its improvement				
	PASSIVE	LQR-EX		LQR-IGA	
	State value	State value	Improvement	State value	Improvement
Sprung mass acceleration	1.468	0.815	44.48%	0.696	52.66%
Dynamic tyre deformation	0.00298	0.00317	-6.38%	0.00281	5.70%
Dynamic deformation of suspension	0.00676	0.00702	-3.85%	0.00783	-15.78%

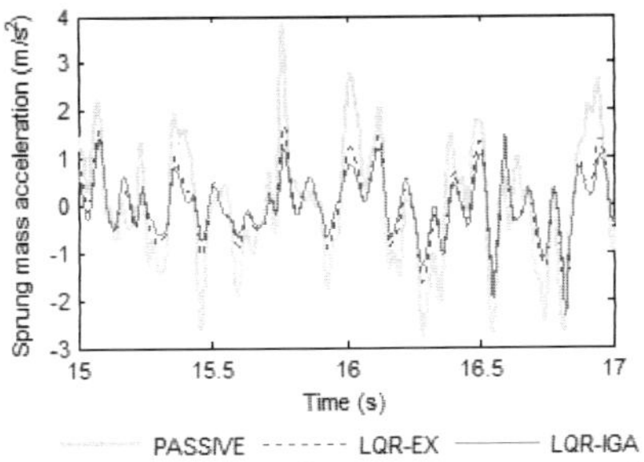

Figure 6. Comparison of sprung mass acceleration.

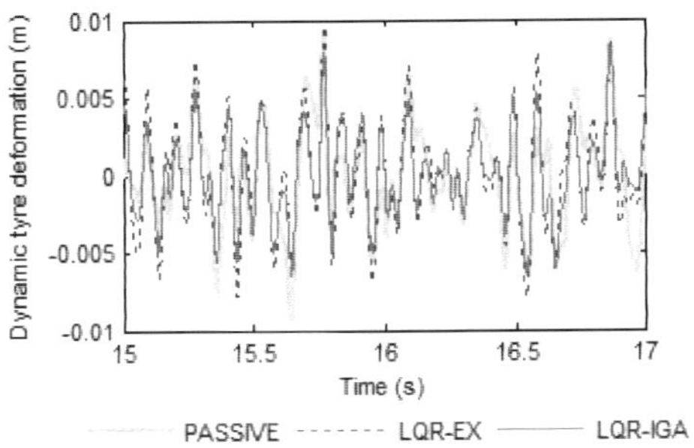

Figure 7. Comparison of dynamic tyre deformation.

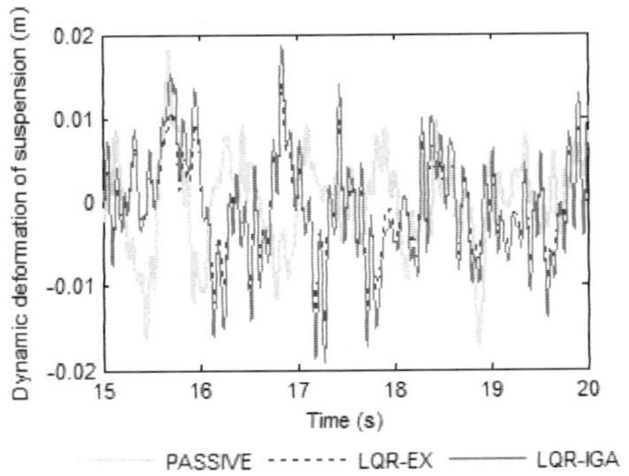

Figure 8. Comparison of dynamic deformation of suspension.

5.2. Simulation in frequency domain

To make a further understanding about the effect of LQR-IGA active control algorithm on suspension system of vehicle, the comparisons of the PSD (power spectrum density) of sprung mass acceleration, dynamic tyre deformation, dynamic deformation of suspension in frequency domain are discussed and analyzed. With the adoption of the scenario of vehicle in the former section, the PSD of sprung mass acceleration, dynamic tyre deformation, dynamic deformation of suspension on LQR-IGA with the comparisons of PASSIVE and LQR-EX are revealed in figure 9 figure 10 and figure 11. The figures illustrate that: 1) in all but high resonance frequency range, the PSD of sprung mass acceleration of

IOP Conf. Series: Materials Science and Engineering **787** (2020) 012011 doi:10.1088/1757-899X/787/1/012011

LQR-EX and LQR-IGA are smaller than that of PASSIVE, the PSD of sprung mass acceleration of LQR-IGA is smaller than that of LQR-EX in low frequency range, low resonance frequency range and mid-frequency range; 2) the riding comfort of both the suspension systems of vehicle with LQR-EX and LQR-IGA is better than that with PASSIVE, the riding comfort of suspension system with LQR-IGA is a little better than that with LQR-EX; 3) the PSD of dynamic tyre deformation of LQR-EX and LQR-IGA are smaller than that of PASSIVE in low frequency range and low resonance frequency range, while that are bigger than PASSIVE in high resonance frequency range, PSD of dynamic tyre deformation of LQR-IGA is bigger than that of LQR-EX in low resonance frequency range and that is smaller than LQR-EX in high resonance frequency range; 4) the PSD of dynamic deformation of suspension system with LQR-EX and LQR-IGA are both improved in low resonance frequency range and deteriorated in low frequency range and high resonance frequency range compared with PASSIVE; the PSD of dynamic deformation of suspension system with LQR-IGA is bigger than that of LQR-EX in low frequency range, low resonance frequency range and high resonance frequency range.

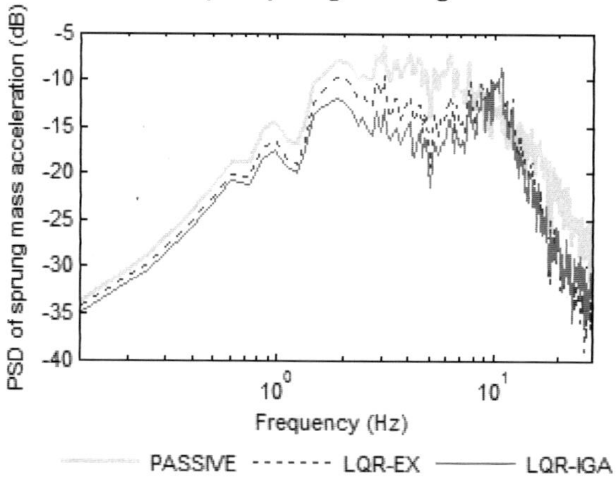

Figure 9. Comparison of the PSD of sprung mass acceleration.

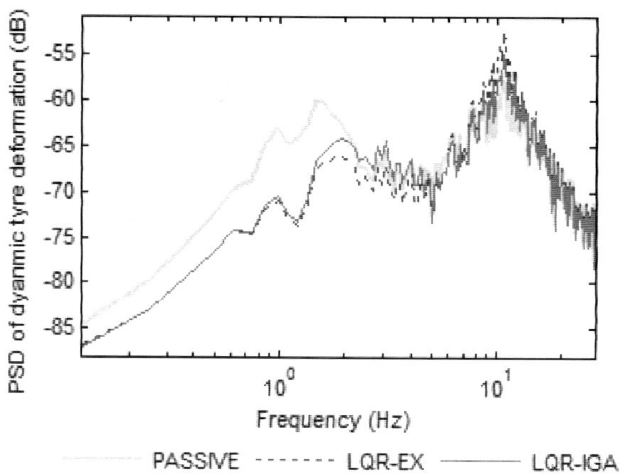

Figure 10. Comparison of the PSD of dynamic tyre deformation.

IOP Conf. Series: Materials Science and Engineering **787** (2020) 012011 doi:10.1088/1757-899X/787/1/012011

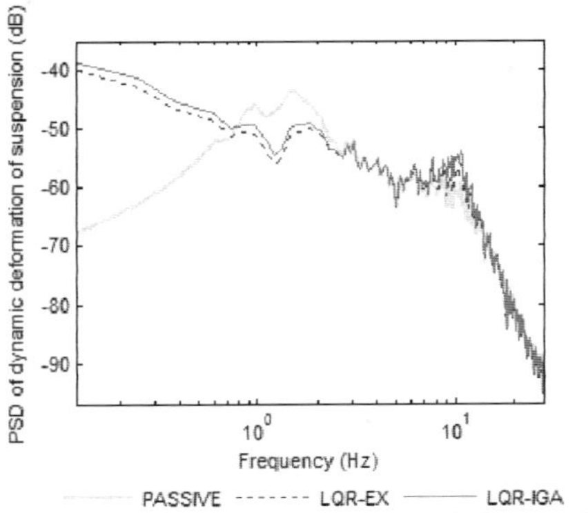

Figure 11. Comparison of the PSD of dynamic deformation of suspension.

6. Simulation and Analysis of Performance of Energy Harvesting

Two types of energy-regenerative suspension system, which are novel energy-regenerative passive suspension (passive energy-harvesting mode of NEAS) and novel energy-regenerative active suspension (active-control mode of NEAS), were presented to analysis the performance of energy harvesting.

6.1. Performance of energy harvesting with standard road excitation

The bar charts of PER, EER, self-power coefficient and average output power of energy harvesting of NEAS on D level standard road with different vehicle speed are shown in figure 12. The PER of NEAS remains constant with the varying of vehicle speed is revealed in figure 12(a), which indicates that the PER is only associated with the suspension system itself; figure 12(b) illustrates that the EER decrease gradually as the vehicle speed rising; the EER is greater than 1 as the recovered energy of suspension system in active-control mode is more than that in passive energy-harvesting mode, it is because the vibration of sprung mass is reduced while the active control algorithm working, then the relative movement between sprung mass and unsprung mass is getting more fierce, thus more energy are recovered as the suspension system operating in active-control mode; figure 12(c) suggests that the self-power coefficient increase gradually as the vehicle speed rising; the self-power coefficient is greater than 1 as the recovered energy of suspension system in active-control mode is more than the required energy of active control, the entire self-sustaining of power supply is realized; figure 12(d) shows that the average output power of energy harvesting increase gradually as the vehicle speed rising.

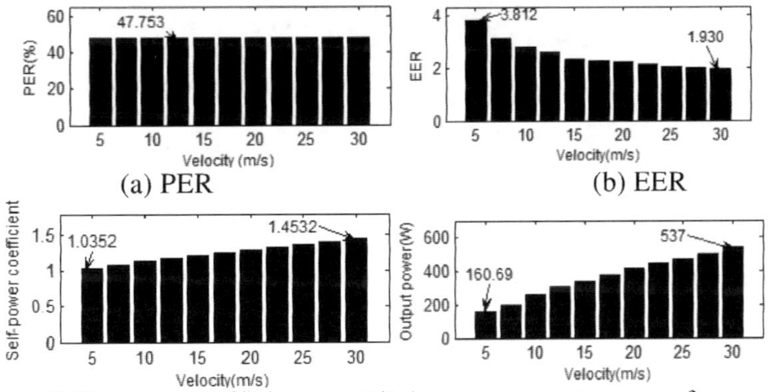

(c) Self-power coefficient (d) Average output power of energy harvesting

Figure 12. The performance of energy harvesting of NEAS on D level standard road with different speed.

6.2. Performance of energy harvesting with sine road excitation

The variation tendency of PER of NEAS under sine road excitation with frequency from 0.5Hz to 15Hz and amplitude from 0.005m to 0.05m is shown in figure 13. The PER is invariable as the alternation of the frequency and amplitude of road excitation, which indicates that the PER of NEAS is only associated with the suspension system itself, independent of road excitation. The effects of parameters of suspension system on PER are discussed in reference [22-23].

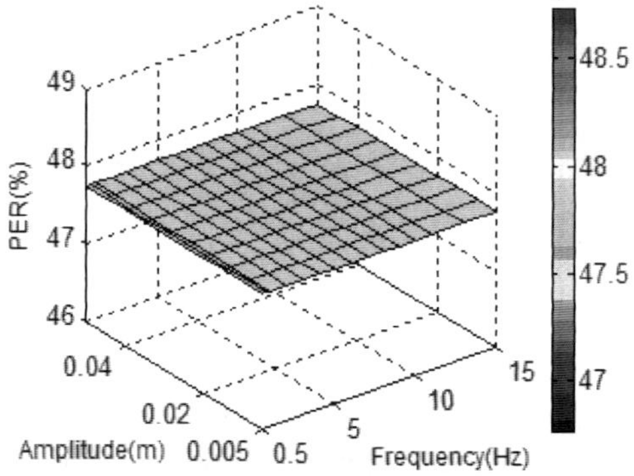

Figure 13. The PER of NEAS under sine road excitation.

The variation tendency of EER of NEAS under sine road excitation with different frequency and amplitude is shown in figure 14, a reference plane, on which the EER is 1, is also presented in the drawing. It manifests that the satisfactory control effectiveness could be obtained with the LQR control algorithm in low frequency range, where high EER could be derived correspondingly; the case that the EER is lower than 1 occurred in medium frequency range for the reason that the variation of relative speed between sprung mass and unsprung mass of NEAS in active-control mode is similar with that in passive energy-harvesting mode, while only the energy-regenerator could recover energy in active-control mode, hence the EER is getting lower.

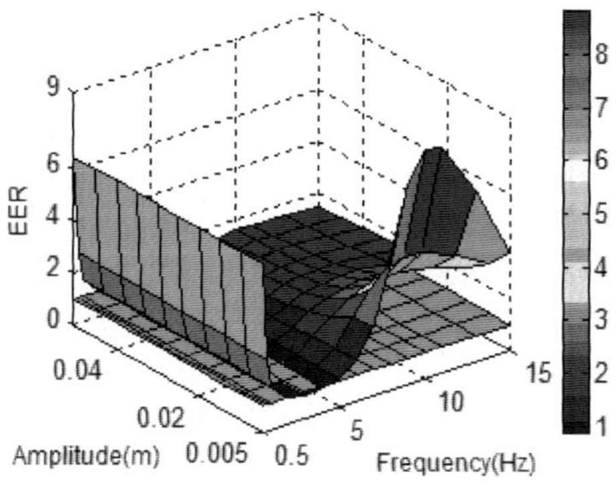

Figure 14. The EER of NEAS under sine road excitation.

The variation tendency of self-power coefficient of NEAS under sine road excitation with different frequency and amplitude is shown in figure 15, a reference plane, on which the self-power coefficient is 1, is also presented in figure 15. It illustrates that the self-power coefficient is independent of the frequency and amplitude of road excitation in low frequency range, which comes out to be smaller than 1, the system needs the supplement of external power; the self-power coefficient increases as the amplitude rises in the medium and high frequency range, the tendency of the increasing becomes more obvious as the frequency gets higher, moreover, the entire self-sustaining of power supply is realized; the self-power coefficient increases as the frequency rises, reaches to its maxima in the higher resonance frequency range, then getting lower, the tendency of rise then fall of self-power coefficient with frequency is more obvious as the amplitude gets higher; the entire self-sustaining of power supply could be realized as the frequency of road excitation higher than 1.72Hz when compared to the reference plane; the self-power coefficient becomes smaller than 1 in a small range before it rising notably.

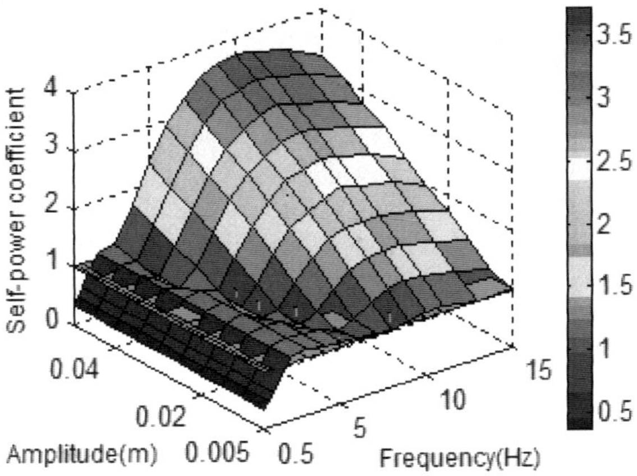

Figure 15. The self-power coefficient of NEAS under sine road excitation.

The variation tendency of average output power of energy harvesting of NEAS under sine road excitation with different frequency and amplitude is shown in figure 16. Conclusions could be made on this figure that the average output power of energy harvesting increase as the amplitude rise, the tendency of the increasing becomes more obvious as the frequency gets higher; the average output power of energy harvesting increases as the frequency rises, reaches to its maxima in the higher resonance frequency range, then getting lower, the tendency of rise then fall of average output power of energy harvesting with frequency is more obvious as the amplitude gets higher. The effects of other parameters of suspension system on the average output power of energy harvesting are discussed in reference [24].

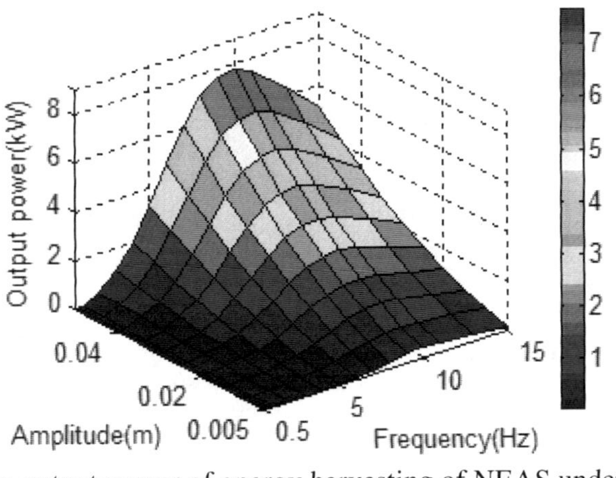

Figure 16. The average output power of energy harvesting of NEAS under sine road excitation.

7. Conclusions

A NEAS system was designed with an electro-magnetic actuator, which includes actuator and energy-regenerator, to realize the simultaneous implementation of energy harvesting and active control. A LQR controller with an improved genetic algorithm in optimizing the weight coefficients of LQR active control algorithm was designed to attenuate and reduce the vibration of suspension system of vehicle. The flowing and transforming processes of power with the corresponding computational formula were provided based on the power flow method. Four evaluation criteria of performance of energy harvesting of suspension system were presented to establish the uniform standard on quantify and analyse the performance of energy harvesting of suspension system.

The running performance of suspension system of vehicle with genetic-algorithm-optimized LQR controller are improved notably, of which the riding comfort is improved by 52.66% and the handling stability is improved by 5.70% in comparison of the passive suspension system. The frequency analysis shows that the improvement on riding comfort is effective in all but high resonance frequency range, the improvement on handling stability is valid in low frequency range and low resonance frequency range.

The PER only depends on the suspension system itself, it's exactly determined by the performance of energy harvesting device in suspension system; The EER, the self-power coefficient, the average output power of energy harvesting are related to the suspension system and the active control algorithm worked on it. High-efficiency active control algorithm, therefore, should be designed to improve the three evaluation criteria effectively. The simulation shows the variation tendencies of the evaluation criteria, but some basic data, such as the average output power of energy harvesting in the higher resonance frequency range, get too high with the ignorance of the limits of the ride comfort and the dynamic deflection of the suspension. The analyses of the areas, that the EER is lower than 1 and the self-power coefficient is smaller than 1, are insufficient. All the three problems will be studied in the future research.

References

[1] Karnopp D. (1978) Power requirements for traversing uneven roadways. Vehicle System Dynamics, 7: 135-152.
[2] Karnopp D. (1992) Power requirement for vehicle suspension system. Vehicle System Dynamics, 21: 65-71.
[3] Nakano K., Suda Y., Nakadai S., et al. (1999) Self-powered active control applied to a truck cab suspension. JSAE Review, 20: 511-516.
[4] Nakano K., Suda Y., Yamaguchi M., et al. (2004) Application of combined type self-powered active suspensions to rubber-tired vehicles. The Japan Society of Mechanical Engineers, Series C, 70: 152-158.
[5] Jones W. (2005) Easy ride: Bose Corp. uses speaker technology to give cars adaptive suspensions. IEEE Spectrum, 42: 12-14.
[6] Fan Y., Min C., Xuechun Z. (2005) Research on the feasibility of vehicle active suspension with energy regeneration. Journal of Vibration and Shock, 24: 27-30.
[7] Kun H., Fan Y., Yongchao Z. (2011) Active control of energy-regenerative electromagnetic suspension based on energy flow analysis. Journal of Shanghai Jiaotong University, 45: 1068-1073.
[8] Changmiao Y., Weihua W., Qingnian W. (2010) Damping characteristic and its influence factors in energy regenerative suspension. Journal of Jilin University (Engineering and Technology Edition), 40: 1482-1486.
[9] Changmiao Y., Weihua W., Qingnian W. (2009) Analysis of energy-saving potential of energy regenerative suspension system on hybrid vehicle. Journal of Jilin University (Engineering and Technology Edition), 39: 841-845.
[10] Pengyun S., Jiye Z., Yanqiu C., et al. (2013) Self-powered active control system design based on energy analysis. Journal of Dynamics and Control, 11: 82-88.
[11] Dongyun C., Likang Y., Minglong C. (2014) Analysis of energy-harvesting active suspension based on certain riding comfort. Journal of Mechanical & Electrical Engineering, 31: 289-294.
[12] Kun H., Yongchao Z., Fan Y., et al. (2009) Coordinate optimization for synthetical performance of electrical energy-regenerative active suspension. Journal of Shanghai Jiaotong University, 43: 226-230.

[13] Jinping O. (2003) Structural vibration control, active, semi-active and intelligent control. Science Press, pp. 61-68.
[14] Fei L. (2010) Investigation of vehicle active suspension based upon electromagnetic actuator. Chongqing University.
[15] Andries P. (2009) Fundamentals of computational swarm intelligence. TAN Y., et al, Translation. Tsinghua University Press, pp. 42-44.
[16] Weizhong H., Guoqin G. (2003) Design of weighting matrix for optimal controller based on genetic algorithm. Computer Measurement & Control, 11: 761-762.
[17] Jie Y., Jinqiu Z., Zhizhao P., et al. (2015) Optimization of electromagnetic suspension LQR control for military tracked vehicle using genetic algorithm. Machinery Design & Manufacture, 288: 229-232.
[18] Guosheng Z., Zongde F., Aimin L., et al. (2007) Optimal control of the active suspension based on the genetic algorithm. China Mechanical Engineering, 18: 1491-1495.
[19] Kehu X., Dashan H., Tianzhao W. (2013) Improved artificial immune algorithm for solving weapon-target assignment problems. Systems Engineering and Electronics, 35: 2121-2127.
[20] Xu L., Ming H. (2011) Modern intelligent optimization mixed algorithm and application. Publishing House of Electronics Industry, pp. 39-43.
[21] Fan Y., Yonchao Z. (2010) Technology of regenerative vehicle active suspensions. Transactions of the Chinese society for agricultural machinery, 41: 1-6.
[22] Zhigang F., Xuexun G., Lei Z., et al. (2013) Potential study and sensitivity analysis of energy-regenerative suspension. Journal of Jiangsu University (Natural Science Edition), 34: 373-377.
[23] Zhigang F. (2013) Energy harvesting theory and damping characteristic research of vehicle hydraulic electromagnetic shock absorber. Wuhan University of Technology.
[24] Songshan L. (2013) Research on damping characteristics of electromagnetic regenerative suspension. Jilin University.

IOP Conf. Series: Materials Science and Engineering **787** (2020) 012012 doi:10.1088/1757-899X/787/1/012012

A decomposition algorithm to solve dynamic city-scale ridesharing problem

H Wang, W X Wu* and L N Qin

School of electrical and control engineering, north China University of technology, Beijing, 100144, China

*Corresponding author's e-mail: wuwenxiang@ncut.edu.cn

Abstract. This paper proposes a computational efficient, rapid, dynamic ridesharing algorithm. Firstly, a designed pre-processing procedure reduces size of the input date of potential nodes for vehicle to pass through. The proposed decomposition algorithm decomposes the large-scale problem with pre-processing date into independent sub-problems, and the method of parallel computing is adopted to solve the sub-problems. We call the dynamic programming algorithm to deal with the vehicle routing problem involved in sub-problems. Greedy random adaptive search algorithm is compared with the decomposition algorithm to verify the algorithm. Finally, a mathematical example shows that the algorithm can effectively solve large-scale dynamic ridesharing problems.

1. Introduction

With the worsening of environmental problems, road congestion, and the high cost of vehicle owners, people recently focus on the empty load rate of vehicles [1]. Ride-sharing can provide travelers with a low-cost and convenient way to travel, leading to reduce emissions, congestion, etc. [2][3]

There is a variety of literature concerning ridesharing problems. This problem is an NP problem, so it cannot be solved by precise algorithm [4]. Therefore, it is a challenging problem to design an effective algorithm that can solve such a large-scale, real-time and complex NP problem in order to realize the real-time requirement of dynamic ridesharing [5~18]. Thus, the goal of this paper is to design an algorithm to solve the problem.

When the scale of a problem is huge, an approximate solution or a heuristic solution can get the desired result [19]. For example, Bian (2019) has come up with a heuristic algorithm to solve the first-mile ridesharing, called the solution pooling approach (SPA). The idea is to build a model for every passenger, and solve the model in a pool. However, because of many models to solve, the algorithm takes a long time and is not applicable in large-scale networks. Masoud (2017) proposed a decomposition algorithm to solve the Multi-Hop peer-to-peer ride-sharing problem. This algorithm needs to calculate the possible routing of each participant. For large-scale network, the routing of participant is extremely huge, so it cannot solve the ridesharing problem of large-scale road network. At the same time, this paper could not plan the routing of the vehicles.

The current research has not made a good solution for large-scale dynamic ridesharing problem. How to make reasonable matching between vehicles and passengers under the condition of large-scale network and make effective routing planning for vehicles is the key point of this paper. The core of our decomposition algorithm is to decompose large-scale problems into small sub-problems, which are independent of each other and can be solved in parallel. In the planning routing stage of decomposition algorithm, we use dynamic programming algorithm to solve. In order to reduce the number of nodes that

IOP Conf. Series: Materials Science and Engineering **787** (2020) 012012 doi:10.1088/1757-899X/787/1/012012

vehicle may pass through, we preprocess the data before decomposition algorithm. Mathematical example shows that our algorithm can solve the ridesharing problem of large-scale networks.

This paper is structured as follows. In section 2, the model of dynamic multiplication in a time interval is established. Section 3 preprocesses the input data to reduce the size of the model. We design a decomposition algorithm to decompose large-scale problems into small sub-problems in section 4. In section 5, the dynamic programming algorithm is applied to solve the vehicle routing problem, and as a comparison, we design another algorithm called greedy random adaptive search problem (GRASP) algorithm in section 6. Finally, a numerical example is proposed in section 7 to verify the algorithm.

2. ridesharing system

We consider a ride-sharing system that including matching and routing planning. We build a model to pursue the highest matching success rate. We have noticed that different objective functions can be set to achieve the different objectives, including minimizing the system cost, minimizing the travel time and so on. In order to show the effectiveness of the algorithm, we set the maximum matching rate.

After discretization of continuous time, in each time interval, the ridesharing is a static problem, and we build a static model for each time interval. At the same time, we know that there is no absolute advantage in the long run between one vehicle matching one passenger and one vehicle matching many passengers in a time interval [20]. In this article, we address the first case, where one vehicle matches one passenger in a time interval.

Table 1. Notations

Notations	Descriptions
G	Road network, $G=(N,A)$
N	Set of points in the road network
A	Set of links in the road network
i	Passenger requests
I	Set of passenger requests, $I=\{1,2,...,i,...,B\}$
p_i	Number of passenger request i
$\overline{s}_i$	Departure point of passenger request i, $\overline{r}_i \in N$
SI	Set of departure points of all passenger requests
$\overline{r}_i$	Arrival point of passenger request i, $\overline{r}_i \in N$
RI	Set of arrival points of passenger points
$\overline{e}_i$	The earliest departure time for passenger request, i
$\overline{l}_i$	The latest arrival time for passenger request, i
k	vehicle
K	Set of vehicles, $K=\{1,2,...,k,...,M\}$
$\underline{s}_k$	Departure point of vehicle k, $\underline{s}_k \in N$
SK	Set of departure points of all vehicles
$\underline{r}_k$	Arrival point of vehicle k, $\underline{r}_k \in N$
RK	Set of arrival point of all vehicles
$\underline{e}_k$	The earliest departure time of vehicle k
$\underline{l}_k$	The latest arrival time of vehicle k
q_k	The seat capacity of vehicle k
VI	The union of the departure points and arrival points set of all passenger requests, $VI=SI \cup RI$, $VI \subset N$
VK	The union of the departure points and arrival points set of all passenger request, $VK=SK \cup RK$, $VK \subset N$
W	The union of the departure points and arrival points set of all passenger request and vehicles, $W=VI \cup VK$, $W \subset N$

IOP Conf. Series: Materials Science and Engineering **787** (2020) 012012 doi:10.1088/1757-899X/787/1/012012

d_{ab}	The shortest distance form point $a \in N$ to point $b \in N$
t_{ab}	The shortest travel time for vehicle form point $a \in N$ to point $b \in N$

Table 2. Notations of variables

Notations	Description
L_k	The routing of vehicle k
x_{kab}	$x_{kab} = 1$,if vehicle k travels to location b after location a immediately, otherwise, $x_{kab} = 0$, $a,b \in SI$
LN_{kb}	The vehicle just left the capacity of the node, $b \in W$
T_{ka}	The time when the vehicle arrives at the node a, $a \in N$

The problem can be formulated as the following model. For the notations, please refer to table 1 and 2.

$$\max \sum_{i \in I} \sum_{k \in K} x_{k\bar{s}_i b} \tag{1}$$

Subject to

$$\sum_{w \in W} x_{kw\underline{r}_k} = 1 , \quad \forall k \in K \tag{2}$$

$$\sum_{w \in W} x_{k\underline{s}_k w} = 1 , \quad \forall k \in K \tag{3}$$

$$\sum_{k \in K} \sum_{w \in W} x_{k\bar{s}_i w} \leq 1 , \quad \forall i \in I \tag{4}$$

$$\sum_{i \in I} \sum_{w \in W} x_{k\bar{s}_i w} \leq 1 \tag{5}$$

$$\sum_{a \in W} x_{kaw} - \sum_{b \in W} x_{kwb} = 0 , \quad \forall w \in W / (\underline{s}_k \cup \underline{r}_k) , \quad \forall k \in K \tag{6}$$

$$x_{kab} \in \{0,1\} , \quad \forall k \in K , \quad \forall a \in W , \quad \forall b \in W \tag{7}$$

$$x_{kab} = 1 \Rightarrow T_{kb} = T_{ka} + t_{ab} , \quad \forall k \in K , \quad \forall a \in W , \quad \forall b \in W \tag{8}$$

$$(\bar{l}_i - T_{k\bar{r}_i}) x_{kw\bar{r}_i} \geq 0 , \quad \forall i \in I , \quad \forall k \in K \tag{9}$$

$$\underline{e}_k \leq T_{k\underline{s}_k} \leq T_{k\underline{r}_k} \leq \underline{l}_k , \quad \forall k \in K \tag{10}$$

$$x_{ka\bar{s}_i} = 1 \Rightarrow LN_{k\bar{s}_i} = LN_{ka} + p_i , \quad \forall k \in K , \quad \forall a \in W , \quad \forall i \in I \tag{11}$$

$$x_{ka\underline{r}_i} = 1 \Rightarrow LN_{k\underline{r}_i} = LN_{ka} - p_i , \quad \forall k \in K , \quad \forall a \in W , \quad \forall i \in I \tag{12}$$

$$0 \leq LN_{kb} \leq q_k , \quad \forall k \in K , \quad b \in W \tag{13}$$

$$LN_{k\underline{s}_k} = LN_{k\underline{r}_k} = 0 , \quad \forall k \in K \tag{14}$$

Formula (1) is the objective function that maximum matching rate. Formulas (2) and (3) represent that the passenger's departure point and arrival point must be in the vehicle routing. Formula (4) represents that one passenger can only be matched to one car, i.e., no transfer. Formula (5) represents that one vehicle can only match one passenger at most. Equation (6) represents the flow balance constraint. Formula (7) represents that the decision variables can only take 0 and 1. Formulas (8), (9) and (10) represent time window constraints. Formulas (11), (12), (13) and (14) represent vehicle load constraints.

3. pre-processing procedure

The goal of the pre-processing procedure is to build a set, G_k, for each vehicle, $k \in K$, which is a set of points that the vehicle can potentially arrive. It must be noted that the potential arrive point of the vehicle not only satisfies the time window constraint, but also satisfies the time window constraint of the passengers inside of the vehicle.

As explained before, we present two points in time, e and l. e is the earliest departure time of vehicle k from point a. l is the latest arrive time of vehicle k to point b. Our detailed definitions of e and l are shown in the following table.

Table 3. The definition of e and l

a is the departure point of vehicle k. b is the arrival point of vehicle k.	$e = \underline{e}_k$	$l = \underline{l}_k$
a is the departure point of vehicle k. b is the arrival point of passenger i.	$e = \underline{e}_k$	$l = \overline{l}_i$
a is the departure point of vehicle k. b is the departure point of passenger i.	$e = \underline{e}_k$	$l = \overline{l}_i - t_{ab}$
a is the departure point of passenger i. b is the arrival point of passenger i.	$e = \overline{e}_i$	$l = \overline{l}_i$
a is the arrival point of passenger i. b is the arrival point of vehicle k.	$e = \overline{e}_i$	$l = \underline{l}_k$

We define a ellipse in the network. The departure and arrival points, (a, b), travel time windows of participants can be used as the focal of the ellipse and long axis of the ellipse. According to the definition, the sum of the distances from two focal points of the ellipse, that is, the departure point and arrival point of the vehicle in the section, to any point on the ellipse satisfies the corresponding time windows constraint.

Table 4. Reduced graph algorithm

algorithm 1 reduced graph algorithm

01 **step1. initialize**
02 $W = \{\overline{s}_i, \overline{r}_i\}$, $\forall i \in I$
03 $G_k = W$, $\forall k \in K$
04 $x_{k\underline{s}_k\underline{r}_k} \leftarrow 1$, $\forall k \in K'$
05 **step2. update or initialize reduced graph for vehicles**
06 for each vehicle $k \in K$
07 $\quad G'_k = \varnothing$
08 $\quad$ for each $x_{kab} = 1$, $\forall a \in L_k$ $\forall b \in L_k$
09 $\quad\quad$ e is the earliest departure time of vehicle k from point a
10 $\quad\quad$ l is the latest arrive time of vehicle k to point b
11 $\quad\quad$ for $w \in G_k$
12 $\quad\quad\quad$ if $t_{wa} + t_{wb} \le l - e$
13 $\quad\quad\quad\quad$ $G'_k = G'_k \cup \{w\}$
14 $\quad\quad\quad$ end if
15 $\quad\quad$ end for
16 $\quad$ end for
17 $\quad G_k = G'_k$
18 end for

As shown in the figure 1, all points in the shaded part are potential points that can be reached by vehicles. The set of these points is G_k.

We start the algorithm by defining a set W that contains all passenger's departure and arrival points (line 02), and initializing the set G_k that each vehicle can potentially reach as W (line 03). For each vehicle $k \in K$, initialize G'_k as an empty set (line 05-06). According to the results of the last interval iteration, for each $x_{kab} = 1$ (line 07), We look for points that satisfy the formula, $t_{wa} + t_{wb} < l - e$, which are potentially reachable by vehicles(line 08-14). Finally, for each vehicle, we set $G_k = G'_k$.

4. decomposition algorithm

The essence of decomposition algorithm shown table 4 is to decompose a large-scale network into small sub-problems, each of which needs less time and space and is easier to solve.

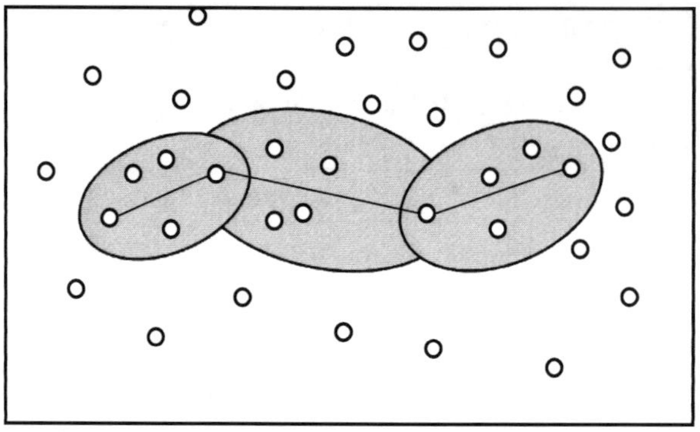

Figure 1. The reduced graph algorithm

We start to decompose the algorithm by setting sub-problem counters, *n*, to 0, initializing the set of sub-problems as an empty set (line 02). We set each sub-problem consists of two parts, the first part is a passenger, the second is a collection of vehicles, the collection of potential arrival points of the vehicle includes the departure point and arrival point of the passenger. For each passenger, we put that passenger in the first part of the sub-question (line 05). For each vehicle, we look for whether the passenger's departure point and arrival point are in the vehicle's set of potential arrival points (line 06-09). We create a sub-problem for that passenger (line 11). If the set of vehicles corresponding to this sub-problem is not an empty set, set act=1 (line 12-14). For the routing planning problem of a sub-problem with passengers and vehicles selected in turn, we adopt the method of dynamic programming, which is introduced in detail in the following text (line 17-20). We notice that the elements of set R_n are the routings that satisfy the constraints in the sub-problem. We set the set Z_m that records the best matching vehicles for each sub-problem to be an empty set (line 22). We set the function obj (L) to be the travel time of routing L. for each sub-problem SP_n with $act(SP_n)=1$, We select the vehicle whose travel time is the minimum travel time of the elements of routing set R_n corresponding to each sub-problem as the solution of the sub-problem (line 24-26). Let Z_m be the set of solutions to all sub-problems. After solving all the sub-problems, we determine whether there is any conflict between the solutions. Since we stipulate that one vehicle can only match one passenger in each time interval, the criterion for judging the conflict is whether a vehicle matches two passengers, that is, whether a vehicle is the solution of two sub-problems. We select the sub-problem whose solution is the shortest travel time of the routing corresponding to the sub-problem of k, take the vehicle as the final solution of this sub-problem, and delete the routing corresponding to the vehicle from the routing solution set of other sub-problems (line 31-33). We set $act(SP_n)=0$ when the vehicle set corresponding to the sub-problem is empty (line 34-36). Output the decision variable according to the routing corresponding to the solution of each sub-problem (line 37-39). The termination criterion is that all $act(SP_n)$ are 0 (line 41-43). If the termination criteria is not met, then go back to step 3.

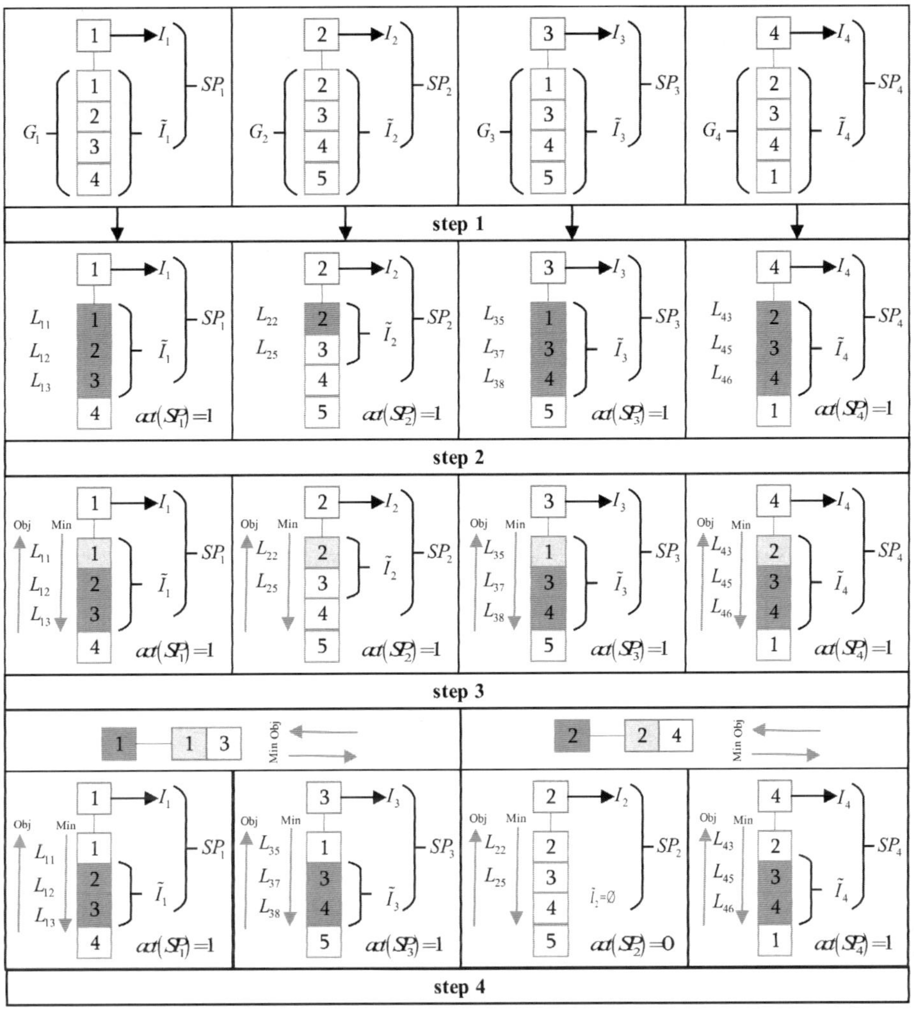

Figure 2. the decomposition algorithm

In the example shown in figure 2, we set 4 sub-problems. The first step establishes four sub-problems. The red box represents the passenger and the green represents the vehicle. Each sub-problem includes a passenger and numbers of passenger-related vehicles. The second step applies dynamic programming to solve the routing problem of passengers and each vehicle. The shaded part represents the vehicle that meets the constraint conditions. In the third step, the solution of each sub-problem is obtained, and the vehicle with the minimum travel time of the routing is selected as the solution of the travel time. The green shade indicates the solution to the sub-problem. In step 4, determine whether there is a conflict. Vehicle 1 and 2 are matched to two passengers. The best passenger is selected to match the vehicle.

5. Dynamic programming algorithm

Dynamic programming algorithm solves the problem of route planning for vehicles and passengers. We define set V to represent the set of points that the vehicle has passed, a to represent the point where the vehicle is in the stage, and V' to represent the node that the vehicle has not passed. Initialization V is a empty set, a is the departure point of the vehicle, V' is the set of the node of the original routing of the vehicle removing the departure point and arrival point of vehicle, add departure point and arrival point of passenger (line 02). Define D_s as the state of the s phase. State D_s of phase 1 is initialized as $\{V, a, V'\}$. The state of the other phases is an empty set (line 03). The purpose of the second step is to find out all the state of each stage. Meanwhile, in the same stage, if the set of nodes through which the vehicles in more than two states have passed and the current node where the vehicles are, they are merged into one state. At the same time, if two or more states have the same set V and the current node a, they are merged into one state. For the status of each s-1 stage, the current node of the vehicle is updated successively. If the current node is the arrival point of the passenger, the departure point of the passenger must not be in the collection of the unarrived points of the vehicle (line 05-14). We define function $D(V_2)$ as the travel time in order of set V. The state corresponding to the smallest set of function values is retained, and other states are deleted (line 15-21). From the last stage, the state transition equation is calculated (line 24-34). Output planning results (line 36-41).

Table 5a. The decomposition algorithm

algorithm 3 the decomposition algorithm

Step 1. initialize and construct sub-problems
$n \leftarrow 0$, $SP=\varnothing$
for each rider $i \in I$
    $n=n+1$
    set $I_n=\{i\}$, $\tilde{I}_n=\varnothing$, $R_n=\varnothing$
    for each driver $k \in K$
        if $(\overline{s}_i,\overline{r}_i) \in G_k^{\;2}$
            $\tilde{I}_n=\tilde{I}_n \cup \{k\}$
        end if
    end for
    $SP_n=\{I_n,\tilde{I}_n\}$, $SP=SP \cup \{SP_n\}$
    if $\tilde{I}_n \neq \varnothing$
        $act(SP_n)=1$
    end if
end for
Step 2. solve the routing problem for each the sub-problem
for each driver $k \in \tilde{I}_n$
    call the dynamic programming algorithm and output L'_k
    $L'_{nk}=L'_k$, $R_n=R_n \cup \{L'_k\}$
end for
Step 3. solve sub-problems
$Z_m=\varnothing$
for each sub-problem SP_n with $act(SP_n)=1$
    if $obj(L'_{nk1})=\min(obj(L'_{nk}))$, $\forall L'_{nk} \in R_n$
        $z_n \leftarrow k1$
    end if
    $Z_m=Z_m \cup \{z_n\}$
end for

IOP Conf. Series: Materials Science and Engineering **787** (2020) 012012 doi:10.1088/1757-899X/787/1/012012

Table 5b. The decomposition algorithm

Step 4. termination criterion and output results

for each driver $k \in Z_m$

$obj(L'_{n1k}) = \min(obj(L'_{nk}))$, $act(SP_n) = 1$

$R_n = R_n / \{L'_{nk}\}$, $\forall n \neq n1$

$L_k = L'_{n1k}$

if $\tilde{I}_n = \varnothing$, $\forall n \neq n1$

$act(SP_n) = 0$

end if

for $j = 1 : length(L'_{n1k}) + 1$

let $x_{kL'_{n1k}(j)L'_{n1k}(j+1)} = 1$

end for

end for

if $\sum_{SP_n \in SP} act(SP_n) = 0$

stop

else

$m = m + 1$, go back to step 3

end

Table 6. Dynamic programming algorithm

algorithm 3 dynamic programming algorithm

Step 1. initialize for rider *i* and driver *k*

$V = \varnothing$, $a = \underline{s}_k$, $V' = L_k \cup \{\bar{s}_i, \bar{r}_i\} / \{\underline{s}_k, \underline{r}_k\}$

$D_1 = \{V, a, V'\}$, $D_s = \varnothing$, $s = 2,3,...,length(L_k) + 2$

Step 2. forward movement

for $s = 2 : length(L_k) + 1$

for each $\{V, a, V'\} \in D_{s-1}$

for $b \in V'_1$

if b is rider $i's$ departure point, or $\bar{s}_i \in V_1$ such that b is rider $i's$ arrival point

if $T_{kb} \leq \bar{l}_i$ and $LN_{kb} \leq q_k$

$V_1 = V \cup \{a\}$, $a \leftarrow b$, $V'_1 = V' / \{b\}$, $D_s = D_s \cup \{(V_1, b, V'_1)\}$

end if

end if

end for

end for

for each $\{V, a, V'\} \in D_s$

for each $\{V_1, a, V'\} \in D_s / \{V, a, V'\}$

if $V_1 = V$ and $D(V_2) = \min(D(V_1), D(V))$, $V_2 \in \{V, V_1\}$

$D_s = D_s / \{V_2, a, V'\}$, $V = V_2$

end if

        end for
    end for
end for
Step 3. backward movement
$s = length(L_k) + 1$
while $s \neq 0$
        for each element $\{V, a, V'\} \in D_s$
            if $s = length(L_k) + 1$
                $d(\{V, a, V'\}) = d_{ar_k}$
            else
                $d(\{V, a, V'\}) = \min\{d_{ab} + d(\{V \cup \{a\}, b, V' / \{b\}\})\}, \ \forall b \in V'$
            end if
        end for
        $s = s - 1$
end while
Step 4. output results
for $j = 1 : length(L_k) + 2$
    if $d(\{V, a, V'\}) = d_{ab} + d(\{V \cup \{a\}, b, V' / \{b\}\})$
        $L'_k(j) = b$
    end if
    $V = V \cup \{a\}$, $V' = V' / \{b\}$, $a \leftarrow b$
end for

6. Greedy random adaptive search algorithm

In order to verify the effectiveness of the algorithm, we designed the greedy random adaptive search algorithm. As shown in the table 7, the greedy random adaptive search algorithm consists of two parts: the first part is to find an initial solution, and the second is searched locally and the exact solution is obtained.

First, we initialize set P to passengers (line 02). When the set P is not empty, we do the following (line 03). Route planning is carried out for each vehicle take turns for each randomly selected passenger (line 04-06). We select two points in the routing of vehicle k in turn, a and b. We insert the passenger departure point after point a and the passenger arrival point after point b. Noting that if a and b are the same point, the passenger departure point is inserted after a and the passenger arrival point is inserted after the passenger starting point (line 06). Determine whether the time window constraint and load constraint are satisfied, and we record the planned routing if constraints are satisfied. Delete the passenger from set P (line 11).

In the local search stage, we adopt the classical exchange operator algorithm. Remove two passengers from their original routing and insert them into other's routing, if greedy function get better, exchange of match (line 16-24).

IOP Conf. Series: Materials Science and Engineering **787** (2020) 012012 doi:10.1088/1757-899X/787/1/012012

Table 7. Greedy random adaptive search algorithm

algorithm 1 Greedy random adaptive search algorithm

Step 1. structural initial solution
$P = I$
while $P = \varnothing$
    for $i \in I$
        for $k \in K$
            $L'_{ki} = \{..a, \overline{s}_i, ..b, \overline{r}_{i1}, ..\}$
            If $T_{kb} \le \overline{l}_i$ and $LN_{kb} \le q_k$
                $L_{ki} = L'_{ki}$
            end
            $D(L_{k1i}) = \min(D(L_{ki}))$
            $P = P / \{i\}$
        end
    end
end
Step 2. Local search
for $i1 \in I$
    for $i2 \in I$
        $L'_{k2i1} = \{..a, \overline{s}_{i1}, ..b, \overline{r}_{i1}, ..\}$, if $L'_{k1i1} \ne \varnothing$
        $L'_{k1i2} = \{..a, \overline{s}_{i2}, ..b, \overline{r}_{i2}, ..\}$, if $L'_{k2i2} \ne \varnothing$
        If $D(L_{k1i2}) + D(L_{k2i1}) < D(L_{k1i1}) + D(L_{k2i2})$
            $L_{k2} = L'_{k2i1}$, $L_{k1} = L'_{k1i2}$
        end if
    end for
end for

7. Numerical examples

In order to verify the effectiveness of the algorithm, we designed a simulation experiment. Since we only verify the validity of the algorithm, we do not have too much requirement on the authenticity of the data. We built a simulated road network consisting of 10 longitudinal roads and 20 transverse roads with 200 nodes. Starting point, arrival point, departure time and arrival time of passengers and drivers are generated by a random generator.

The ridesharing problem instances are solved on a PC with core i5 and 8GB of RAM. We use MATLAB to program the algorithm.

From the figure 3, we can see clearly that the running time of the decomposition algorithm is far less than that of the greedy random adaptive search algorithm. It is because that the decomposition algorithm does not search globally and applies parallel computing

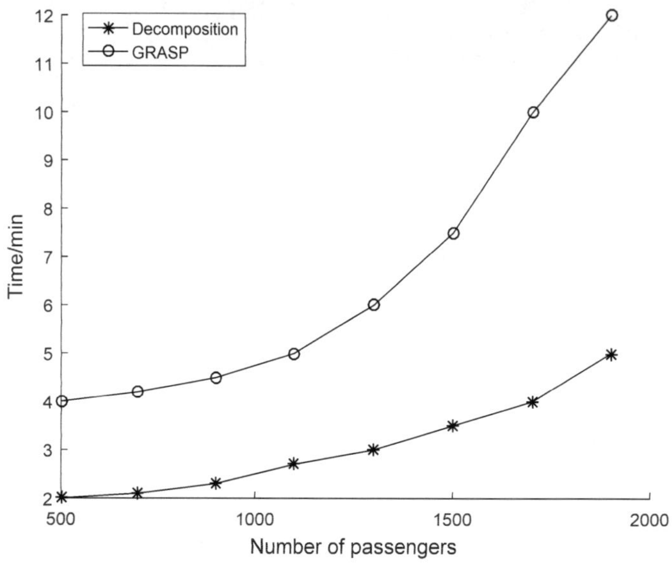

Figure 3. The running time of different algorithms

8. Conclusions

In this paper, we designed a decomposition algorithm for dynamic ridesharing problem. Compared with greedy adaptive search algorithm, it can effectively solve large-scale ridesharing problem and greatly reduce the running time. The central idea of this proposed algorithm is to turn large-scale problems into a number of independent small-scale sub-problems. We pursue parallel computation of these sub-problems. In order to solve the problem of vehicle path planning in the decomposition algorithm, we adopt the method of dynamic programming. Before the decomposition algorithm starts to run, we pre-process the input data in order to reduce the size of the points that the vehicle may pass through.

Finally, numerical examples are used to verify the effectiveness of our decomposition algorithm.

References

[1] Masoud, N. (2017) A decomposition algorithm to solve the multi-hop peer-to-peer ride-matching problem. Transportation Research Part B, 99: 1-29.

[2] Furuhata, M. et al. (2019) Ridesharing: the state-of-the-art and future directions. Transportation Research Part B. 57: 28-46.

[3] Cici, B. et al. (2014) assessing the potential of ride-sharing using mobile and social data: a tale of four cities, in: proceedings of the 2014 ACM international joint conference on pervasive and ubiquitous computing. ACM.

[4] Lenstra, J.K., Kan. (1981) complexity of vehicle routing scheduling problems. Networks 11(2): 221-227.

[5] Kamar, E., Horvitz, E. (2009). Collaboration and shared plans in the open world: studies of ridesharing. In: Proceedings of the International Joint Conference on Artificial Intelligence (IJCAI), pp. 187–194.

[6] Cheng, S.F., Nguyen, D.T., Lau, H.C. (2014) Mechanisms for arranging ride sharing and fare splitting for last-mile travel demands. In: Proceedings of the 2014 International Conference on Autonomous Agents and Multi-Agent Systems. International Foundation for Autonomous Agents and Multivalent Systems.

[7] Zhao, D. et al. (2014) Incentives in ridesharing with deficit control. In: Proceedings of the 2014 International Conference on Autonomous Agents and Multi--Agent Systems. International Foundation for Autonomous Agents and Multiagent Systems.

[8] Zhao, D. Ramchurn, S.D., Jennings, N.R. (2015) "Incentive design for ridesharing with

uncertainty." arXiv preprint arXiv:1505.01617
[9] Asghari, M. et al. (2016) Price-aware real-time ride-sharing at scale: an auction-based approach. In: Proceedings of the 24th ACM SIGSPATIAL International Conference on Advances in Geographic Information Systems. ACM.
[10] Asghari, M., Shahabi, C. (2017) An On-line truthful and individually rational pricing mechanism for ride-sharing. In: Proceedings of the 25th ACM SIGSPATIAL International Conference on Advances in Geographic Information Systems. ACM.
[11] Shen, W. Lopes, C.V., Crandall J.W. (2016) An online mechanism for ridesharing in autonomous mobility-on-demand systems. arXiv: 1603.02208
[12] Nguyen, D.T. (2013). Fair Cost Sharing Auction Mechanisms in Last Mile Ridesharing. Singapore Management University (Singapore).
[13] Zhang, J., Wen, D., Zeng, S. (2016) A discounted trade reduction mechanism for dynamic ridesharing pricing. IEEE Trans. Intell. Transp. Syst. 17 (6): 1586–1595.
[14] Kleiner, A., Nebel, B., Ziparo, V.A. (2011) A mechanism for dynamic ride sharing based on parallel auctions. Twenty-Second International Joint Conferences on Artificial Intelligence.
[15] Lloret B., Masoud, R.N., Nam D. (2017) P2p Ridesharing with Ride-Back on Hov Lanes: Towards a Practical Alternative Mode for Daily Commuting. No. 17-06253.
[16] Masoud, N., Lloret, B.R., Jayakrishnan, R. (2017b) Using bilateral trading to increase ridership and user permanence in ridesharing systems. Transp. Res. E 102: 60–77.
[17] Masoud, N., Lloret B.R. (2016). 95th Annual Meeting of the Transportation Research Board. Washington, DC.
[18] Ma, H., Fang, F., and DC. (2018) Parkes Spatio-temporal pricing for ridesharing platforms. arXiv preprint arXiv:1801.04015.
[19] Bian, Z.Y., Liu, X. (2019) Mechanism design for first-mile ridesharing based on personalized requirements part II: solution algorithm for large-scale problems. Transp. Res. B 120:172–192.
[20] Simonetto, A., Mnteil, J., Gambella, C. (2019). Real-time city-scale ridesharing via linear assignment problems. Transportation Research Part C. 101:208-232.

IOP Conf. Series: Materials Science and Engineering **787** (2020) 012013 doi:10.1088/1757-899X/787/1/012013

Reinforce-Optimization of regional network seismic based on connectivity reliability

You Tingting[1,*], Meng Weiwei[1]

[1]North China Municipal Engineering Design & Research Institute Co. Ltd, TianJin, China

[*]Corresponding author's e-mail: youtingting1993@163.com

Abstract: Road traffic is an important access for emergency service and disaster relief after the earthquake. In this paper, a reinforce-optimization model for road network seismic is established in combination with the analysis results of the seismic reliability of road segment and the importance of road segment. The model is aimed at the lowest cost of reinforcement and optimization, ensuring the aseismic reliability of the road segment and the system to reach the lowest limited value after the reinforcement and optimization, and the highest reliability of the system is mainly considered after the reinforcement and optimization. The orthogonal enumeration method is used to solve the scheme which satisfies the model condition, and a design method for seismic reinforcement and optimization of road network is given. Finally, it confirms the feasibility of this method with an example.

1. Introduction

The probability of the earthquake, especially the destructive earthquake, is low, but it is devastating. It not only has a certain influence on the national economy, but also has a profound influence on the social stability and human psychology, which has attracted wide attention from all over the world. The data shows that China is one of the most serious countries which has ever suffered from earthquakes. Since2013, there are 184 earthquakes of more than 5.0 degree in our country.

Although most cities have developed road networks in China, they lack relatively fast response capabilities in earthquake emergency response. Thus, through the evaluation of the seismic reliability of the traffic system composition unit, estimating the post-earthquake state of the traffic system scientifically, putting forward improvements of the reinforcement and optimization measures of various composition unit, improving the overall connectivity of the traffic system after earthquakes, and transforming the emergency repair of the unit after the earthquake into the reinforcement and optimization before earthquakes are urgent task for us at present and have certainly guiding significance in engineering.

2. Literature review

Reliability was regarded as a subject for professional research. It first appeared during World War II. Since 1970s, the research of reliability has entered a mature era and has made rapid development. Reliability has infiltrated into all aspects of people's life. The information and complication of civil products have greatly promoted the improvement of reliability research.

The reliability research of road network system started relatively late, and it has only been several decades since the concept was put forward. Road network reliability is an important index to measure the performance of urban road network, especially for traffic system under unexpected events. The reliability evaluation of road network is one of the effective methods to analyze the uncertainty and randomicity of the road network. Its function is to evaluate the probability of the road traffic network to reach a certain level of service[1], that is, to reach a certain level of traffic service. Researches of Evaluation index and evaluation method are key technology of road network reliability evaluation. It has been developing rapidly since 1990s.

Mine and Kawai first proposed the concept of connectivity, which reflected the probability of maintaining connectivity between two nodes traffic network. Generally, only 0, 1, two states of a section or node are studied , that is, connectivity or disconnection[2]; Iida and Wakabayashi further expanded connectivity from two points to k points and road network[3]; Nicholson used graph theory and the Floyd algorithm to calculate terminal reliability[4]. Asakura and Kashiwadani first put forward the concept of travel time reliability, pointing out that the volatility of travel time can be used to represent the reliability of traffic network, that is, the reliability of travel time[5]; Lida[6] defined the travel time reliability as the probability that travellers can reach the destination within a specified time range. On this basis, Yin, et al.[7] estimated the travel time change caused by traffic demand change by sensitivity analysis; Chen[8] proposed the concept of road network capacity reliability for the first time, and then put forward a method to evaluate the reliability of urban road network capacity[9]; Yang and Bell[10] put forward a two-dimensional graphical reliability analysis method based on the overall consideration of road network capacity reliability and travel time reliability; Sumalee and Kurauchi[11] calculated the reliability of urban road network capacity under different conditions by changing traffic management control measures, analyzed it with Monte Carlo method and verified it in a small experimental network.

Zhang Yong and Yang Xiaoguang analyzed the reliability of urban road network based on complex network theory. The results show that the road network has different robustness under the two attacks of randomicity and selectivity[12]; Liu Sifeng et al. put forward the corresponding matrix algorithm and used it to build the Graph Evaluation and Review Technique model[13]; By combining the reliability theory of series system and parallel system, Zhang Jieanalyzed the reliability of the road between OD points, the reliability of the emergency rescue road network, the reliability of the transportation network of the rescue materials and the seismic reliability of the medical rescue traffic network[14]; He Xuan took the influence factors into the network topology model, and further defined the relationship between the influencing factors, so as to analyze the reliability of the emergency rescue system[15]; Tian Fangshuai et al. analyzed the main factors that affect the connectivity reliability of the traffic system, offered calculation methods for the connectivity reliability of the transportation system components and the traffic network and established the seismic reliability model of the urban transportation system[16]. Based on Monte Carlo simulation method, Hou Benwei et al. established an post-earthquake connectivity reliability and traffic volume transit time analysis model, and analyzed the influence of the seismic that effect the road network support capability from three aspects of the road and bridge structure (unit), the section (line) and the network system, and compared the results characteristics and applicability of the two models[17].

3. Aseismic reinforcement and optimization of road traffic system

3.1 Reliability analysis of the road network

The road traffic network is large and complex, it has so many miscellaneous data that it is difficult to calculate accurately. In this paper, Monte Carlo simulation method is used to analyze connectivity reliability of post-earthquake traffic network. Based on the Monte Carlo simulation method for seismic reliability evaluation of road traffic system, the following assumptions are made:

①Only road sections may fail in the road network, and nodes will not fail;

②There are only two states in the road network, connectivity or failure;

③The road segments are independent of each other, and the two traffic flow reliability of road segment is consistent.

The basic idea of the seismic connectivity reliability evaluation of road traffic system based on

Monte Carlo simulation method is based on obtaining the seismic reliability of post-earthquake road segment. Through a large number of simulation experiments, this paper obtains the probability of reliable connectivity of each path, and the system seismic connectivity reliability. The process is as follows:

①Mark the node of the road network and the road segment in the study area with m nodes (i, j) and M road segment (expressed in U), input the seismic reliability value $P_{ij}$of each road segment (i is the starting mark of a road unit, j is the end point mark of a road unit).

②Set up the adjacency matrix U (m, m) between each node.

$$\text{Which: } u_{ii} = 0, u_{ij} = \begin{cases} 0 & \text{Node } i\text{ , }j \text{ do not connect} \\ 1 & \text{Node } i\text{ , }j \text{ connect} \end{cases} \tag{3.1}$$

③At each road segment, the random number X_{ij} on the M [0,1] interval is generated and compared with the aseismic reliability P_{ij} of the road segments, and the adjacency matrix is dealt with correspondingly.

$$\begin{cases} \text{When } X_{ij} \le P_{ij} & \text{The analog unit is connected, letbe } u_{ij} = 1,\ u_{ji} = 1 \\ \text{When } X_{ij} > P_{ij} & \text{This analog unit is not connected, letbe } u_{ij} = 0,\ u_{ji} = 0 \end{cases} \tag{3.2}$$

④By using Breadth-First Search, searching out the connected nodes gradually from the source point, and the connection path of the source point to the target node i denoted by "1", and the shortest path denoted by "2", and each unit u (u=1,2,... M) in the shortest path denoted by 3u (u=1,2,... M). At the same time, we record the simulation times of i from source point to different node.

⑤Repeat steps ③—④, statistic the times of "1" in all links and denote as t_i, the shortest path number in the connection case denoted by d, and each unit u (u=1,2,...M) in the shortest path occurrence number denoted by d_u, the number of simulations time is T, thus the connectivity reliability R_i of source point O to node$i$and the connected reliability edge betweennessλc are:

$$R_i = \frac{t_i}{T} \tag{3.3}$$

$$\lambda_c(e_u) = \frac{d_u}{d} \tag{3.4}$$

λc refers to the proportion of the number of specified edges in the shortest path of the network to all the shortest path. The higher the proportion, the greater the edge betweenness, indicating that this edge plays a more important role in the road network.

⑥To some extent, the connectivity reliability of nodes in the road network reflects the connectivity reliability. Therefore, the average value of the connected reliability of the source point to the nodes can be used as the evaluation value of the whole connectivity degree of the road network. The seismic connectivity reliability R_S of road network system is as follows:

$$R_S = \frac{\sum_{i=1}^{m} R_i}{m} \tag{3.5}$$

⑦Improve each unit u (u=1,2,...M) with the original seismic reliability Pu' and repeat the above steps, a new connection reliability Ri' from the source point to the node i and the seismic connectivity reliability R_S of the road network system will be obtainned. The corresponding key importanceI_c is:

$$I_c = \frac{1 - P_u}{R_S} \cdot \frac{R_S' - R_S}{P_u' - P_u} \tag{3.6}$$

The key importance is to consider the influence of the failure probability of the road segment on the connectivity reliability of the system while considering the change of the system connectivity reliability caused by the change of the reliability of the road segment[18].

3.2 Reinforcement and optimization modeling

Considering many factors such as post-earthquake road network reliability and economic input etc., on the basis of ensuring the post-earthquake reliability of the road network, a optimization model for the lowest cost of the road network reinforcement is established in this paper. The optimization model is as follows:

$$\begin{aligned}&\min C = \sum_{i\in U} C(i)\\ &s.t.\ P_S(x_i) \geq P_{S_0}\\ &P_i \geq P_{i_0}\\ &I_{g_i} \geq I_{g_j}\ (\exists j \in A \neq i)\\ &\text{When } I_{g_i} = I_{g_j},\ \lambda(x_i) \geq \lambda(x_j)(\exists j \in A \neq i)\end{aligned} \tag{3.7}$$

Where: C is the total cost of strengthening and optimizing road network, including the cost and equipment of installation engineering, the purchase cost of tools and other costs of engineering construction, all of which are determined according to the estimate rating and the national standard; U is the set of all the units in the road network; x_i is the unit, which is taken strengthening measures in the road network; $C(i)$ is the cost of strengthen measures for unit i; $P_S(x)$ is the connectivity reliability of road traffic system after road segment is strengthened and optimized; P_{S0} is the connecting reliability limit value of road traffic system after the road segment is strengthened and optimized; P_i is the seismic reliability after the road segment i is strengthened and optimized; P_{i0} is the seismic reliability limit value of unit i; I_{gi} is the key importance of unit i; I_{gj} is the key importance of unit j; $\lambda(x_i)$ is connected reliable edge betweenness of unit i; $\lambda(x_j)$ is connected reliable edge betweenness of unit j.

For the constraints of key importance, when the existing scheme cannot satisfy the constraints, we can choose the suboptimal key importance to choose the plan. For connected edge betweenness, the higher edge betweenness can be considered only at the same key importance.

4. Illustrative example

4.1 Reliability of regional road network

In this virtual network, it contains 15 nodes and 20 road segments. Select node 1 as the source point. In the earthquake, the source point can be considered as the abstract node of the disaster area, and calculate the seismic connectivity reliability with different nodes, as shown in Figure 1.

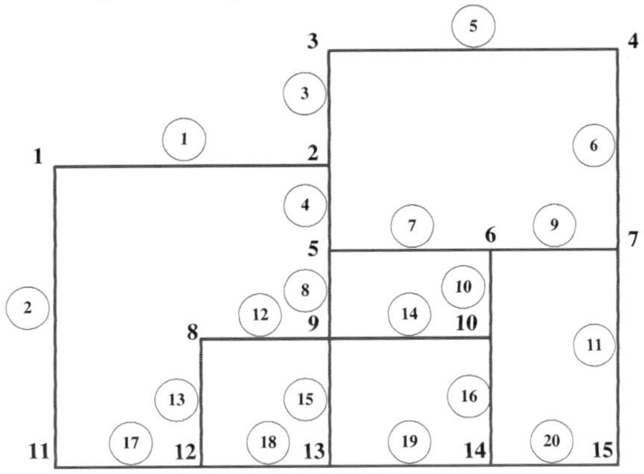

Figure1. Virtual road network diagram

The aseismic reliability of road segment under different intensities is calculated by seismic reliability calculation method, as shown in Table 1.

Table 1. Analysis result of seismic reliability of road section

Number	Road segment	Sidework width (m)	Lane width (m)	Road length (m)	Five degree	Six degree	Seven degree	Eight degree	Nine degree
1	1-2	9.0	24.0	1300	1.000	1.000	1.000	0.999	0.899
2	1-11	7.0	9.1	1100	1.000	1.000	0.999	0.998	0.895
3	2-3	4.0	9.0	380	1.000	1.000	1.000	0.996	0.880
4	2-5	4.0	9.0	230	1.000	1.000	0.933	0.789	0.459
5	3-4	9.0	12.0	1500	0.999	0.891	0.821	0.413	0.236
6	4-7	9.0	12.0	695	0.937	0.879	0.802	0.407	0.103
7	5-6	4.5	14.0	420	1.000	0.984	0.934	0.910	0.806
8	5-9	9.0	12.0	265	1.000	1.000	0.999	0.927	0.899
9	6-7	4.5	14.0	240	1.000	1.000	0.998	0.926	0.896
10	6-10	8.9	11.5	460	1.000	1.000	0.986	0.950	0.801
11	7-15	4.5	14.0	405	1.000	0.998	0.954	0.882	0.800
12	8-9	6.0	11.0	455	1.000	1.000	0.906	0.731	0.400
13	8-12	11.0	28.0	385	1.000	1.000	0.928	0.726	0.413
14	9-10	5.0	10.0	240	1.000	1.000	0.932	0.768	0.401
15	9-13	6.3	21.0	343	1.000	1.000	0.984	0.935	0.908
16	10-14	9.0	12.0	466	1.000	1.000	0.996	0.799	0.496
17	11-12	6.3	21.0	1200	1.000	1.000	0.996	0.910	0.898
18	12-13	6.3	21.0	600	1.000	1.000	0.997	0.978	0.832
19	13-14	10.0	12.0	580	1.000	1.000	0.991	0.781	0.430
20	14-15	6.3	21.0	740	1.000	1.000	0.989	0.913	0.832

The seismic reliability of road segment under different intensities is statistically analyzed, and the reliability percentage of each section under different intensities can be obtained by analyzing the data, as shown in Table2.

Table 2. Percentage of reliability under each intensity

Intensity	Very reliable (0.950,1.000]	Basic reliable (0.800,0.950]	Medium reliable (0.500,0.800]	Basically unreliable (0.200,0.500]	Serious unreliable (0.000,0.200]
5	95%	5%	0	0	0
6	90%	10%	0	0	0
7	65%	35%	0	0	0
8	20%	40%	30%	10%	0
9	0	55%	5%	35%	5%

Based on the above analysis, when the seismic intensity of the road segment is 5 degree and 6 degree, the proportion of the road segment that remains very reliable is more than 90%. The simulation process is aimed at the earthquake intensity above 7 degree or more. Monte Carlo simulation method was used to simulate the path selection from source point to different node under different intensities, and their seismic connectivity reliability was obtained, as shown in Table 3.

Table 3. The connectivity reliability of the path under different intensity

Node	Seven degree	Eight degree	Nine degree	Node	Seven degree	Eight degree	Nine degree
2	1.000	1.000	0.897	9	0.991	0.818	0.627
3	1.000	0.996	0.711	10	0.974	0.832	0.508
4	0.964	0.889	0.407	11	0.961	0.830	0.807
5	0.995	0.861	0.690	12	0.963	0.843	0.608
6	0.994	0.811	0.735	13	0.964	0.860	0.714
7	0.997	0.873	0.881	14	0.997	0.875	0.591
8	0.959	0.833	0.406	15	0.993	0.810	0.758
R_S	0.982	0.866	0.667				

According to the analysis of the results, the road segments are less affected when the earthquake intensity is 5, 6 and 7 degrees, all of which can remain basically reliable. Therefore, we only consider the road units under the seismic intensity of 8 and 9 degrees when analyzing the importance of the units.

4.2 Research on reinforcement and optimization

In the process of reinforcement and optimization, when the seismic reliability Pu≥0.8 of the road segment, such units are not considered for reinforcement. When Pu<0.8, you can choose to reinforce it. In summary, the road units that need to be reinforced at 8 degree earthquake intensity are numbered 4, 5, 6, 12, 13, 14, 16 and 19 (the number of road segments to be reinforced and the 8 degree are same as 9 degree).

For each element under different states, the cost of reinforcement required for each increase 0.1 seismic reliability is different. It is assumed that the reinforcement cost of medium reliability is a yuan/m2, the reinforcement cost of the basic unreliable state is 2a yuan/m^2, and the reinforcement cost of the serious unreliable state is 5a yuan/m^2. The seismic reliability of the strengthened segment needs to reach 0.8 or more. If the unit is reconstructed, the seismic reliability of the reconstructed unit will be 1.000, the reinforcement cost will be 10a yuan/m^2, as shown in Table 4.

Table 4. The road segment information and the reinforcement cost

(a)

Number	Original aseismic reliability	Aseismic reliability after reinforcement	Sidework width	Lane width	Road length	Reinforcement cost (yuan)	
						Original base	Reconstruct
4	0.789	0.933	4.0	9.0	230	1192*a*	11923*a*
5	0.413	0.821	9.0	12.0	1500	132192*a*	660960*a*
6	0.407	0.802	9.0	12.0	695	59297*a*	296487*a*
12	0.731	0.906	6.0	11.0	455	5255*a*	52553*a*
13	0.726	0.928	11.0	28.0	385	23953*a*	239532*a*

(b)

Number	Original aseismic reliability	Aseismic reliability after reinforcement	Sidework width	Lane width	Road length	Reinforcement cost (yuan)	
						Original base	Reconstruct
14	0.768	0.932	5.0	10.0	240	1968*a*	19680*a*
16	0.799	0.996	9.0	12.0	466	9915*a*	99146*a*
19	0.781	0.991	10.0	12.0	580	14616*a*	146160*a*

Reinforce the road segment on the basis of the importance degree and make the scheme test number in order, the seismic connectivity reliability of the original unit reinforcement and reconstruction unit are obtained after multiple simulation based on the Monte Carlo simulation method, as shown in Table 5.

Table 5. Reinforcement optimization index value of road segment

Number	Reinforcement on the original unit			Reconstruction unit optimization		
	Aseismic connectivity reliability	Variety	Cost (yuan)	Aseismic connectivity reliability	Variety	Cost (yuan)
4	0.866	0.00%	1192*a*	0.873	0.66%	11923*a*
5	0.923	5.65%	132192*a*	0.967	10.10%	660960*a*
6	0.923	5.64%	59297*a*	0.951	8.50%	296487*a*
12	0.867	0.08%	5255*a*	0.868	0.16%	52553*a*
13	0.867	0.07%	23953*a*	0.868	0.14%	239532*a*
14	0.872	0.59%	1968*a*	0.880	1.33%	19680*a*
16	0.877	1.04%	9915*a*	0.886	1.99%	99146*a*
19	0.920	5.35%	14616*a*	0.921	5.49%	146160*a*

The reliability of system aseismic connectivity calculated by the reinforcement scheme is compared, and the 4, 5, 6, 12, 14, 16 and 19 factors are selected, and the reinforcement optimization of the road network is selected based on the orthogonal enumeration method. The concrete scheme is shown in Table 6. Among them, the level figure is 2, and the "1" represents the reinforcement on the basis of the original unit, "2" represents the reconstruction of the unit.

Table 6. Orthogonal table L8(27)

(a)

Test number	Factor（Element number）							Test number	Factor（Element number）						
	4	5	6	12	14	16	19		4	5	6	12	14	16	19
1	1	1	1	1	1	1	1	5	2	1	2	1	2	1	2
2	1	1	1	2	2	2	2	6	2	1	2	2	1	2	1
3	1	2	2	1	1	2	2	7	2	2	1	1	2	2	1

(b)

Test number	Factor（Element number）							Test number	Factor（Element number）						
	4	5	6	12	14	16	19		4	5	6	12	14	16	19
4	1	2	2	2	2	1	1	8	2	2	1	2	1	1	2

According to the scheme determined by orthogonal enumeration method to reinforce and optimize, and the result of optimization is shown in Table 7. When the limit value of the seismic connectivity reliability is 0.800, all the schemes are satisfied. Therefore, considering the cost and the change amplitude of the network seismic connectivity reliability, the maximum increase of the connectivity reliability is No.3 and No.4. By comparison, No.4 costs the least; therefore, choose No.4 plan as the best plan and No.3 plan as suboptimal plan. So according to plan No.4, reinforce units 4, 16 and 19 on the original basis, while reconstructing the rest units.

Table 7.Reinforcement optimization index value of road segment

Test number	Aseismatic connectivity reliability of road network	Seismic Connectivity Reliability Change of Road Network	Reinforcement cost (yuan)
1	0.874	0.75%	224435*a*
2	0.870	0.39%	510220*a*
3	0.979	11.27%	1211168*a*
4	0.981	11.45%	1055403*a*
5	0.918	5.14%	621612*a*
6	0.918	5.19%	608885*a*
7	0.910	4.34%	870877*a*
8	0.914	4.72%	942776*a*

5. Conclusion

Aimmed at researching the reliability of the post-earthquake network, we calculated the seismic reliability of the study area under different intensities, used the Monte Carlo simulation method to analyze the system connectivity reliability under different intensities and the units with high degree of importance. The reinforcement optimization model of network is calculated by orthogonal enumeration method with the goal of strengthening and optimizing cost at least, the condition of aseismic connectivity reliability is higher, which provides a theoretical basis for seismic network reinforcement optimization. The conclusions are as follows:

(1) Based on Monte Carlo simulation method, we analyzed the seismic connectivity reliability of simple road network and it was proved to be feasible.

(2) Established the aseismic reinforcement optimization model of road network by minimizing the cost of reinforcement optimization of road segment as the objective function, the minimum value of the overall road network connectivity reliability that meets the set criteria after optimized, and importance height of road segment optimization as constraints. Orthogonal enumeration method is used to solve the reinforcement optimization scheme, and the optimal reinforcement optimization scheme is selected by analyzing the example.

Acknowledgments

Thanks Meng Weiwei for his help in writing the paper, and thanks for the company's strong support in scientific research, which enabled me to finish the paper with the support of abundant resources.

References

[1] Hou Liwen, Jiang Fu (2000),"Research on the reliability of urban road network," Systems engineering, 18(5): 44-48.
[2] Mine H, Kawai H (1982),"Mathematics for Reliability Analysis," Asakurashoten, 12-14.
[3] Lida Y and Wakabayashi H (1989),"An Approximation Method of Terminal Reliability of a Road Network Using Partial Minimal Path and Cut Set[A], Proceedings of the Fifth WCTR," Japan: Yokohama, 367-380.
[4] Nicholson A (2003),"Assessing Transport Reliability: Malevolence And User Knowledge[A]. The network reliability of transport: proceedings of the 1st international symposium on transportation network reliability (INSTR)," Kyoto: Pergamon Press.
[5] Asakura Y, Kashiwadani M (1991),"Road Network Reliability Caused by Daily Fluctuation of Traffic Flow," // Proceedings of the 19th PTRC Summer Annual Meeting. Brighton: 73- 84.
[6] Lida Y (1999),"Basic Concepts and Future Directions of Road Network Reliability Analysis," Journal of Advanced Transportation, 33(2) : 125-134.
[7] Yin Y, Ieda H (2001),"Assessing Performance Reliability of Road Network under Nonrecurrent Congestion," Transportation Research Record, , 1771: 148-155.

[8] Chen A, Yang H, Lo HK, et al (1999),"Capacity related reliability for transportation networks," Journal of Advanced Transportation, 33(2) :183-200.
[9] Chen A, Yang H, Lo H K and Tang W (2002),"Capacity Reliability of a Road Network: an assessment methodology and numerical results," Transportation Research, 36B, 225-252.
[10] Yang H, Bell MGH (1998),"Models and algorithms for road network design: a review and some new developments. Transp Rev," Transport Reviews,18(3) :257-278.
[11] Sumalee A, Kurauchi F (2006),"Network Capacity Reliability Analysis Considering Traffic Regulation after a Major Disaster,"Networks and Spatial Economics, 6(3) : 205-219.
[12] Zhang Yong, Yang Xiaoguang (2008),"Simulation and analysis of complex network characteristics and reliability of urban road network," system simulation journal, 22 (2): 464-467.
[13] Liu Sifeng, "Wan Shouqing, Lu Zhipeng, et al (2009), Research on the link importance evaluation model between rescue points and accident points in complex traffic network" China management science, 17 (1): 119-124.
[14] Zhang Jie (2012). "Composition and seismic reliability of urban emergency road network," Qingdao: Ocean University of China.
[15] He Xuan (2012). "Reliability study of major emergency rescue system," Wuhan: Huazhong University of Science and Technology.
[16] Tian Fang Shuai (2013). "Earthquake reliability research and application of urban emergency road network," Qingdao: Ocean University of China.
[17] Hou Benwei, Li Xiaojun, et al (2017), "Post-earthquake connectivity and traffic time analysis of highway network based on Monte Carlo simulation," Chinese Journal of highway engineering, 30 (6): 287-296.
[18] Han Ran Ran (2011), "Study on connectivity reliability analysis of highway network in Western China," Beijing Jiaotong University.

IOP Conf. Series: Materials Science and Engineering **787** (2020) 012014 doi:10.1088/1757-899X/787/1/012014

Intermittent instability and its control in parallel converters

Wang Lili[1,2*], Chen Junning[2]

[1]School of Electronic and Information Engineering, Anhui Jianzhu University, Hefei Anhui Province, 230601, China

[2]School of Electronic and Information Engineering, Anhui University, Hefei Anhui Province, 230601, China

[*]Corresponding author's e-mail: ll_wang@yeah.net

Abstract. This paper studies intermittency instability and its control in parallel converters. By numerical simulation, the intermittent subharmonic and chaos are captured with coupled interference signal , their characteristics are described by bifurcation diagrams as parameter varies. Then the paper proposes control methods and it is proved effective by the comparison of power spectrums. The conclusion can help engineer to explain the inherent irregular motion in vehicle, and design more stable power electronic supplies.

1. Introduction

DC-DC converter is the core part of power supply[1]. Its stability is very important for the performance of power system. Parallel converters are suitable for the needs of low voltage, large current and distributed modular power system[2-4]. It is widely used invehicleengineering to provide power to air condition and other auxiliary equipments. Intermittency is an abnormal working state in converters for its unpredictability, which will affect the stability of the vehicle. Intermittency is characterized by sporadicnoise[5]. Further study can help engineer to explain the inherent irregular motion in DC-DC converter, and design more stable power electronic supplies.

2. Parallel Buck converters with coupled interference signal

2.1 Overview of system operation

In practice, the system is often disturbed by external signals. Interference signals are transmitted or radiated into the circuit through the coupling channel, which affects the normal operation of converters. For that, we introduce interference signals (the circuit is often interfered by such periodic sinusoidal signals in the RF environment) and add them to the input voltage,

$$E^{*} = E + v_s = E + \hat{v}_s \sin(2\pi f_0 t) = E[1 + \alpha_v(2\pi f_0 t)] \tag{1}$$

where E is input voltage, $\hat{v}_s$ is the voltage amplitude of the interference signal, f_0 is the frequency of interference signal. α_v is the amplitude ratio of the interference voltage to the input voltage, represents the intensity of the interference source, is given by

$$\alpha_v = \hat{v}_s / E \tag{2}$$

Figure 1. shows parallel Buck converters with coupled interference signals.

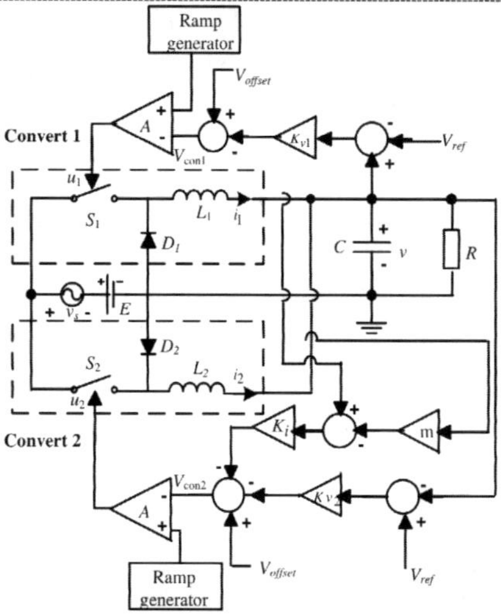

Figure 1.Parallel buck converters with coupled interference signal

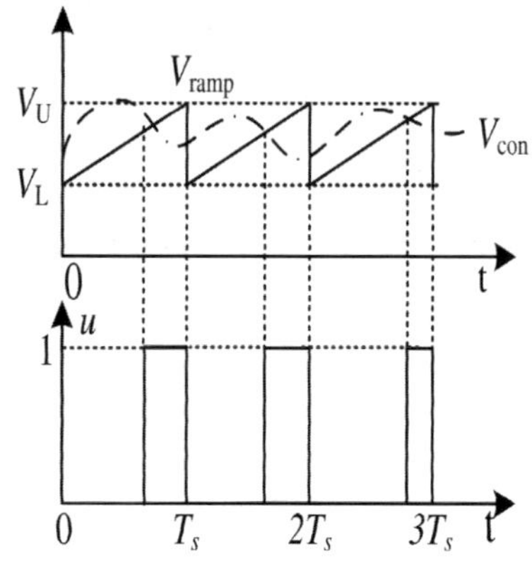

Figure 2.Pulse-width modulation

The state equation of parallel bucks is listed as follows:

$$\begin{cases} \dot{v} = -\dfrac{v}{(R+r_c)C} + \dfrac{R}{(R+r_c)C} i_1 + \dfrac{R}{(R+r_c)C} i_2 \\ \dot{i}_1 = -\dfrac{Rv}{(R+r_c)L_1} - \dfrac{1}{L_1}\left(\dfrac{Rr_c}{R+r_c} + r_{L1}\right) i_1 - \dfrac{Rr_c}{(R+r_c)L_1} i_2 + \dfrac{E^* u_1}{L_1} \\ \dot{i}_2 = -\dfrac{Rv}{(R+r_c)L_2} - \dfrac{Rr_c}{(R+r_c)L_2} i_1 - \dfrac{1}{L_1}\left(\dfrac{Rr_c}{(R+r_c)L_1} + r_{L2}\right) i_2 + \dfrac{E^* u_2}{L_2} \end{cases} \tag{3}$$

2.2. Simulation results

According to the state equation, the simulation model is established by MATLAB, circuit parameters are chosen as follows:

Table 1 Component values.

Circuit Components	Value
Capacitor C, ESR r_C	$47\mu F$, 0.01 Ω
Inductor L_1, ESR r_{L1}	$20mH$, 0.05Ω
DC offset voltage	$5V$
Input voltage E	$22V$
Current gain K_i	5
Lower limit voltag V_L	3.8 V
Switching period T	$400\mu S$
Load resistor R	10Ω
Inductor L_2, ESR r_{L2}	$20mH$, 0.2Ω
Reference voltage V_{ref}	$11.3V$
Voltage gain K_{v1}, K_{v2}	3.5,3.5
Split ratio m	1
Upper limit voltage V_U	8.2 V

The waveforms of the state variables as time varies can be observed directly. Without interference signal, inductor current and capacitor voltage are periodic in time domain as shown in figure 3, indicating that parallel converters present period-1, the system is in a stable state[6].

Coupling with intruding interference signal, the parallel converters present different operations due to different interference frequency.

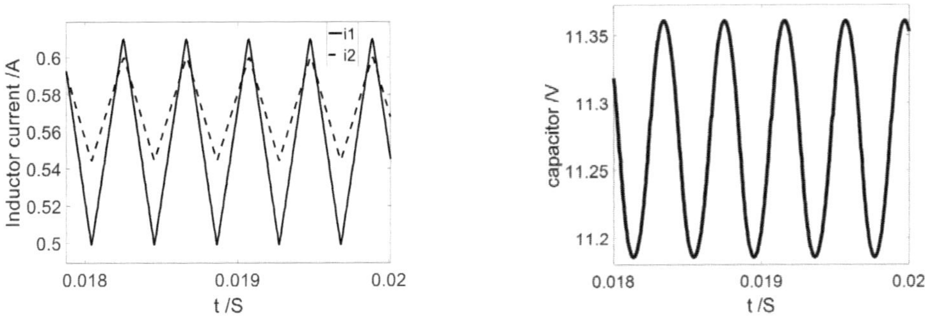

(a) inductor current of time domain (b) capacitor voltage of time domain

Figure 3. System state without interference signals

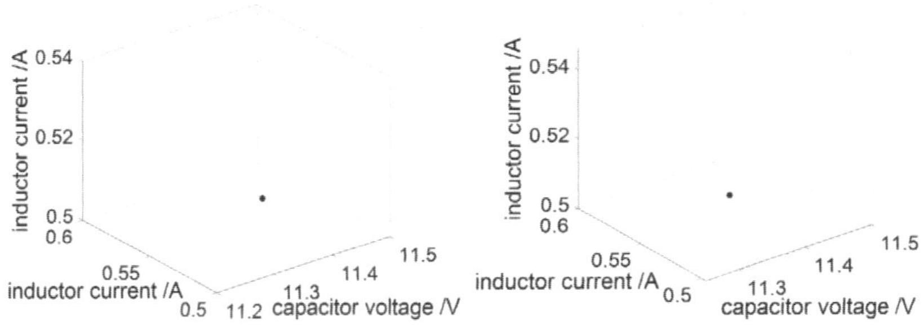

(a) $f_0 = f_s$ (b) $f_0 = 2f_s$

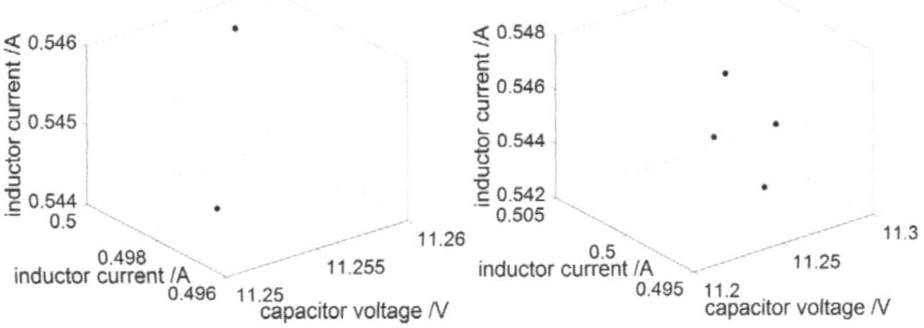

(c) $f_0 = \frac{1}{2} f_s$ (d) $f_0 = \frac{1}{4} f_s$

Figure 4. Poincare section

If the frequency of interference signal is n times of the switching frequency of the converters, n is a positive integer. Taking n=1 and n=2 as examples, the Poincare section of the system is observed and exhibitioned as figure 4(a) and (b). Poincare section describes the sample state at each switching point. It is a useful way to describe the operation of the system.

When $f_0 = f_s$ and $f_0 = 2f_s$, where f_0 is interference frequency, they are both characterized by only one intersection, showing the converters is in period-1, indicating that system is stable. If the frequency of interference signal is 1/n times of the switching frequency of the converters, Figure

4(c)and(d) show two examples with n=2 and n=4, parallel converters are in period-2 and period-4,characterized by 2 and 4 intersections on the Poincare sections.

However, in most case, the interference signal is random, and its frequency is not just the rational multiples of the switching frequency exactly. Generally, the frequency of the interference signal is close to the switching frequency. Therefore, assuming $f_0 = 2501Hz$, taking the interference intensity of intruding interference signal as the parameter. Figure 5. shows the time-bifurcation diagram of inductance current under different interference intensities.

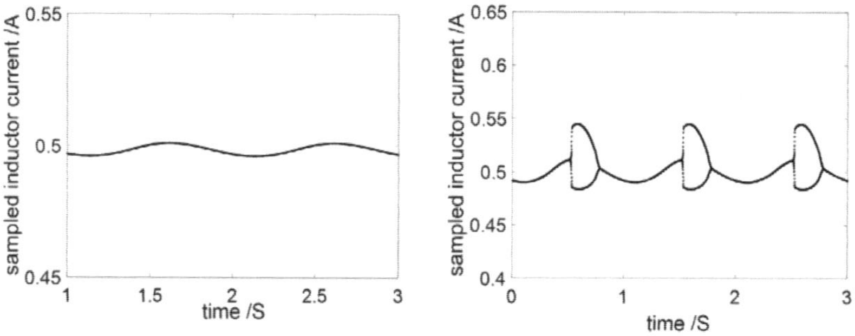

(a) $\alpha_v = 0.1$ (b) $\alpha_v = 0.35$

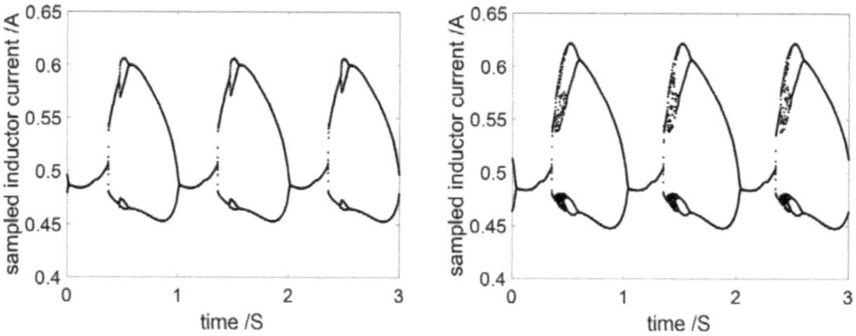

(c) $\alpha_v = 0.62$ (d) $\alpha_v = 0.67$.

Figure5.Time-bifurcation diagrams with varied intensity of the interference signal.

From these four figures, we have the following observations:

- When the interference signal intensity is weak, as shown in Figure5. (a), the converters are still in period-1 although the sampled inductance current fluctuates around 0.5A, indicating that the interference signal has little impact on the converters.
- As the intensity of the interference signal increases gradually, intermittency can be captured intermittency in the converters. The converter alternates between stable period-n and subharmonic states regularly. When $\alpha_v = 0.35$, converters experiences period-1 and period-2 subharmonic alternately, and when $\alpha_v = 0.62$ the converter experiences period-1 and period-2 subharmonic alternately. Figure. 5 (b) and (c) give the corresponding time-bifurcation diagrams.
- Further increase the intensity of the coupling interference signal can cause the converter experiences period-1 and 2i subharmonic alternately. Until the intensity of the coupling interference signal reaches a critical value, intermittent chaos appears in converters, period 1 and chaos present alternately. Figure.5(d) gives the corresponding time bifurcation diagrams as $\alpha_v = 0.67$.
- Intermittency is periodic. Intermittent period is defined as $T_{in} = 1/|f_0 - f_s|$ after summary and analysis. When the frequency of the interference signal is close to the switching frequency of the converter, converters take a long time to finish an intermittent process. In this example, there is $1Hz$ between two frequencies, so the intermittent period $T_{in} = 1s$.

IOP Conf. Series: Materials Science and Engineering **787** (2020) 012014 doi:10.1088/1757-899X/787/1/012014

3. Suppression of Intermittent instability by Resonant Parametric Perturbation

In actual systems, chaos is destructive and should be avoided. But structure and parameters of designed systemare determined, and they cannot be changed. Resonance parameter perturbation is an effective method to transform converters from chaos to regular state or cause the system to move away from its original periodic orbit by perturbating the frequency or amplitude of appropriate parameters. It is one of non-feedback chaotic control methods.

There are many parameters in parallel converters. The choice of parameters is related to the control effect. After careful comparison, we choose the reference voltage which has great influence on converters and is easily disturbed as the resonance perturbation parameter[7]. When $\alpha_v = 0.67$, we replace $V_{ref}[1+\alpha_c \sin(2\pi f_s t+\psi)]$ with V_{ref}, where α_c is the amplitude of the perturbing signal, ψ is the phase shift between the perturbing signal and the switching frequency of converters. The perturbing frequency is as same as the frequency of converters.

Setting $\psi = 0$, α_c is varies from 0 to 0.02 .Lots of simulation show when the system is stable at period 1, the minimum effective perturbing amplitude is 0.017. Figure6(a) gives the bifurcation diagram of the state variable of with different perturbing signal amplitude.

Furthermore, the operation with phase shift as parameter can be observed when the perturbing amplitude is 0.017. Figure 6 (b) clearly illustrates the period-1 region of converters is centered at around $\psi = 1.5$. Finally the operation with phase shift as parameter can be observed when the $\psi = 1.5$. It is concluded that $\alpha_c = 0.01$ 和 $\psi = 1.5$ are the optimal amplitude and phase in parallel converters.

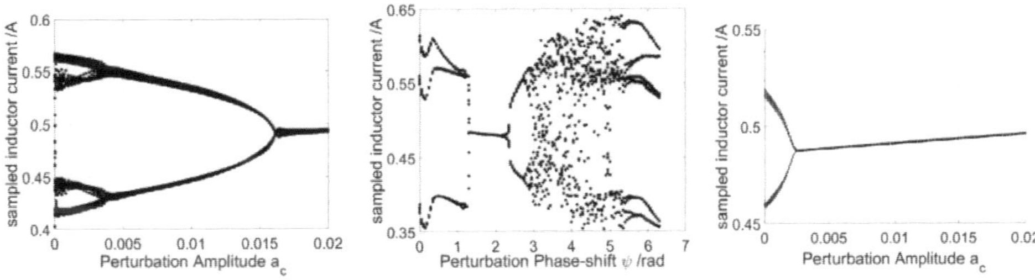

(a) α_c as parameter($\psi = 0$) (b) ψ as parameter($\alpha_c = 0.017$)(c) α_c as parameter($\psi = 1.5$)

Figure 6. Parameter bifurcation diagrams

To illustrate the effect of perturbation, we observe the phase portraitsof system by experiments. The phase portrait without perturbation contains countless orbits as shown in Figure7 (a), indicating the system is unstable. But the phase portrait with perturbation is depicted by a closed orbit, as shown in Figure 7 (b), indicating the system is in a stable period-1.By comparing the phase portraits, resonance parameters perturbation transforms convertersto the regular state effectively.

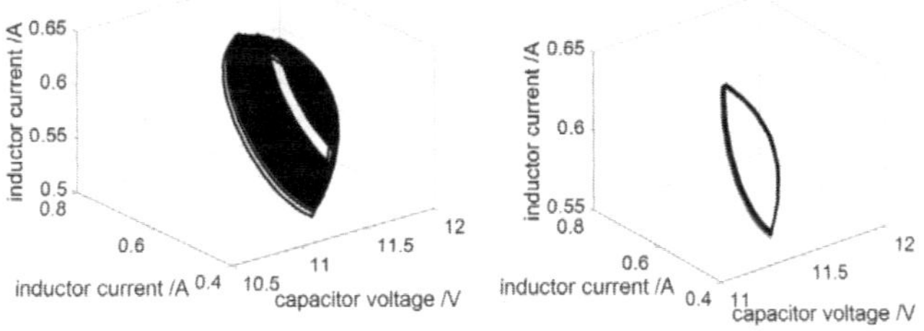

(a) $\alpha_c = 0$, $\psi = 0$ (b) $\alpha_c = 0.017$, $\psi = 1.5$

Figure 7.Phase portraits

4. Conclusions

In this paper, parallel Buck converters with coupled interference signal are constructed and investigated. It is found that the intermittent instability of converters is characterized by intermittent harmonics or intermittent chaos. The research results show that when the system is disturbed by coupled parasitic signals, the intermittent harmonics and intermittent chaos can be captured. The specific mode of operation relates to the intensity of the interference. Coupling circuit conduction and radiation interference are the source of intermittency. The resonant parameter perturbation method can control the system in a stable state.

Acknowledgments
This work was supported by National Natural Science Foundation of China, Grant No.60672023 and No. 61071023 and the Key research projects of Natural Science in Universities in Anhui, Grant No. KJ2017A489 and No.KJ2015A031.

References

[1] Subashini, M.(2017)Bifurcation and chaos in converter interfaces in solar PV systems-A review.Journal of Engineering Science and Technology Review,10:90-97.
[2] Zhang, F.Y.,Hu,W.(2015) Chaos control and anti-control in Boost converter based on altering correlation.Acta Physica Sinica,64:DOI:10.7498/aps.64.048401
[3] Harb, Ahmad M., Harb, Souhib M., Batarseh, Issa E.(2010) Chaos and bifurcation of voltage-mode-controlled buck dc-dc converter with multi control parameters.International Journal of Modelling and Simulation,30: 472-478.
[4] Wang, L.B.,Meng, Z., Sun,Y.G,.(2017)Bifurcation and chaos analysis of power converter for switched reluctance motor drive. International Journal of Electronics.104: 157-173.
[5] He, J., Zheng, X., Wang, W., Ren, Y.(2013)Terminal sliding mode control of Boost converter with chaos.Transactions of China Electrotechnical Society,28:104-108.
[6] Li, G.L., Li, ChunY.,Chen, X., Zhang, X.W.(2013)Chaos control of SEPIC converter based on resonant parametric perturbation method.Acta Physica Sinica,62:DOI: 10.7498/aps.62.210505
[7] Jia, M.M., Zhang, G.S.(2016)Chaos control for the voltage-controlled buck converter.Journal of the University of Electronic Science and Technology of China,45:208-214.
[8] Vergara P.Z., Dario D.C. Trujillo, S.C. (2018)Period addition phenomenon and chaos control in a zad controlled boost converterr.International Journal of Bifurcation and Chaos,28:DOI:10.1142/S0218127418501572.

IOP Conf. Series: Materials Science and Engineering **787** (2020) 012015 doi:10.1088/1757-899X/787/1/012015

Finite Element Analysis of Stress Distribution Characteristics of Vehicle Retreaded Tire Composite Layer

Wang Qiang, Jiang Li*, Wang Yunlong, Qi Xiaojie and Wang Guotian

School of Automobile and Traffic Engineering, Heilongjiang Institute of Technology, Harbin, Heilongjiang, 150050, China

*Corresponding author's e-mail: 630702666@qq.com

Abstract. In order to further reveal the failure and damage mechanism of retreaded tires in use, such as tread delamination and pressure burst, we constructed a retreaded tire material distribution model, a stress constraint model for each layer, and a finite element model. Stress distribution characteristics of the tread, breaker, belt, carcass, sidewall, toe rubber, etc. under free condition and static grounding condition were analyzed by numerical simulation, and the following main conclusions were obtained: Under the free-inflation condition, the tire carcass of the retreaded tire is subjected to the most stress, the belt is next to it, the breaker and the tread are the least, and the stress on the tread and tire shoulder is relatively great. Under the static grounding condition, the belt of the retreaded tire of a vehicle is subjected to the most stress, the carcass next to it, the stress distributions of the belt and the carcass are similar, and the stress on the tread and the breaker is relatively small.

1. Introduction

There are various methods for the disposal of used vehicle tires. In accordance with global development trends and environmental awareness, tire retreading has been promoted. Due to its unique low-carbon environmental protecting characteristic and advantages, it has developed rapidly in the vehicle tire industry and has accelerated the overall development of the tire industry [1]. Therefore, increasing the retread rate of used vehicle tires can effectively increase the utilization rate of used vehicle tires, which will greatly save rubber resources and promote environmental protection. In the process of using retreading tires, the tires often experience increased tread wear, chipping, delamination, and even failure modes such as pressing burst and puncturing burst. At present, researches mainly focus on improving the design of the tread rubber matrix formula and improving the retreading technology to solve the above problems, and the research on the comprehensive mechanical properties and internal micro-mechanical characteristics of retreaded tires of vehicles are few, resulting in uneven quality of retreaded tires and low service life [2-3]. To this end, by constructing a retreaded tire material distribution model, a stress constraint model for each layer, and a finite element analysis model, this paper qualitatively and quantitatively describes and evaluates the stress distribution characteristics of each layer of the retreaded tire, and further reveals the problems appearing when the retreaded tire are in use, such as tread delamination, pressing burst and other failure and damage mechanisms, so as to provide theoretical guidance for improving the performance of retreaded tires and optimizing the retreading technology.

2. Material Distribution Model of Retreaded Tire Composite Layer

This article takes 26.5R25 vehicle retreaded tire as the main research object, and its composite layer

IOP Conf. Series: Materials Science and Engineering **787** (2020) 012015 doi:10.1088/1757-899X/787/1/012015

material distribution model is shown in Figure 1. It is mainly composed of 7 parts: tread, breaker, breaker belt, old carcass, sidewall, toe rubber and bead ring [4-6].

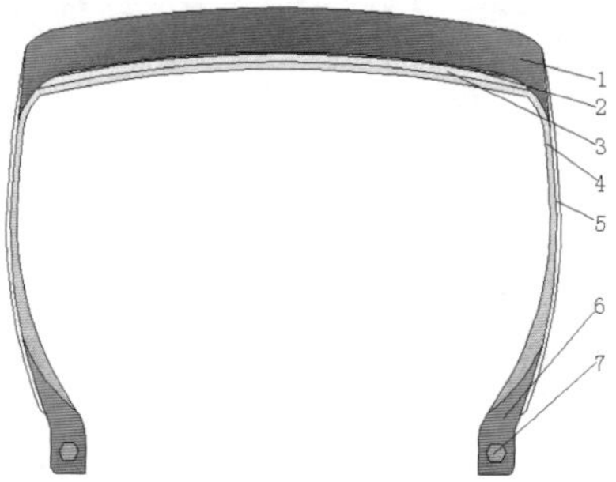

1-Tread 2-breaker 3-Belt 4-Old Carcass 5-Sidewall 6-Bead Rubber 7-Bead Ring
Figure 1 Material Distribution Model of Retreaded Vehicle Tires

3. Stress Constrained Models of Each Layer of Vehicle Retread Tires

The strength and stiffness of the retreaded tires of a vehicle mainly depend on the maximum stress that various layers can withstand without being damaged due to external forces. Among them, the stress constraint model of the carcass layer of the retreaded tire [7-10] is

$$\sigma_t \le \frac{10^3 P(R^2{}_k - R^2{}_0)}{2R_k n i_k} \cdot K \tag{1}$$

where σ_t—stress on the carcass, MPa; P—tire pressure, kPa; R_k—radius of the tread point, m; R_0—radius of the widest point of the section, m; n—wire of carcass layer-cord, layer; i_k—wire of the tread point-cord density, piece/ m; K—design safety factor of carcass.

The stress constrained model of belt [7-10] is

$$\sigma_d \le \frac{10^3 P R_k}{n i_k \cos^2 \alpha_k} \cdot K \tag{2}$$

Where, σ_d—stress of the belt, MPa; P—inflated internal pressure, kPa; R_k—crown point radius, m; n—number of layers of wire ply working layers in the belt, layer; i_k—density of the belt at the tread point , piece/ m; α_k—the wire on the belt-cord angle, (°); K— design safety factor of the belt.

The constraint model of the stress on the tread, breaker and sidewall [7-10] is

$$\sigma_m \le [\sigma_{mr}] \cdot K \tag{3}$$

$$\sigma_h \le [\sigma_{hr}] \cdot K \tag{4}$$

$$\sigma_c \le [\sigma_{cr}] \cdot K \tag{5}$$

where σ_m—stress on the tread, MPa; σ_h—stress on the breaker, MPa; σ_c—stress on the sidewall, MPa; K—design safety factor of the tread, breaker, and sidewall; $[\sigma_{mr}]$—maximum allowable stress on tread rubber material, MPa; $[\sigma_{hr}]$—maximum allowable stress on breaker rubber material, MPa; $[\sigma_{cr}]$—maximum allowable stress on sidewall rubber material, MPa.

4. Finite Element Analysis Model of Vehicle Retreading Tire

The finite element model constructed by using ANSYS Workbench software is shown in Figure 2. The grid is divided into tetrahedral elements and some parts are refined. The model consists of 218,954 degrees of freedom, 79,478 nodes, and 32,666 elements. The material parameters of tread, breaker, sidewall, toe rubber and bead ring are shown in Table 1. The material parameters of old carcass and belt are shown in Table 2. Among them, the tread, breaker, sidewall and toe rubber are simulated by Mooney-Rivlin model, the bead ring is simulated by Solid element, and the old carcass and breaker are simulated by Layer element [11-12].

Figure 2 Finite Element Model

Table 1.Material Parameters of Tread, etc.

Material	Elastic Modulus/MPa	Poisson's Ratio	Density/kg/m^3
Tread	8.14	0.48	1810
Breaker	6.12	0.48	1120
Sidewall	9.96	0.48	1260
Toe Rubber	12.16	0.48	1470
Bead Ring	2.11 e5	0.29	7850

Table 2.Material Parameters of the Old Carcass and Belt

Material	Elastic Modulus/MPa			Shear Modulus/MPa			Poisson's Ratio			Density/ kg/m^3
	E_1	E_2	E_3	G_{12}	G_{23}	G_{31}	μ_{12}	μ_{23}	μ_{31}	
Old carcass	9.8e4	6.86	6.86	6.14	3.26	3.26	0.388	0.388	0.48	4650
Belt	1.6e5	2.82	2.82	9.68	2.46	2.46	0.366	0.366	0.48	6570

5. Finite Element Analysis of Stress in Each Layer Under Free Inflation Condition

The comprehensive stress cloud diagram of the engineering retreaded tire at an inflation pressure of 600 kPa is shown in Figure 3. The maximum stress is 38.05 MPa, which occurs at the bead ring. The stress distribution along the radial direction of the tire is shown in Figure 4. From Figure 4, it can be seen that the stress on the carcass is the maximum, which is 0.81 MPa, followed by the breaker, which is 0.42 MPa, the stress decreases gradually along the breaker and the tread, and there is a large value at the shoulder of the carcass, which is 2.23 MPa. The results show that under the condition of free inflation of the retreaded tire of a vehicle, the main stressed component is the carcass, which expands outward under air pressure, which causes the belt, the breaker and the tread to expand outward and form a tight hoop. The effect indicates that the quality of the old carcass of used car retreaded tires will have a direct impact on the radial deformation of the free-inflated retreaded tires. The stress distribution curve of each layer along the tire width direction is shown in Figure 5. Among them, the carcass layer is subjected to the most stress, followed by the belt, the breaker, and the tread. The stress change law of each layer is basically the same. A larger value appears at the mid-line of the tread, but the carcass is different from other layers that the stress reaches a maximum at the shoulder. The results show that the shoulder of the old carcass of the retreaded vehicle must be carefully inspected before retreading. If there are damage defects, it will directly affect the mechanical properties of the retreaded tire.

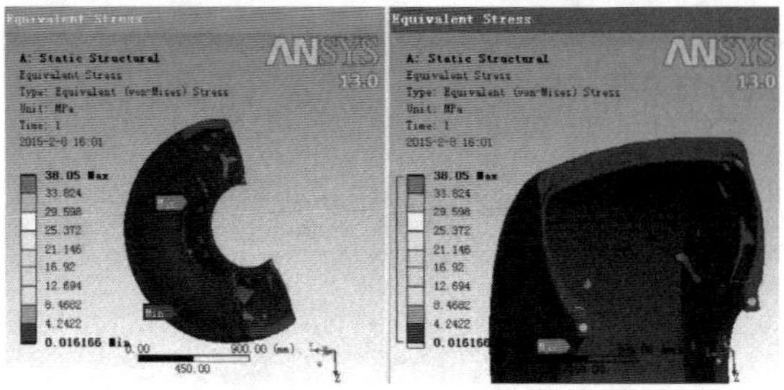

Figure 3 Comprehensive Stress Cloud Diagram

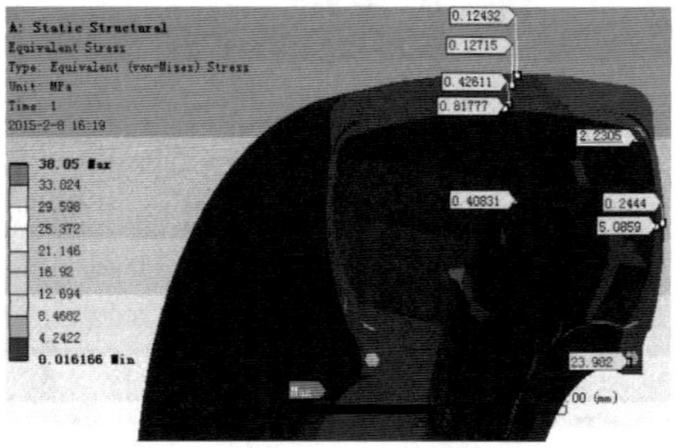

Figure 4 Stress Cloud Diagram in the Radial Direction of the Tire

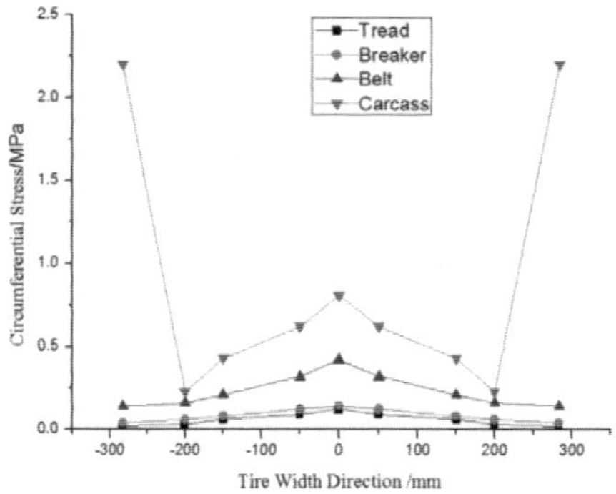

Figure 5 Stress of Each Layer in the Tire Width Direction

6. Finite Element Analysis of Stress in Each Layer Under Static Ground Contact Condition

Figure 6 shows the comprehensive stress cloud diagram of the tire when the inflation pressure is 600kPa, the load is 135kN, and the interface friction coefficient is 0.9. As shown in Figure 6, the maximum stress is 15.11 MPa, which occurs at the belt.

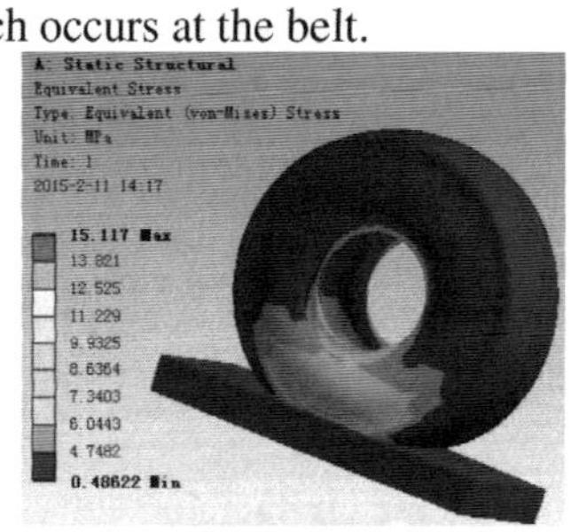

Figure 6 Comprehensive Stress Cloud Diagram

The stress cloud diagram of each layer is shown in Figure 7, and the stress distribution curve of each layer along the tire width direction is shown in Figure 8. It can be seen from Figures 7 and 8 that the stress distribution in the tread ranges from 0.97 to 4.25 MPa, and the maximum stress value of 4.25

MPa occurs at the shoulder, gradually decreases toward the center of the ground contact along the width direction, and then becomes larger when approaching the center of the tread. The stress distribution range in the breaker is 0.49-1.65MPa, the maximum stress value of 1.65MPa occurs at the edge of the breaker, and the stress gradually decreases in the width direction toward the ground contact center. The stress distribution range in the belt is 3.40-14.90 MPa, the maximum stress value of 14.90 MPa appears at the center of the ground contact area between the belt and the breaker, and gradually decreases to the two sides along the width direction, and then increases at the edge of the belt. The stress distribution range in the carcass. It is 2.43-10.64 MPa, and its maximum stress value of 10.64 MPa appears in the center of the ground contact area between the carcass and the belt, and gradually decreases to the two sides along the width direction, and then increases at the edge of the carcass, which is similar to the changing trend of the belt. The stress distribution range in the sidewall is 1.45-6.38 MPa, and the maximum stress value of 6.38 MPa appears at the place where the sidewall contacts the shoulder. The stress distribution range in the toe is 1.45-9.58 MPa, its maximum stress value of 9.58M Pa appears inside the toe rubber.

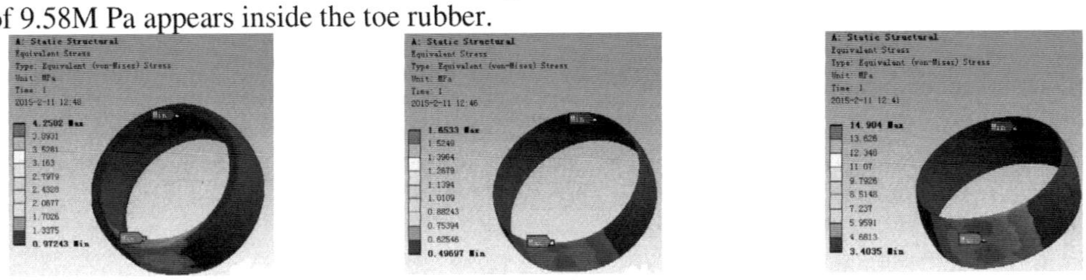

(a) Tread Stress Cloud Diagram (b) Breaker Stress Cloud Diagram (c) Belt Stress Cloud Diagram

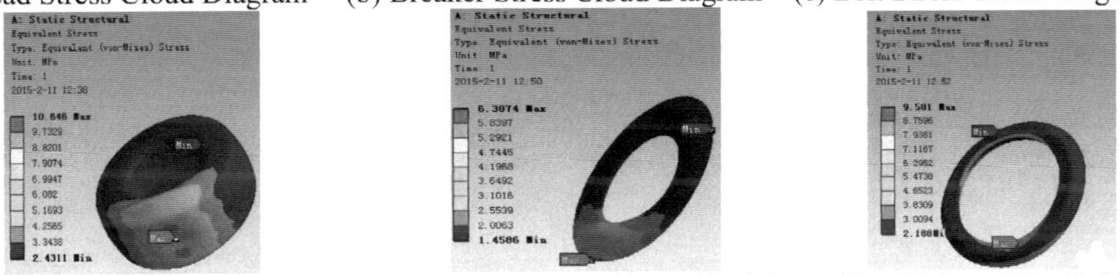

(d)Stress Cloud Map of Carcass (e)Stress Cloud Map of Sidewall (f)Stress Cloud Map of Toe Rubber

Figure 7 Stress Cloud Diagram of Each Layer

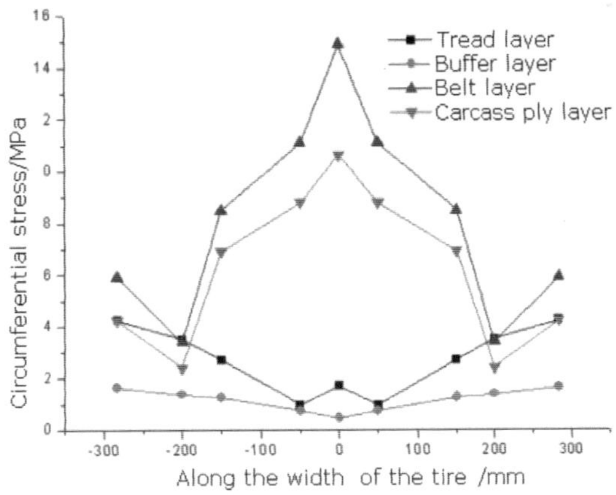

Figure 8 Stress Distribution Curve of Each Layer Along the Tire Width

7. Conclusion

(1) The composite layer material distribution model, stress constraint model of each layer, and infinite element analysis model of the vehicle retreaded tire were constructed, and the stress distribution characteristics of the vehicle retreaded tire were obtained.

(2) Under the condition of free inflation, the carcass of the retreaded tire is subjected to the most stress, followed by the belt, the stress on the breaker and the tread is smaller, and the stress on the shoulder is relatively large.

(3) The breaker belt of the retreaded tire of a vehicle is the most stressed, followed by the carcass, the stress distribution of the belt and the carcass is similar, and the stress on the tread and the breaker is relatively small.

(4) The results of finite element analysis show that the old carcass and belt of the retreaded tire of the vehicle are the main load-carrying components. Therefore, it is recommended that the quality inspection of the old carcass and belt should be strengthened before retreading, especially the quality of the bundle cord in the old carcass must be strictly tested.

Acknowledgments

Fund Project: Project supported by basic scientific research business expenses of Heilongjiang Provincial Undergraduate Universities (2018CX07); Heilongjiang Natural Science Foundation Project (LH2019E115); Heilongjiang Institute of Technology PhD Research Startup Fund Project (2016BJ02)
Author: Wang Qiang (1981-), male, Harbin, associate professor, doctor of engineering, postdoctor. Research interests: vehicle tire technology, vehicle driving safety.

References

[1] Zhao Li. The Current Status, Problems and Countermeasures of Waste Tire Resources in China [J]. Comprehensive Utilization of Tire Resources in China, 2018 (12): 41-43.

[2] Zhong Junxiong. Necessity Analysis of Quality Monitoring of Retreaded Tire Products [J]. China New Technology & Products, 2016, (11): 161.

[3] Liu Daochun. Tire Retreading Industry Leads the Future of Low Carbon and Energy Saving [J]. Modern Rubber Technology, 2016, (02): 9-14.

[4] Wang Wenfeng. Finite Element Simulation and Optimization Design of Pneumatic Tires [D]. Qingdao University of Science and Technology, 2019.

[5] Wang Zhiping, Pei Quanhua, Song Chaoxing, Wang Caihong, Wang Ruofei. Finite Element Three-dimensional Modeling Method of Tires with Complex Tread Pattern [J]. Rubber Science and Technology, 2019, 17 (05): 254-256.

[6] Liu Yunpeng, Zhou Tao, Yang Xiaoguang. Application of Finite Element Analysis in Tire Structure Design [J]. Tire Industry, 2019, 39 (05): 263-267.

[7] Su Xiaofeng, Fu Ping, Ding Zhongjun, Li Dewei, Yang Mingfei. Finite Element Model Establishment and Simulation Analysis of the Tire Based on Abaqus Software [J]. Rubber Industry, 2019, 66 (02): 121-127.

[8] Wang Jun, Li Liang, Sun Lin, Sha Changxin. Finite Element Analysis of 245 / 70R16 Tires [J]. Tire Industry, 2016, (09): 520-528.

[9] Jing Yu, Jiang Zhanghua, Tan Lin. Design of 11.00R2018PR All-Steel Light Load Radial Ply Tire [J]. Rubber Science and Technology, 2016, (12): 37-39.

[10] Liu Shenglin. Design of 385 / 55R19.5 Wide Base Tubeless All-steel Radial Tire [J]. Tire Industry, 2016, (09): 529-532.

[11] Wang Fengchi, Tian Peipei, Feng Xuning, Wang Xinbo. Finite Element Analysis of the Structure Foundation of Waste Rubber Tires [J]. Journal of Shenyang Architecture University (Natural Science Edition), 2019, 35 (05): 867-874. [2] Wang Zhiping, Pei Quanhua, Wang Caihong, Chen Yu. Finite Element Analysis of Bead Undulation Mechanism of All-Steel Radial Ply Tires [J] .Tire Industry, 2019,39 (08): 463-466.

IOP Conf. Series: Materials Science and Engineering **787** (2020) 012015 doi:10.1088/1757-899X/787/1/012015

[12] Wei Yintao, Liu Zhe, Zhou Fuqiang, Zhao Chonglei. Three-Dimensional Ring Model of Tire Considering Out-Of-Plane Vibration [J]. Journal of Vibration Engineering, 2016, (05): 795-803.

An optimized Algorithm for broadcast beacon frequency and transmission power in VANETs

Guoyan Yang, Xiangfeng Li and Haiyang Shi

School of Data Science and Technology Heilongjiang University, Harbin, Heilongjiang, 150080, China

*Corresponding author's e-mail: fengxue1995101@163.com

Abstract. Power control has been an important branch of the research field for VANETs. In order to meet the safety of vehicle driving, Vehicle-to-Vehicle(V2V) communication has become a research hotpot simultaneously. In the internet of vehicles, most of the messages generated by V2V communication are implemented by broadcasting. This paper proposed a joint optimization algorithm for power control and broadcast beacon frequency based on the LTE-V2V communication protocol. This algorithm applies deep reinforcement learning (DRL) and designs a reward function that can represent the communication effect of LTE-V2V. The experimental results show that the proposed algorithm can better configure the transmitting power and broadcast beacon frequency of the vehicle terminal, and obtain more reliable communication effects in the dynamic traffic flow environment.

1. Introduction

Mobile communication of vehicles is a vital part of the Intelligent Transportation System (ITS). LTE-V2V (long-term-evolution vehicle-to-vehicle) communication, as a kind of in-vehicle communication technology, is achieved by the device-to-device (D2D) communication between cars which are approaching each other in geographical location [1]. On the basis of LTE-V2V, the in-vehicle communication terminal can multiplex the frequency domain and the time domain communication resource broadcast beacon of the cellular communication user so that simultaneous improving the transmission rate of V2V data and the resource utilization efficiency of the cellular communication.

However, the delay, reliability, and data transmission rate of V2V wireless communication, in view of the application scenario about LTE-V2V vehicle cooperative sensing, are mainly affected by the following two aspects:

On the one hand, the transmission power of the vehicle communication terminal has an impact on Vehicle cooperative sensing performance. Literature [2] has put the delay and reliability of the vehicle link as a constraint with a consider of slow time-varying channel state information and has used the power control and resource allocation algorithm to maximize the total rate of cellular users and to take into account the fairness between users in the meantime. On the other hand, Vehicle cooperative sensing performance depends on the frequency of broadcasting beacons of in-vehicle communication terminals. Literature [3] has suggested a broadcast relay algorithm that selects the next hop node relay broadcast message when adaptively estimating the link quality and the distance between the source broadcast and the potential transponder. The key point in this method is selecting a certain number of relay nodes to improve the reliability of packets. Secondly, literature [4] has proposed an algorithm that based on the average speed difference between the vehicle and the neighbouring vehicle to form a cluster could relay the message through the cluster head.

The existing LTE-V2V vehicle wireless communication technology mainly adopts the broadcast beacon frequency and transmit power of a fixed configuration of the vehicle communication terminal as raw materials for independent design. Of course, the constraint condition that meeting the QoS requirements of vehicle collaboration awareness related applications in a specific traffic flow environment needs to be satisfy during the design process [5,6]. However, this independent configuration design method has not considered the potential coupling effect of the broadcast beacon frequency and transmits power of the in-vehicle communication terminal on the communication performance of the LTE-V2V in-vehicle communication terminal.

Aiming at the above two aspects affecting the communication effect of LTE-V2V, this paper has proposed a joint optimization algorithm of LTE-V2V power control and broadcast beacon frequency based on DDPG, which is suitable for LTE-V2V by constructing a reasonable reward function. The communication environment is trained by the deep reinforcement learning algorithm to obtain the optimal control strategy of the power of the vehicle communication terminal and the broadcast beacon frequency, and the effectiveness is verified by experiments.

2. System model

In the geographic range covered by the signal coverage of the LTE cellular communication base station, the number of different types of traffic flow states that have appeared in the road traffic environment and the average LTE-V2V vehicle density at the m-th level are defined as M （M =0,1,2, … n） and ρ_m respectively. The definitions of the following parameters used in this paper are shown below:

P_{Tx}: the allowable power of the vehicle communication terminal;P_{max}: the maximum amount of P_{Tx};

R_{CA}: the range of expected LTE-V2V vehicle collaborative sensing;α : the path fading index of TE-V2V wireless communication;γ_{min} : the minimum SINR that can be reliably guaranteed for LTE-V2V vehicle cooperative sensing application QoS;P_{noise} : the average noise power of LTE-V2V wireless communication channel;d_0: the path fading coefficient of LTE-V2V unit vehicle relative to geographical distance LTE-V2V wireless communication;G_{Rx} : the antenna gain of LTE-V2V vehicle communication terminal receiving wireless signal; θ : the coefficient of self-interference power cancellation strength of LTE-V2V vehicle communication terminal;L_b: the load of each broadcast beacon of LTE-V2V vehicle communication terminal; $N_{max}^{subframe}$: the amount of resource blocks corresponding to each sub-frame wireless signal of the LTE-V2V wireless communication channel;$N_{avg}^{subframe}$: the average amount of resource blocks occupied by each sub-frame radio signal in the LTE-V2V wireless communication channel used for sharing;N_{beacon} : the amount of resource blocks required for each broadcast beacon of the LTE-V2V in-vehicle communication terminal;

This paper has stated the following functions according to these definitions:

$$N_{subframe-beacon} = \min\left\{n \in z \mid n \geq N_{beacon} / N_{avg}^{subframe}\right\} \quad (1)$$

which represents the amount of wireless signal sub-frames corresponding to each broadcast beacon of the LTE-V2V vehicle communication terminal;

$$\eta = \min\left\{n \in z \mid n \geq N_{beacon} / N_{subframe-beacon}\right\} / N_{\max}^{subframe} \quad (2)$$

which represents the bandwidth resource occupancy rate of the broadcast beacon of the LTE-V2V vehicle communication terminal(η) ;

Therefore, using the equation (1) and (2) to get the minimum value that the LTE-V2V vehicle communication terminal needs to meet the transmit power ($P_{\min}$):

$$P_{\min} = \max\left\{0, \frac{R_{CA}^{\alpha}\lambda_{\min}d_0\eta P_{noise}}{G_{Rx} - R_{CA}^{\alpha}\lambda_{\min}d_0\eta}\right\} \quad (3)$$

Then, constructing the constraints of LTE-V2V vehicle communication terminal transmit power:

$$P_{\min} \le P_{Tx} \le P_{\max} \quad (4)$$

Under the condition of maintaining the communication link between LTE-V2V in-vehicle wireless communication terminals without interruption reliably, we construct the function relationship between the average multiplexing distance supported by LTE-V2V resources ($R_{LTE-V2V}$) and the transmission power of LTE-V2V vehicle wireless communication terminal (P_{Tx}),between LTE-V2V in-vehicle wireless communication terminals without interruption reliably, in order to achieve the function relationship between the average multiplexing distance supported by LTE-V2V resources ($R_{LTE-V2V}$) and the transmission power of LTE-V2V vehicle wireless communication terminal (P_{Tx}):

$$R_{LTE-V2V}\left(P_{Tx}\right)=\max\left\{2,1+\left[\frac{1}{\lambda_{\min}}-\frac{\eta P_{noise} d_0 R_{CA}^{\alpha}}{G_{Rx} P_{Tx}}\right]^{-\frac{1}{\alpha}}\right\} R_{CA} \quad (5)$$

Simultaneously, this paper considers the relative geographic distance error between LTE-V2V in-vehicle wireless communication terminals acquired by LTE base stations and assumes that the error is $\delta_{LTE-V2V}$. As a result, the function relationship between the distance supported by LTE-V2V ($\overline{R}_{LTE-V2V}$) and the transmission power of LTE-V2V vehicle wireless communication terminal (P_{Tx}) is

$$\overline{R}_{LTE-V2V}\left(P_{Tx}\right)=\max\left\{2,1+\left[\frac{1}{\lambda_{\min}}-\frac{\eta P_{noise} d_0 R_{CA}^{\alpha}}{G_{Rx} P_{Tx}}\right]^{-\frac{1}{\alpha}}\right\} R_{CA}+\delta_{\text{LTE-V2V}} \quad (6)$$

The definition allows the maximum and minimum values of the broadcast beacon frequency of the in-vehicle communication terminal configuring the LTE-V2V is $f_{\max}$ and $f_{\min}$ respectively. Then, the constraints of LTE-V2V vehicle communication terminal broadcast beacon frequency is constructed a $f_{\min} \le f_b \le f_{\max}$.

The average duration of wireless signals per sub-frame in LTE-V2V wireless communication channel is defined as T_{sbuframe}. As a consequence, the relationship between the maximum number of LTE-V2V vehicle terminals that could simultaneously support multiplexing resources supported by LTE-V2V resources () and the broadcast beacon frequency of LTE-V2V vehicle wireless communication terminal ($N_{\text{LTE-V2V}}(f_b)$) is shown below:

$$N_{LTE-V2V}\left(f_b\right)=\max\left\{n\in z \mid \text{n} \le \frac{\max\left\{n' \le \frac{1}{f_b T_{sbuframe}}\right\}}{N_{subframe-beacon}}\right\} \quad (7)$$

The density of LTE-V2V vehicle in traffic flow environment is ρ_m . With this condition, the probability function that represents he number of LTE-V2V vehicle terminals that are simultaneously resource multiplexed (φ) does not exceed $N_{LTE-V2V}\left(f_b\right)$ is constructed as

$$p_{\lambda_j}\left(f_b, P_{Tx}\right)=\sum_{\varphi=0}^{N_{LTE-V2V(f_b)}} \frac{\left(\lambda_j \overline{R}_{LTE-V2V}\left(P_{Tx}\right)\right)^{\varphi} e^{-\lambda_j \overline{R}_{LTE-V2V}\left(P_{Tx}\right)}}{\varphi!} \quad (8)$$

Meanwhile, the estimation function of LTE-V2V vehicle collaborative sensing energy efficiency indicator ($E\left(f_b, \mathrm{P}_{Tx} \mid \lambda\right)$) is constructed as

$$E(f_b, P_{Tx} \mid \lambda) = \frac{p_\lambda(f_b, P_{Tx}) f_b L_b}{P_{Tx}} \quad (9)$$

Table 1 shows the parameters and values used in the model, as follows:

Table 1 Model Parameter and Value.

Parameter	Value
Vehicle collaboration perception range (R_{CA})	420m
Wireless communication path fading index (α)	2.75db
Noise power (P_{noise})	-125db
Antenna gain at the receiver (G_{Rx})	3db
Lowest SINR ($\gamma_{\min}$)	20.88db
Maximum Power (P_{Tx})	10db
Path loss at 1 m (d_0)	47.86 dB
Beacon size (L_b)	1600bit
Number of resource blocks corresponding to each sub-frame wireless signal ($N_{\max}^{\text{subframe}}$)	100
$N_{\max}^{\text{subframe}}$ for shared ($N_{\text{avg}}^{\text{subframe}}$)	81
Number of resource blocks required for each beacon (N_{beacon})	100

3. A Joint optimized algorithm for power control and broadcast beacon frequency based on DDPG in VANETs

Reinforcement Learning (RL) is a kind of learning from environmental state mapping to action and the learning aim of it is to maximize the cumulative rewards of agents in their interactions with the environment [7]. The Markov Decision Process (MDP) can usually be used to model RL. MDP is generally defined as a five-tuple $< S, A, P, R, \gamma >$. And S is a collection of all environmental states. A is a set of executable actions of the agent.P is a state transition probability. R is represented as a transient reward. γ is a discount factor ($\gamma \in [0,1]$),which is used to calculate a bonus value for the next state. In the process of reinforcement learning, the agent observes that it is currently in a state s_t, and uses this action a_t to interact with the environment. The environment switches to the next state s_{t+1} according to the current state and the action selected by the agent. The purpose of the Agent is to learn a strategy μ to maximize the formula: $R_{total} = R_1 + \lambda R_2 + \ldots = \sum_{k=0}^{\infty} \gamma^k R_{k+1}$.

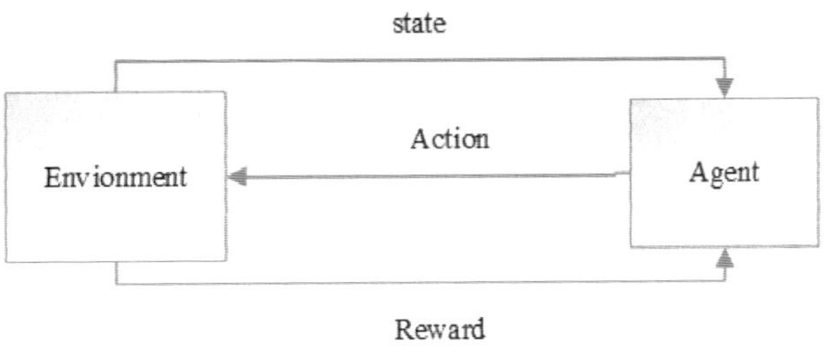

Figure 1 DRL working principle

In our model, the agent is first defined as the communication device of the vehicle. The interaction with the agent is the environment, and the state is the basis for the agent to make the current decision.

IOP Conf. Series: Materials Science and Engineering **787** (2020) 012016 doi:10.1088/1757-899X/787/1/012016

The state is defined in our model. For vehicle density, the vehicle decides which strategy to adopt based on the current state. According to formula (9), the reward function is set to:

$$R_t = E\left(a_t \mid s_t\right) \tag{10}$$

s_t is the current environment and a_t is the action that the agent makes based on the current environment.

4. Simulation Results

In order to evaluate the feasibility of the proposed algorithm, the simulation experiment simulates an environment with 9 traffic flows on the GYM platform, and each change of traffic flow follows the Markov probability transfer matrix. This paper considers the LTE-V2V communication protocol scenario in [8, 9]

In the DDPG experiment, the Actor network is designed to have two fully connected hidden layers, each of which has 200 neurons, the activation function is the *Relu* function and the activation function of the output layer is the *Tanh* function. There are three hidden layers in the Critic network, and the number of neurons in the hidden layer is 200, 50, and 200, respectively. The rest of the neural network hyper parameters involved are listed in Table 2.

Table 2. Hyper Parameter Setting

Parameter	Value
learning rate for actor（LR_A）	0.005
learning rate for critic（LR_C）	0.005
soft replacement（TAU）	0.01
reward discount（GAMMA）	0.9
Replaybuffer MEMORY_SIZE	10000
BATCH_SIZE	32

In Figure.2, we can observe that under the same transmission power and broadcast beacon frequency based on LTE-V2V protocol, the probability of normal communication between vehicles becomes greater when the density of traffic becomes larger and the communication distance becomes longer. In Fig.3, the simulation results show that under the same transmission power (33dbm), the density of each type of vehicle will have a greater impact on the reward function as the broadcast beacon frequency increases.

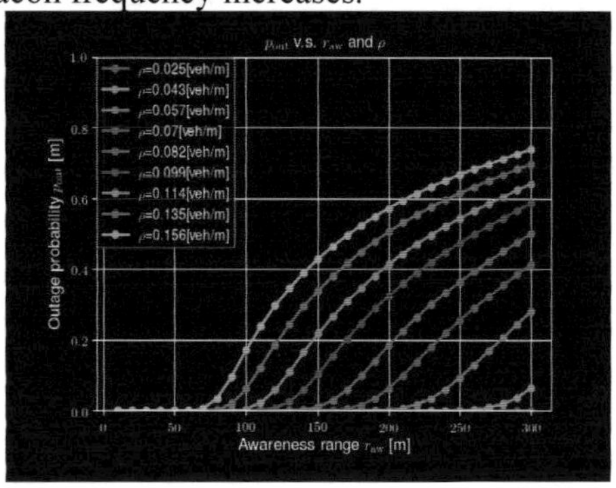

Figure.2 The probability of communication Disconnection changes with distance Between vechicles increasing.

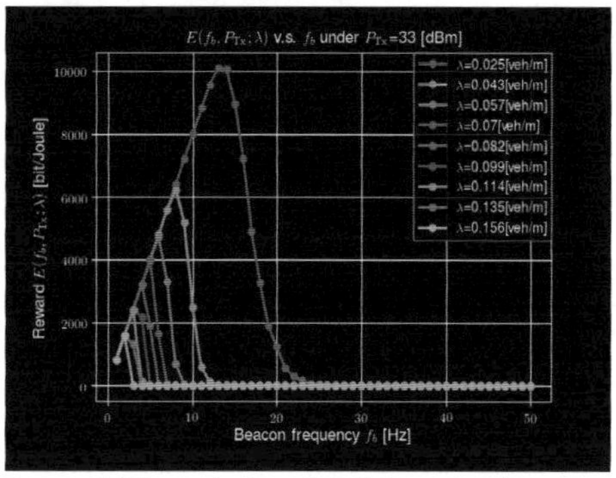

Figure.3 The relationship between the reward function and the broadcast beacon frequency under the specific power value.

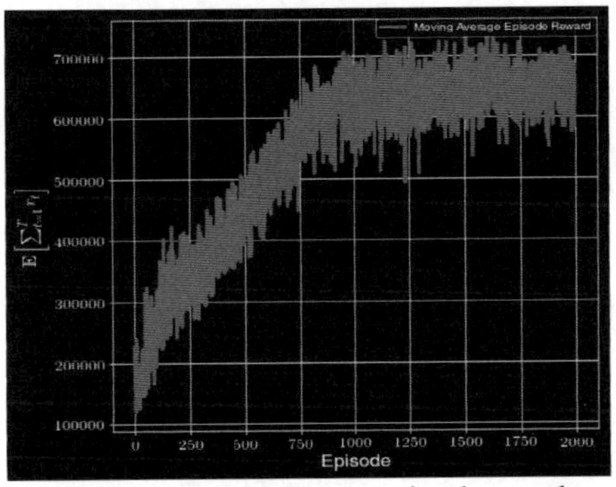

Figure.4 Rewards per episode graph

In Fig.4, it can be clearly seen from the simulation results that the algorithm proposed in this paper can realize that the reward value does not change greatly with the change of the current environment when the number of iteration steps increases.

5. Conclusion

In order to obtain the largest LTE-V2V communication cooperative sensing energy efficiency target, this paper proposed a joint optimization algorithm based on deep reinforcement learning for broadcast beacon frequency and transmit power. On the GYM platform under Open-AI, the simulation results show that the proposed algorithm can accurately make better configuration of the power of the vehicle communication terminal and the broadcast beacon transmission frequency in the dynamic traffic flow environment based on the LTE-V2V communication protocol. At the same time the proposed algorithm can get more reliable communication quality.

References

[1] Chen S , Hu J , Shi Y , et al. (2016). LTE-V: A TD-LTE based V2X Solution for Future Vehicular Network. IEEE Internet of Things Journal, 3:997-1005.

[2] Sun W L,Strom E G,Branteron F,et al.(2016). Radio resource management for D2D-based V2V communication.IEEE Trans. On Vehicular Technology,65:6636 - 6650.

[3] Rehman O, Ould-Khaoua M, Bourdoucen H.(2016) An adaptive relay nodes selection scheme for multi-hop broadcast in VANETs. Computer Communication,87:76-90.

[4] Hosmani S, Mathpati B.(2017) Survey on cluster based routing protocol in VANET. In 2017 International Conference on Electrical, Electronics, Communication, Computer, and OPtimization Techniques(ICEECCOT) Mysuru,India, pp:1-6.

[5] Sattari M R J, Noor R M, Ghahremani S. (2012) A taxonomy for congestion control alggorithms in Vehicular Ad Hoc Networks. In 2012 IEEE International Conference on Communication , Networks and Satelite (ComNetSat). Bali, Indonesia.pp:95-108.

[6] Naja A, Boulmalf M. (2017) Performance analysis of an improved probability-based and counter-based broadcast protocols for VANETs. In 2017 International Conference on Elictrical and Information Technologies(ICEIT).Rabat Morocco.pp:1-5.

[7] Liu Q , Zhai J W, Zhang Z C.(2018) A Survey on Deep Reinforcement Learning.Chinese Journal of Computers. 41:1-27.

[8] A. Bazzi, B. M. Masini, and A. Zanella. (2015) Performance analysis of V2V beaconing using LTE in direct mode with full duplex radios. IEEE Wireless Communication.4:685-688.

[9] A. Bazzi, B. M. Masini, and A. Zanella. (2017) On the performance of IEEE 802.11p and LTE-V2V for the Cooperative Awareness of Connected Vehicles. IEEE Transactions on Vehicular Technology. 66:10419-10432.

Greenhouse gas emission of corn butanol in life-cycle

Jiang Li, Wang Qiang* and Zhu Rongfu

School of Automobile and Traffic Engineering, Heilongjiang Institute of Technology, Harbin, Heilongjiang, 150050, China

*Corresponding author's e-mail:630702666@qq.com

Abstract. Since advances in the ABE (acetone-butanol-ethanol) fermentation process in recent years have led to significant increases in its productivity and yields, the production of butanol and its use in motor vehicles have become an option worth evaluating. It employs a well-to-wheels (WTW) analysis tool, and ascertains Bio-butanol-related WTW activities including corn farming, corn transportation, butanol production, butanol transportation, and vehicle operation. According to the 3 kinds of by-product of different methods of distribution, calculation and comparison of the greenhouse gas emissions in the life cycle of corn alcohol. The results show that 1) life cycle GHG emission of the corn butanol reduced 26%—49.5% according to the energy value method and product substitution method;2) bio-acetone that was coproduced with bio-butanol as an alternative to petroleum-based acetone could significantly reduce the life cycle GHG emission of butanol; 3) In the life cycle of the corn butanol, the main source of GHG emission was the use of natural gas, diesel and nitrogen, , accounting for the total GHG emission by 53%, 21.4% and 13.38% respectively.

1. Introduction

2010, crude oil consumed in transportation accounted for about 68% of the total crude oil consumption, of which vehicle oil consumption accounted for about 57%. There is a shortage of domestic oil resources, and more than half of the oil is dependent on imports. The NDRC forecasts that by 2015, China may import 370 million tons of crude oil, equivalent to 7.4 million barrels a day, and the dependence on oil imports will reach 65 percent. The rapid increase of energy consumption has not only brought serious threats to energy security, especially oil security, but also led to high environmental emissions. The development of traffic alternative fuels has become an important measure for our country to get rid of the energy, especially oil predicament [1]. As an alternative fuel for vehicles, butanol has more advantages than ethanol and has become one of the most concerned fuels in recent years.

From the whole life cycle of traffic fuel, the greenhouse gas (greenhouse gas, GHG) emitted by the engine during the driving process of the motor vehicle accounts for only a part of the total carbon emissions, and the rest is derived from the fuel production link, which needs to consume a variety of energy and resources, and greenhouse gas emissions . At present, the domestic research on fuel butanol is focused on the use of vehicles, and there is little research on the life cycle of butanol. In this paper, the life cycle of fuel butanol is evaluated by using corn as raw material.

2. Life cycle inventory analysis of butanol

Taking the national average corn planting area (8500 kg / ha corn yield) as the research object, taking the 150000 t / ha butanol production system as an example, taking the GHG emission index from the raw material production to the vehicle use in the life cycle of ordinary gasoline as the evaluation basis, the GHG emission in the life cycle of corn butanol is evaluated.

2.1. Formatting the title

The title is set 17 point Times Bold, flush left, unjustified. The first letter of the title should be capitalized with the rest in lower case. It should not be indented. Leave 28 mm of space above the title and 10 mm after the title.

2.2. Life cycle systems

The Butyl alcohol life cycle system includes five stages: raw material production, raw material transportation, butanol production, butanol transport and the use of butanol on vehicles. The raw material production stage refers to the "sowing, growing, harvesting and processing" process of corn. The main material inputs include seed, machinery, nitrogen fertilizer, phosphate fertilizer, potassium fertilizer, electricity, diesel oil, pesticide and so on. The input of substances per hectare of corn production and processing is shown in Table 1. The production stage of butanol is the process of converting the solar energy absorbed by corn production into chemical energy. The main processes include crushing, liquefaction, saccharification, ABE fermentation, gas stripping, distillation, drying, etc. In addition to butanol, there are acetone, ethanol and dry distillerundefineds grains soluble (DDGS) [2].

Table 1. Material input of every hectare corn production and processing

seeds/kg	machinery /kg	powe /kwh	diesel /L	pesticide /kg
32	10.8	270	97	1.25

In the Butyl alcohol production phase, natural gas is mainly used to provide steam, and the required electricity comes from the national power grid. The demand for natural gas and electricity during the butanol production phase is shown in Table 2.

Table2. Natural gas and electricity requirement for the corn butanol production process

Production	Natural gas（MJ/kg）	Electric energy （kWh/kg）
smash	5.1	0.04
dry	5.5	
ABE fermentation and gas Stripping, distillation	24.05	0.04
amount to	34.65	0.08

2.3. Evaluation indicators

Greenhouse gases (GHG) are natural and man-made gases in the atmosphere that absorb and re-emerge infrared radiation, such as carbon dioxide ($C0_2$), methane (CH_4), nitrous oxide (N_20) and chlorofluorocarbons (CFCs), and are related to vehicle fuels in the first three terms. The greenhouse gases calculated in this paper include CO_2, CH_4 and N_20.

Global warming potential (global warming potential, GWP) refers to the measured values of radiation effects of different greenhouse gases relative to CO_2 in a certain period of time [4]. In this paper, the global warming potential of various GHG issued by the United Nations Intergovernmental Panel on Climate change in 2006 is used, as shown in Table 3. The GHG emission factor EF of various terminal energy or agricultural materials is calculated by three greenhouse gas emissions E, and the unit is carbon dioxide equivalent, as shown in formula (1).

$$EF = E_{CO_2} + 25 \times E_{CH_4} + 298 \times E_{N_2O} \qquad (1)$$

Table3. GWP of three GHGS

Greenhouse gas	carbon dioxide equivalent
CO_2	1
CH_4	25
N_2O	298

In the life cycle of corn butanol, there are three sources of greenhouse gas emissions from the system. Firstly, greenhouse gas emissions from a variety of energy and materials in each stage, including chemical fertilizers, natural gas, electricity, diesel oil, etc.; Secondly, CO_2 produced by fermentation during butanol conversion; Thirdly, CO_2 released by combustion of butanol during vehicle use. In this study, only the majority of CO_2 emissions are considered, and greenhouse gas emissions from pesticides and artificial labor are not included in the systematic calculation. The carbon dioxide produced by the chemical reaction of butanol fermentation and combustion comes from the carbon dioxide absorbed from the atmosphere through photosynthesis during corn growth, forming a closed circuit cycle between the atmosphere and corn, as shown in figure 1. Therefore, CO_2 absorbed by corn can be offset by CO_2 emissions in the second and third items at the planting stage. Only the first item is considered when calculating greenhouse gas emissions for the life cycle of corn butanol.

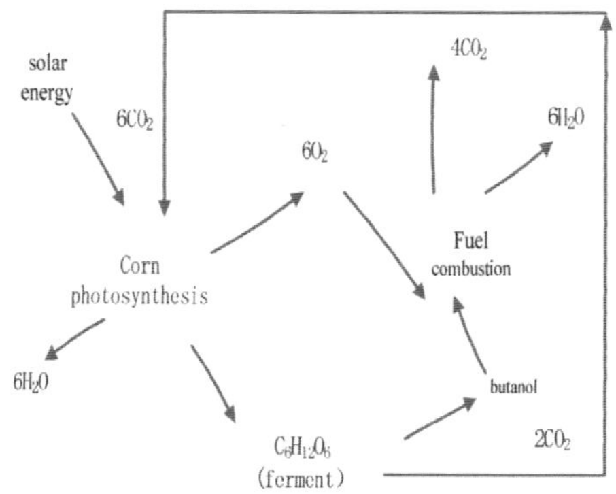

Fig1. CO2 cycle in corn butanol life cycle

2.4. *Data allocation methods for products*

In the analysis of GHG emissions, the data distribution methods of products mainly include weight method, energy value method, product substitution method, market value method and so on. Each allocation method will have an important impact on the evaluation of GHG emission results in the life cycle [3]. Because the yield of acetone and DDGS by-products is higher in the process of producing butanol, different distribution methods will lead to different analysis results, so three distribution methods are selected for analysis in this paper.Method 1: do not consider the treatment and utilization of by-products. Method two: energy distribution method. Acetone, ethanol and DDGS are all regarded as energy products. According to the calorific value and output of the products and by-products, the energy output ratio of various products is calculated to allocate GHG emissions. Method 3: acetone produced by corn fermentation is regarded as renewable energy, which is used to replace acetone produced by traditional oil, and DDGS is used to replace feed.

2.5. *Calculation model*

The concept of functional unit needs to be introduced in the life cycle analysis of vehicle fuel, which is based on "1 kg corn butanol fuel". Define the total greenhouse gas emissions over the life cycle of corn butanol as GHG_{LC}. In the production phase j, the greenhouse gas emissions caused by consumer material I is calculated by the GHG emission factor EF and the consumption Si, j unit corresponding to I, as shown in formula (2).

$$GHG_{LC} = \sum_{j=1}^{4} GHG_{LC.j} = \sum_{j=1}^{4} \left(\sum_{i=1}^{6} EF_i \times S_{i.j} \right) \quad (2)$$

In the formula, i = 1 to 6 denote the nitrogen fertilizer, the phosphate fertilizer, the potassium fertilizer, the natural gas, the electric power and the diesel oil, and j = 1 to 4 are the raw material production stage, the raw material transportation stage, the butanol production stage and the butanol transportation stage, respectively.

2.6. Data

This study mainly obtained the data related to corn butanol production through literature research, expert interviews and field research. the relevant data of the use of various agricultural materials in the raw material planting stage mainly came from the domestic statistical yearbook [4], the literature value;in the butanol production stage, the industrial production related data mainly came from the foreign literature and the enterprise research; the GHG emission factor of each material referred to the domestic literature value .

According to the literature, the annual production of butanol is 150000 tons, the consumption of corn is 834000 tons, the yield of butanol is 18%, and the by-product acetone is 84000 tons, ethanol is 4050 tons, DDGS is about 259500 tons. Diesel freight car is used to transport raw materials and butanol by road, and the consumption of 100 tons of diesel oil is about 5 liters, and the transportation distance is 300 kilometers and 500 kilometers, respectively.

3.Research results

3.1 Greenhouse gas emissions under the three options

According to method 1, the material consumption of 1kg corn butanol is calculated according to the data above, and the greenhouse gas emission of corn butanol is calculated as 3473g CO_2 equivalent / kg butanol according to the formula, as shown in the table4.

Table4. Life cycle GHG emission of the corn butanol

	nitrogenous fertilize /kg	phosphate fertilize /kg	potassic fertilize /kg	natural gas /m^3	powe /kWh	diesel oil /L
EF*	5148	587	811	1980	1071.7	4372.5
consumption	0.105	0.056	0.058	0.93	0.25	0.17
GHG /g	540.5	32.9	47.0	1841.4	267.9	743.3
GHG /%	15.6	0.9	1.4	53	7.7	21.4

* the unit of EF: gCO2 •kg butanol -1

According to method 2, the output and calorific value of products and by-products in the process of butanol production are shown in Table 5 by using energy distribution method. According to formula (3), the life cycle GHG emission of corn butanol can be calculated to be 1962gCO_2 equivalent / kg butanol.

$$GHG_{BU} = GHG_{LC} \times \frac{E_{BU}}{E_{LC}} \quad (3)$$

Table5. Yield and LHV of ABE fermentation products

Product	Output (kg)	Low heating value (MJ/kg)	Energy output ratio
butanol	1	35.1	56.5%
acetone	0.56	29.73	26.8%
alcohol	0.027	26.77	1.1%
DDGS	1.73	5.6*	15.6%

*energy substitution facto of DDGS

In the formula, the life cycle GHG emission of the main product butanol according to the method is the life cycle GHG emission of the method one butanol, and the energy output of the main product butanol is the energy output of all the products. According to method 3, the petroleum-based acetone was replaced by acetone co-produced by biobutanol, and DDGS is used to replace feed by product substitution method. The traditional raw material for the production of acetone is cumene. It is set that 2.21 tons of cumene is needed to produce 1 ton of acetone. The distance between the diesel truck and the user is 500 kilometers, and the consumption of 100 tons of diesel oil is about 5 liters. Using the GREET model [5] and combined with the relevant literature, it is calculated that the GHG emission of the main product in the life cycle is about 38.5% of that of scheme 1, which is 1337gCO_2 equivalent /

kg butatol.The results of the GHG emissions from three programmes are summarized and compared with the equivalent calorific value of the gasoline GHG emissions[6].

3.2 Analysis of results
Compared with the GHG emissions of gasoline with the same calorific value in its life cycle, the GHG emissions of corn butanol in scheme 1 increased by 31%, and the GHG emissions in options II and III decreased by 26% and 49.5%, respectively. If both acetone and DDGS are used as fuel, GHG emissions can be effectively reduced, and biological acetone can be used instead of traditional petroleum acetone. GHG emissions can be significantly reduced. Therefore, it can be seen that acetone is a very critical factor in GHG emission analysis of the corn butanol life cycle. First, traditional acetone production is fossil energy-intensive production, and isopropyl benzene is produced from propylene and benzene, which are extracted from naphtha and natural gas, which is 100%fossil raw material. Secondly, the acetone production of ABE fermentation process is high, and every ton of butanol is produced, 0.56 tons of acetone is produced. When replacing petroleum acetone with bioacetone, renewable energy has replaced fossil energy, which significantly reduces GHG emissions.

4. Conclusion
(1) Using ABE fermentation method to completely replace gasoline with corn as raw material can effectively reduce GHG emission from the point of view of life cycle. According to different distribution schemes, the GHG emission of corn butanol life cycle can be reduced by 26%—49.5% compared with the same calorific value gasoline.

(2) The yield of acetone in the ABE fermentation process is high, and 0.56 ton of acetone is produced when one ton of butanol is produced. As a result, acetone is a very critical factor in the GHG emissions analysis of the corn butanol life cycle. When replacing petroleum acetone with bioacetone, the renewable energy replaces the fossil energy and can significantly reduce GHG emissions.

(3) In the life cycle of the corn, the main source of GHG emissions is the use of natural gas, diesel and nitrogen, accounting for 53%, 21.4% and 13.38% of the total GHG emissions, respectively. If the transport vehicle is replaced with a gasoline truck and corn-butanol is used as a substitute fuel, while minimizing the amount of nitrogen fertilizer used, the GHG emissions can be further reduced.

Acknowledgments
Heilongjiang Institute of Engineering Ph. D. Fund. （2016BJ02）； The Funded Project on Basic Scientific Research of Provincial Universities in Heilongjiang Province（2018CX07）

References
[1] Wu M, Wang M. (2008) Assessment of potential life-cycle energy and greenhouse gas emission effects from using coin·based butanol as a transportation fuel. Biotechnology Progress, 24:1204-1214.
[2] Lucas G, Pereira,Mateus F, Chagas,M. (2014)Life cycle assessment of butanol production in sugarcane biorefineries in Brazil. Journal of Cleaner Production.16:1-12.
[3] Gnansounou E, Dauriat A, Villegas J. (2009) Life cycle assessment of biofuels: energy and greenhouse gas balances. Bioresource Technology, 100: 4919-4930.
[4] Xia X, Zhang Y. (2012) Evaluation and Policy Research of fuel ethanol based on Life cycle . China Environmental Science press, BeiJing ,pp. 60-62.
[5] Ou X. (2010) Life Cycle Energy Use for and GHG Emissions from the Road Transport Sector in China. Tsinghua University.
[6] Gao H, Hu S, Li Y. (2012) Greenhouse gas emission of sweet sorghum ethanol in life-cycle. Transactions of the CSAE, 28:178-183.

Application Study on Urban Road Volume-Time Function Based on RFID Data

Jian Gu[1]*, Linghua Yu[2] and Miaohua Li[3]

[1]Engineering Research Center of Catastrophic Prophylaxis and Treatment of Road & Traffic Safety of Ministry of Education, Changsha University of Science & Technology, Changsha, Hunan, 410114, China

[2]School of Transportation, Southeast University, Nanjing, Jiangsu, 211189, China

[3]Hisense TransTech Co.,Ltd., Changsha, Hunan, 410000, China

* Corresponding Author's email: gujian@csust.edu.cn

Abstract. Radio Frequency Identification (RFID) plays an important role on road traffic management and basic data could be collected from automotive electronic registration identifications, which could be applied for road traffic management and planning. Firstly, road section volume-time function was discussed in the paper, and then urban road RFID data collected from the city of Nanjing were analyzed with mature methods on identifying redundant and wrong data based on license plate number. Then the function from Bureau of Public Road (BPR) was calibrated to get the basic volume-time function with RFID data, and the basic function was improved based on the time periods, and the improved functions were used to describe relationships between volume and travel time in weekdays and weekends, and the fitting effect is compared with the basic function. It was indicated that the fitting effort R^2 was increased by 40% at least. The results can provide reference and support for urban road traffic planning.

1. Introduction

In the process of road traffic data collection and detection, due to the inherent superiority, radio frequency identification (RFID) technology is receiving more and more attention in traffic management. There is a significant advantage that it is not affected by environmental conditions when collecting the identity information of the identified vehicle, which could be applied to performing traffic demand characteristics identification and traffic flow parameter analysis. By further utilizing the detection records of the identified vehicle, the location information and time stamp information can be quickly extracted, thereby there are more and more studies and applications in the travel time and travel speed of the vehicle characteristics [1-2], travel demand analysis [3-4], dangerous road cargo transportation [5], traffic control [6], and so on.

Some cities in China have completed the coverage and application of RFID readers. Under this background, in-depth study of urban traffic operation characteristics and vehicle travel rules based on RFID data is of great significance to traffic planning and management, especially in road networks. In order to achieve a reasonable spatial and temporal distribution of traffic demand, a reasonable route impedance function should be constructed from the perspectives of minimum travel time or minimum energy consumption. Combining with mathematical methods such as graph theory, it can be known that a route in a road network is composed of a set of road segments, and the travel time of a route is obtained by adding the travel time of each road segment element in the set and the travel time of the intersection

IOP Conf. Series: Materials Science and Engineering **787** (2020) 012018 doi:10.1088/1757-899X/787/1/012018

between road segments.

In the application and theoretical studies of road traffic planning, the BPR (Bureau of Public Road) function is a widely used mathematical model to calculate route impedance [7-8]. Considering that the model is established based on the characteristics of the highway condition, it lacks considering and quantifying the delay caused by urban road intersections, there will be differences when using this model to analyze urban road traffic operation. To solve this problem, improvement studies are carried out, and scholars have also separately analyzed the BPR function from the model perspective [9] and the data perspective [10]. In order to describe the relationship between the travel time and urban roads flow volume, this paper starts with the RFID data from urban roads, and the basic BPR function is improved (such as the model and parameters recommended by the US Highway Administration, $\alpha = 0.15, \beta = 4.0$), travel time and flow data between road detection sections are extracted, the BPR function framework is further used to improve the model form and perform parameter calibration, and the improved functional model between travel time and flow is obtained in the analysis area in different periods. The analysis results can be used for urban traffic demand management, and scientific reference could be provided from road operation characteristics analysis.

2. Methodology

According to the basic form of the BPR model, the travel time is proportional to the flow volume, as shown in equation (1).

$$t = t_0\left(1+\alpha\left(\frac{q}{c}\right)^{\beta}\right) \tag{1}$$

The Equation 1 can be transformed as shown in equation (2) and equation (3).

$$\frac{t}{t_0} - 1 = \alpha\left(\frac{q}{c}\right)^{\beta} \tag{2}$$

$$\log\left(\frac{t}{t_0} - 1\right) = \log\alpha + \beta\log\left(\frac{q}{c}\right) \tag{3}$$

Equation 3 can be simplified as shown in equation (5) based on equation (4).

$$y = \log\left(\frac{t}{t_0} - 1\right), x = \log\left(\frac{q}{c}\right) \tag{4}$$

$$y = \beta x + \log\alpha \tag{5}$$

3. Data analysis

3.1 Data quality analysis

The data used in this paper is collected from the city of Nanjing local roads for one month in December 2015. The data fields mainly include the station code, lane number, record time, license plate number, license plate color, and vehicle speed. Before further using the data to calibrate the model, data quality analysis of the original data is needed. There are many researches and applications of data quality analysis and processing methods for RFID data features [11-12]. In this paper, the redundant data and error data are mainly processed, results as shown in table 1.

Table 1. Example of one station record data processed

Record No.	License plate No.	Record time	License plate color	Lane No.	Vehicle speed (km/h)	Type
1	"Su A**129"	2015-12-9 6:39:00	blue	1	35	repeat
2	"Su A**129"	2015-12-9 6:39:00	blue	2	24	
3	"Su A**401"	2015-12-9 6:39:16	yellow	1	32	similar
4	"Su A**401"	2015-12-9 6:39:23	yellow	1	32	
5	"unrecognized"	2015-12-9 6:40:34	-	1	22	unrecognized

In this paper the wrong license plates are classified according to the standards of motor vehicle number plates in the collected data of 21 stations in the study area, mainly three types of wrong license plate data are included. The error of no Chinese characters in provinces, autonomous regions, and municipalities is classified to the first type of wrong license plate. The error of no serial number or digital identification is classified to the second type of wrong license plate. The error of the code including "O" or "I" is classified to the third type of wrong license plate.

By counting the data from 21 stations throughout the day on December 1, 2015, a total of 219,456 entries were obtained, and the statistics were identified according to the wrong license plate classification. The results are shown in table 2.

Table 2. Statistical characteristics of wrong license plate data

Error type	Quantity (items)	The proportion (%)
First type of error	556	0.25%
Second type of error	1746	0.80%
Third type of error	161	0.07%
Correct license plate	216993	98.88%
Total	219456	100%

It is analyzed that the proportion of license plate error rate is low in table 2, but the total amount of error data is 2463, of which the second type of incorrect license plate data is the main error. It can be seen that in the identification process, license numbers are incorrectly recognized as letters. As the license plate is the unique identification, the vehicle cannot be uniquely verified by wrong license plate, so it should be deleted. In addition, when analyzing the rationality of vehicle travel time characteristics between neighbor stations, the spatial distance and speed characteristics of vehicles passing through neighbor base stations are mainly considered.

3.2 Traffic flow characteristics

Based on the erroneous data and redundant data process in the data set, the traffic flow volume characteristics of the stations and the travel time characteristics between the stations can be extracted. The local path (path 1 and path 2) and the distribution characteristics of base stations are shown in figure 1. Taking the characteristics of the base station 6292's all-day traffic volume as an example, the characteristics of the traffic flow volume are shown in figure 2, where solid line stands for weekday and dot line stands for weekend.

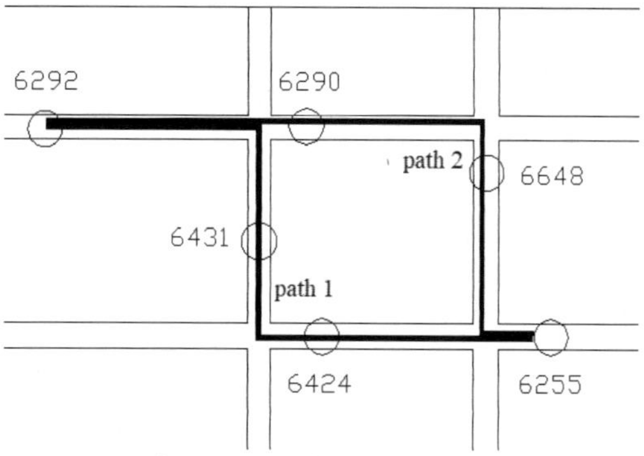

Figure 1. Path and stations

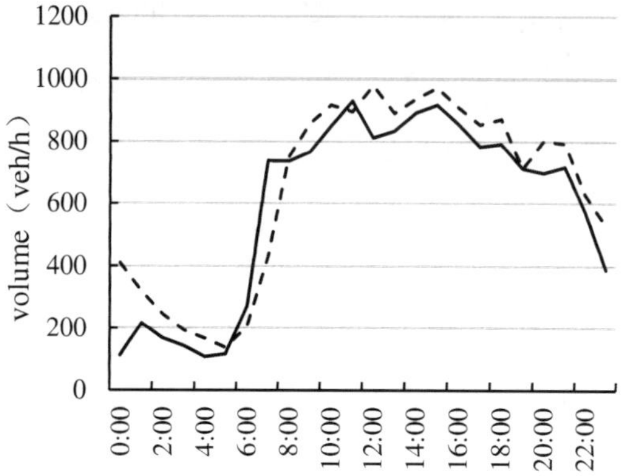

Figure 2. Traffic volume at station 6292

Taking the travel time of two path during peak hours as an example, the route starts at station 6292 and ends at station 6255. The license plate data is analyzed to obtain the traffic flow volume characteristics and average travel time characteristics of the route within a single week, statistical characteristics results are shown in table 3.

Table 3. Statistical characteristics of route volume and travel time

Period	Path	Volume (veh)	Lane No.	Travel time (s)	Ratio
Weekday	1	307	3	496	47%
	2	345	3	325	53%
Weekend	1	94	3	435	54%
	2	82	3	305	46%

4. Calibration and analysis

The data from December 1st to 20th, 2015 is taken as sample data, the scatter diagram of the traffic and travel time of the station pair 6292-6290 is shown in Figure 3. From figure 3 it can be known that the data corresponding to the exponential form does not have a high fitting degree. The BPR function is directly calibrated using sample data, and the result is shown in figure 4.

In order to improve the accuracy, according to the traffic volume analysis requirements at 8-12 a.m. in the area, the traffic flow data is divided into two categories, of which four hours in the 8-12 a.m. time period per day are regarded as a type, and the data of the remaining periods are regarded as a type. The BPR models are constructed and fitted separately, and the fitting conditions for weekend are compared and analyzed.

This paper uses the station pair 6292-6290 as an example. There are many non-motorized vehicles and pedestrians in the surrounding area, which have a large impact on the traffic flow, then the road capacity in the model uses the empirical value of 680 pcu / h for one lane and 2040 pcu / h for three lanes.

IOP Conf. Series: Materials Science and Engineering **787** (2020) 012018 doi:10.1088/1757-899X/787/1/012018

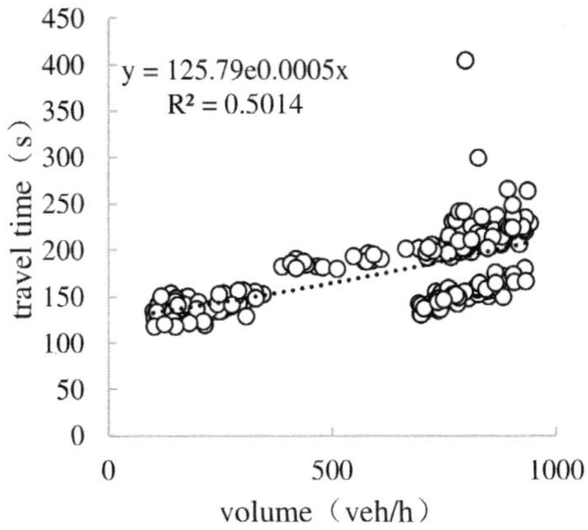

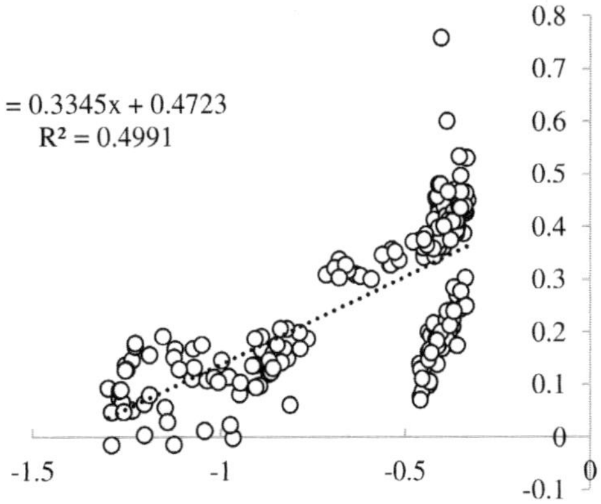

Figure 3. Travel time vs. flow volume for weekday Figure 4. Calibration of BPR function for weekday

(1) The fitting situation on weekdays (not 8-12 a.m.) is shown in figure 5, and the fitting formula is shown as equation (6).

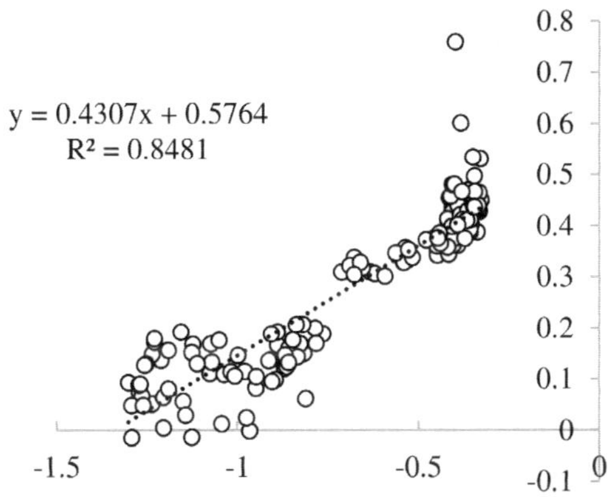

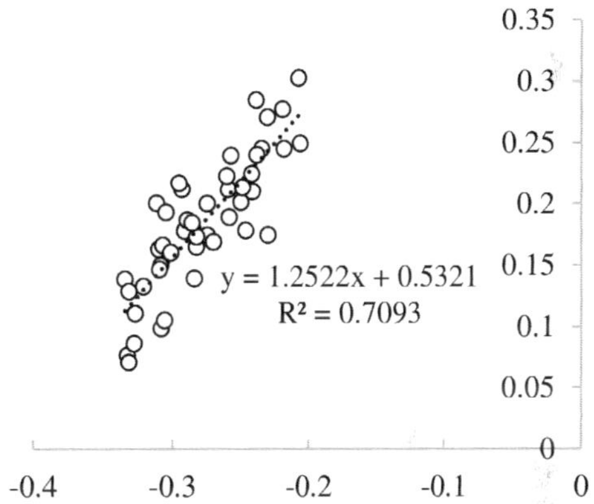

Figure 5. Fitting on weekdays (non 8-12 a.m.) Figure 6. Fitting on weekdays (8-12 a.m.)

$$y = 0.4307x + 0.5764 \tag{6}$$

(2) Function of station on weekday (8-12 a.m.)

Weekday (8-12 a.m.) fitting situation is shown in figure 6, and the fitting formula is shown as equation (7).

$$y = 1.2522x + 0.5321 \tag{7}$$

(3) Weekend function (non-8-12 a.m.)

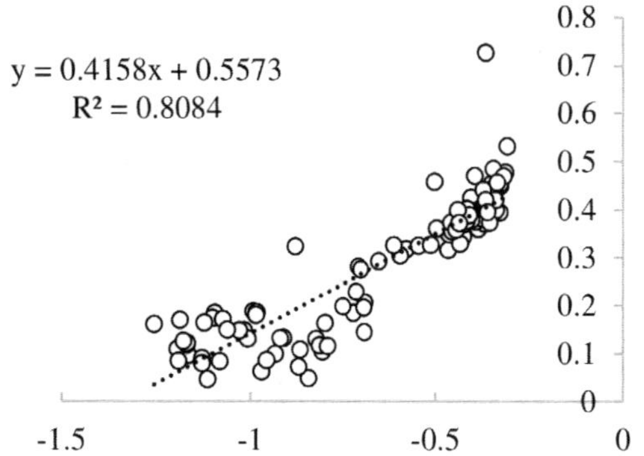

Figure 7. Fitting on weekends (non-8-12 a.m.)

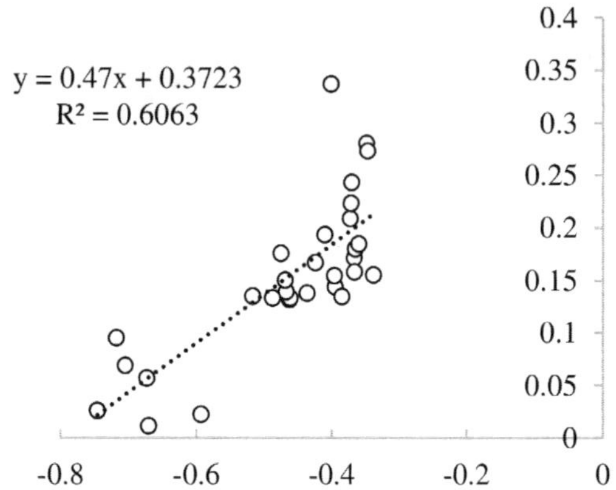

Figure 8. Fitting on weekends (8-12 a.m.)

The weekend (non-8-12 a.m.) fitting situation is shown in figure 7, and the fitting formula is shown as equation (8).

$$y = 0.4158x + 0.5573 \tag{8}$$

(4) Weekend function (8-12 a.m.)

The weekend (8-12 a.m.) fitting situation is shown in figure 8, and the fitting formula is shown as equation (9).

$$y = 0.47x + 0.3722 \tag{9}$$

The fitting of each function is shown as equation (10).

$$\begin{cases} \text{weekday(non 8-12 a.m.)} & y_1 = 0.4307x_1 + 0.5764 \\ \text{weekday(8-12 a.m.)} & y_2 = 1.2522x_2 + 0.5321 \\ \text{weekend(non 8-12 a.m.)} & y_3 = 0.4158x_3 + 0.5573 \\ \text{weekend(8-12 a.m.)} & y_4 = 0.4700x_4 + 0.3722 \end{cases} \tag{10}$$

The impedance function for the four cases is shown as equation (11).

$$\text{weekday(non 8-12a.m.)} \quad t_1 = 60 \times \left(1 + 2.70 \times \left(\frac{q}{c}\right)^{0.4307}\right) \tag{11-1}$$

$$\text{weekday(8-12a.m.)} \quad t_2 = 60 \times \left(1 + 3.41 \times \left(\frac{q}{c}\right)^{1.2522}\right) \tag{11-2}$$

$$\text{weekend(non 8-12a.m.)} \quad t_3 = 60 \times \left(1 + 3.61 \times \left(\frac{q}{c}\right)^{0.4158}\right) \tag{11-3}$$

$$\text{weekend(8-12a.m.)} \quad t_4 = 60 \times \left(1 + 2.36 \times \left(\frac{q}{c}\right)^{0.4700}\right) \tag{11-4}$$

The comparison of function parameter calibrations and the fitting conditions between BPR function and its improvements are shown in table 4 and table 5.

Table 4. Comparison of calibration results of BPR function

BPR model			Improved BPR model		
Data type	α	β	t_0	α	β
weekday（non 8-12）	2.96	0.3345	60	2.70	0.5764
weekday（8-12）			60	3.41	1.2522
weekend（non 8-12）			60	3.61	0.4158
weekend（8-12）			60	2.36	0.47

Table 5. Calibration and fitting

Comparison of fitting		R^2
BPR model		0.4991
Improved BPR model	weekday（non 8-12 a.m.）	0.8481
	weekday（8-12 a.m.）	0.7093
	weekend（non 8-12 a.m.）	0.8084
	weekend（8-12 a.m.）	0.6068

From table 4 and table 5 it could be analyzed that, for the fourth case, due to the relatively small amount of sample data in the weekend 8-12 a.m. conditions, the R^2 in this case is lower than the other three conditions. In the other three cases, the data fitting results of the weekday 8-12 a.m. conditions are higher than the results obtained by direct data calibration.

From the comprehensive analysis, it is illustrated that the impedance model obtained by the data extracted from the RFID in Nanjing is significantly different from the recommended parameters by the U.S. Highway Administration, which is caused by the calibration data collected from urban roads with many intersections and small distance, it will bring greater disturbance to the traffic flow continuity, which is inconsistent with the highway condition. Therefore, applying the actual data to fit the impedance model is more practical for urban road traffic planning and management.

5. Conclusion

This paper starts with urban road traffic impedance model calibration, travel time is selected as route impedance, urban road RFID station data is collected and used. Based on the redundant data, license plate error data, and travel time error data processing, basic BPR model is calibrated for local area of urban road. Combined with the demand of multi-period analysis, the models are calibrated separately, and the fitting effect is compared with the basic model. The comparison results show that the model calibrated by using the period data has better fitting effect, and further comparison and analysis are performed on weekday and weekend respectively. The results can provide effective support for urban road traffic planning.

Acknowledgments

The authors express thanks to all who participated in this research for their cooperation, and we would like to give great thanks to the hard work by the peer reviewers and editors. We also acknowledge the financial support provided by Open Fund of Engineering Research Center of Catastrophic Prophylaxis and Treatment of Road & Traffic Safety of Ministry of Education (Grant No. KFJ170403) (Changsha University of Science & Technology) and Hunan Provincial Key Laboratory of Smart Roadway and Cooperative Vehicle-Infrastructure Systems (Grant No. 2017TP1016).

References

[1] Tam, M. L., Lam, W. H. K. (2011) Application of automatic vehicle identification technology for real-time journey time estimation. Information Fusion, 12:11-19.

[2] Li, M. H., Wu, J. M., Sun, L. H., et. al. (2019) Bus travel time prediction method based on RFID electronic license plate data. China Journal of Highway and Transport, 32:165-173.

[3] Cheng, W., Wang, S. L., Cheng, X. Z. (2014) Virtual track: applications and challenges of the

RFID system on roads. Network, IEEE. 28: 42-47.
[4] Zhao, X. X., Du, W., Zhou, X., et. al. (2016) A method for estimating OD matrix of urban road network using RFID and analysis of its spatio-temporal features. Journal of Transport Information and Safety, 2016:30-36, 50.
[5] Yao, Z. Q., Wang, J., Hu, Y. X., et. al. (2013) An intelligent supervision system for hazardous materials logistics based on RFID/GPRS/GPS/GIS. Journal of Highway and Transportation Research and Development, 2013:147-152.
[6] Chao, K., Chen, P. (2014) An intelligent traffic flow control system based on radio frequency identification and wireless sensor networks. International Journal of Distributed Sensor Networks, 2014:1-10.
[7] Kurzth, D. L., van den Hout, A., Ives, B. (1996) Implementation of highway capacity manual-based volume-delay functions in regional traffic assignment process. Transportation Research Record, 1996:27-36.
[8] Liu, N., Zhao, S. C., He, N. (2013) Further study of impedance function based on BPR function. Journal of Wuhan University of Technology (Transportation Science & Engineering), 37:545-548.
[9] Pan, Y. Y., Yu, T., Ma, J. X. (2017) Improvement of urban road impedance function based on section impedance and node impedance. Journal of Chongqing Jiaotong University (Natural Science), 36:76-81.
[10] Zhou, L., Ma, X. F., Zhong, M. (2017) A study on impedance function of city secondary road based on taxi GPS data. Journal of Transport Information and Safety, 35: 34-42.
[11] Yang, Y. S., Du, W., Ning, D., et. al. (2016) Detection and an analysis of redundant RFID traffic data. Journal of Transport Information and Safety, 34:72-80.
[12] Du, W., Yang, Y. S., Ning, D., et. al. (2016) An analysis and a novel detection method of erroneous traffic data from RFID devices. Journal of Transport Information and Safety, 34:81-87.

IOP Conf. Series: Materials Science and Engineering **787** (2020) 012019 doi:10.1088/1757-899X/787/1/012019

A power shift electro-hydraulic-controlled system of a seven-speed automatic transmission designed by the logic method

Yi Fan

Transportation College, Nanning University, Nanning, Guangxi, 530200, China

E-mail: 64608769@qq.com

Abstract. Aiming at a self-designed seven-speed planetary gear automatic transmission, the electro-hydraulic-controlled system of power shift is designed by logic method. First of all, according to the transmission diagrams and the working condition of each gear shift element, design the connection condition of the shift valves of each forward gears and the reverse gear in turn, and write out the corresponding logical equations. Then, on this basis, carry out the interlock and failure analysis, and complete the diagram of the power shift electro-hydraulic-controlled system. Finally, according to the designed electro-hydraulic-controlled system and shift valve control status table, logical equations are verified. The results show that the function of the system is realized by using the logic combination of six two-position three-way electro-hydraulic-controlled shift valves and one manual control valve, and the design of the power shift electro-hydraulic-controlled system can be carried out clearly, efficiently and accurately.

1. Introduction

The planetary gear automatic transmission has its advantages of compact structure and smooth shifting, therefore, it has become the focus of research and development of major automobile enterprises [1]. The design of power shift electro-hydraulic-controlled system is one of the core work of automatic transmission design [2-3]. The conventional methods like trial and reference to similar design have certain blindness [4-5]. What's worse, the design result isn't often the optimal value, which is difficult to meet the design requirements of complex hydraulic system. When the power shift automatic transmission realizes the normal operation of different gears, it controls the movement of the hydraulic valve spool in the shift valve through the on or off power of the shift valves. While the moving of the hydraulic valve spool makes the shift valve in different working positions, the corresponding oil circuit is connected or closed, and the combination and separation of the corresponding shift elements are realized. Different gear lever positions and different gears require a combination of several shift elements and the separation of other shift elements. It can be seen that the design of the electro-hydraulic-controlled system of the automatic transmission is essentially a logical problem. Aiming at a self-designed seven-speed automatic transmission, under the principle of shift design, this paper uses the logic method to design the shift structure of power shift electro-hydraulic-controlled system, then analyzes its interlock and actual effect, and finally uses logical equations to verify the correctness of each shift.

2. Design principle of logic method

The combination or separation of shift elements has two values, and different gears can be achieved through the combination of several shift valves. The two-position three-way electro-hydraulic-controlled shift valve is shown in Figure 1. "Z" on the left is the control end of shift valve, the spring end is on the right, and the corresponding oil circuits are Y, X_1 and X_2. When the shift valve is energized, the shift valve is on the left position; when the shift valve is switched off, the shift valve is on the right position under the action of the spring. The corresponding logical equation can be expressed as:

$$Y = ZX_2 + \overline{Z}X_1 \quad (1)$$

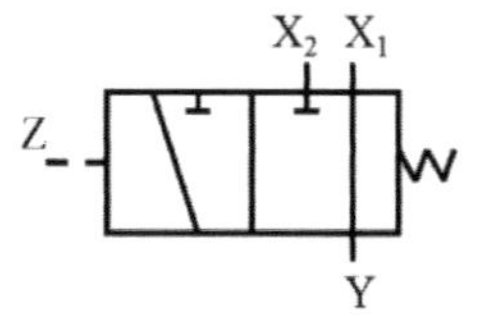

Figure 1. The two-position three-way electro-hydraulic-controlled shift valve.

During the design process of the power shift electro-hydraulic-controlled system, some basic design criteria shall be followed:

Principle 1: achieve basic gears. The designed electro-hydraulic-controlled system can separate or combine the corresponding actuator according to the position of each gear lever and each gear in the working diagram of the shift elements.

Principle 2: no movement interference or shifting action. The designed electro-hydraulic-controlled system cannot interfere with the components in the planetary row. Take one component like the sun gear, the planet frame or the gear ring in the planetary row as an example. If the clutch and brake are arranged on the component, these two shifting components cannot be combined at the same time. In order to ensure the smoothness of gear shifting and to reduce the impact of gear shifting, the phenomenon of gear skipping should not occur in the design process, such as directly gear changing from the 1st gear to the 3rd gear or to the 4th gear.

Principle 3: invalidation protection function. In the designed electro-hydraulic-controlled system, if some or all of the shift valves cannot work, the electro-hydraulic-controlled system should be able to maintain a certain gear to ensure that the automobile can travel a certain distance to the automobile-repair enterprise.

3. Transmission structure and working principle of the seven-speed automatic transmission

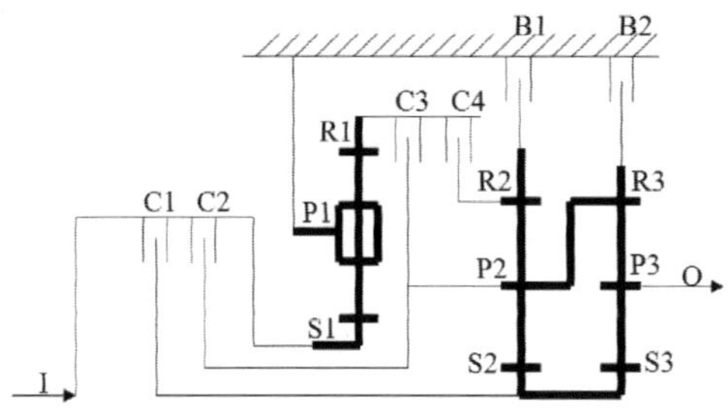

Figure 2. Structural diagram of a seven-speed automatic transmission.

Figure 2 is the structural diagram of a seven-speed planetary gear automatic transmission designed in reference [6], which includes three planetary gear rows and six shift actuators. The corresponding connection relationship is shown in the Figure 2. Table 1 is the working state of corresponding shift elements in different gear lever positions and different gears. It can be seen that each gear consists of 2 combined shift elements and 4 separated shift elements. Gear R can be controlled by the manual valve alone. Among the seven gears of gear D, the adjacent gears are all reserved for the combination of one shift element, and one shift element is added for the combination and one for separation at the same time.

Table 1. Working state of each shift element.

Gear lever positions	Gear	Shift elements					
		C_1	C_2	C_3	C_4	B_1	B_2
D	1	●					●
	2	●				●	
	3	●		●			
	4	●			●		
	5	●	●				
	6		●		●		
	7		●			●	
R	R				●		●

4. Design of the hydraulic system

4.1. Design analysis of different gears

It can be seen from the working state of shift elements of each gear in Table 1 that gear R can directly oil C4 and B2 through the manual valve instead of the shift valve, so that the two shift elements can be combined to realize gear R. In the position of gear lever D, each gear is a combination of two shift elements. C1 is combined from the 1st to the 5th gear, and the corresponding gear is realized by changing the combination of another shift element; C2 is combined from the 5th to the 7th gear, and the corresponding gear is realized by changing the combination of another shift element, too. The 5th to the 6th gear is a shift point, which keeps C2 unchanged and both C1 and C4 changed. Set the oil return port as "E" and the system oil pressure connection port as "P", and design from D1.

Gear D1 (C1, B2) and gear D2 (C1, B1): according to the above analysis, C1 is combined in the first five gears, and there is a transformation between C1 and C4 from the 5th gear to the 6th gear. Therefore, a two-position three-way electro-hydraulic-controlled shift valve H1 is designed to realize the separation and combination between C1 and C4, as shown in Figure 3; the transformation between B2 and B1 can be realized by a two-position three-way electro-hydraulic-controlled shift valve H2, as shown in Figure 4.

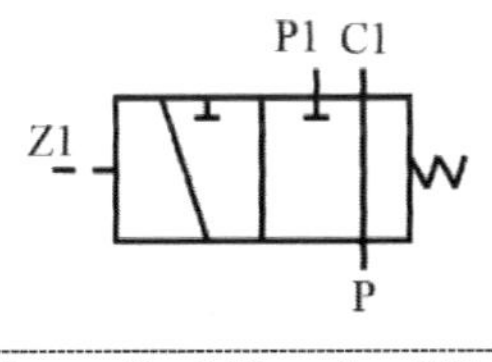

Figure 3. shift valve H1.

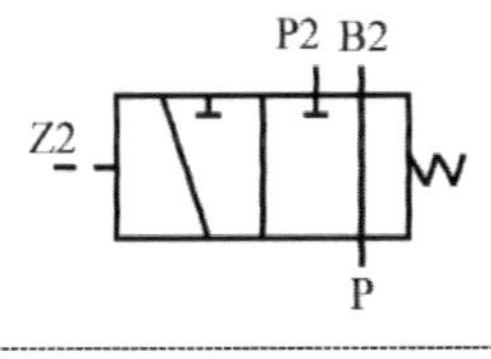

Figure 4. shift valve H2.

The hydraulic logical equations are as shown in equations (2) and (3). When the shift valve H1 is switched off, the valve is on the right position, and P is connected with C1, so that the shift element C1 is combined. When the shift valve H1 is energized, the valve is on the left position, and the system oil pressure P is connected with P1 which is considered as the input oil pressure of C4. When the shift valve H2 is switched off, the valve is on the right position, and P is connected with B2, so that the shift element B2 is combined to achieve gear D1. When the shift valve H2 is energized, the valve is on the left position, and the system oil pressure P is connected with P2 which is considered as the input oil pressure of B1.

$$P = Z_1 P_1 + \overline{Z_1} C_1 \tag{2}$$

$$P = Z_2 P_2 + \overline{Z_2} B_2 \tag{3}$$

Gear D3 (C1, C3): on the basis of the 2nd gear, C1 remains combination unchanged in the 3rd gear. B1 is separated and C3 is combined, so a two-position three-way electro-hydraulic-controlled shift valve H3 is designed to realize the separation and combination between B1 and C3, as shown in Figure 5. The hydraulic logical equation is shown in equation (4). When the shift valve H3 is switched off, the valve is on the right position, and P2 is connected with B1, so that the shift element B1 is combined to realize gear D2. When the shift valve H3 is energized, the valve is on the left position, and P2 is connected with P3, which is considered as the input oil pressure of C3.

$$P_2 = Z_3 P_3 + \overline{Z_3} B_1 \tag{4}$$

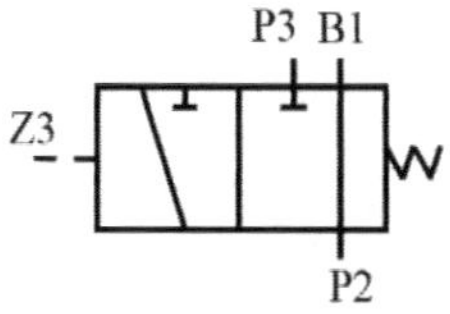

Figure 5. shift valve H3.

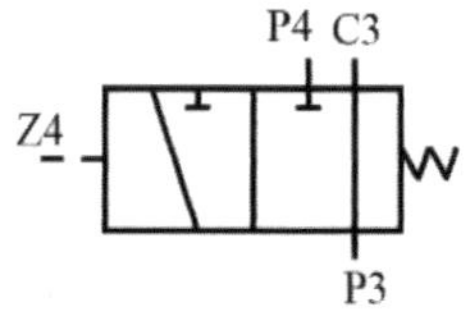

Figure 6. shift valve H4.

Gear D4 (C1, C4): on the basis of the 3rd gear, C1 remains combination unchanged in the 4th gear. C3 is separated and C4 is combined, so a two-position three-way electro-hydraulic-controlled shift valve H4 is designed to realize the separation and combination between C3 and C4, as shown in Figure 6. The hydraulic logical equation is shown in equation (5). When the shift valve H4 is switched off, the valve is on the right position, and P3 is connected with C3, so that the shift element C3 is combined to realize gear D3. When the shift valve H4 is energized, the valve is on the left position, and P3 is connected with P4, which is considered as the input oil pressure of C4.

$$P_3 = Z_4 P_4 + \overline{Z_4} C_3 \tag{5}$$

Gear D5 (C1, C2): on the basis of the 4th gear, C1 remains combination unchanged in the 5th gear. C4 is separated and C2 is combined, so a two-position three-way electro-hydraulic-controlled shift valve H5 is designed to realize the separation and combination between C4 and C2, as shown in Figure 7. The hydraulic logical equation is shown in equation (6). When the shift valve H5 is switched off, the valve is on the right position, and P4 is connected with C4, so that the shift element C4 is combined to realize gear D4. When the shift valve H5 is energized, the valve is on the left position, and P4 is connected with P2, so that the shift element C2 is combined to realize gear D5.

$$P_4 = Z_5 C_2 + \overline{Z_5} C_4 \tag{6}$$

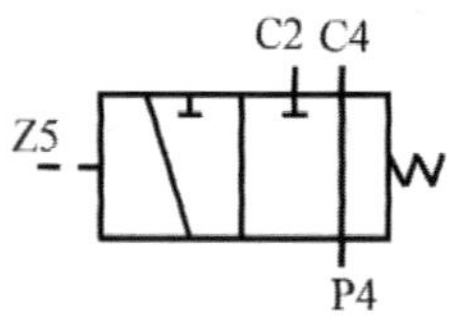

Figure 7. shift valve H5.

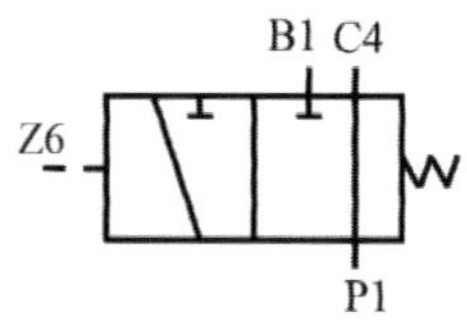

Figure 8. shift valve H6.

Gear D6 (C2, C4) and gear D7 (C2, B1): on the basis of the 5th gear, C2 remains combination unchanged in the 6th gear. C1 is separated and C4 is combined, so a two-position three-way electro-hydraulic-controlled shift valve H6 is designed to realize the separation and combination between C1 and C4, as shown in Figure 8. The hydraulic logical equation is shown in equation (7). When the shift valve H6 is switched off, the valve is on the right position, and P1 is connected with C4, so that the shift element C4 is combined to realize gear D6. When the shift valve H6 is energized, the valve is on the left position, and P1 is connected with B1, so that the shift element B1 is combined to realize gear D7.

$$P_1 = Z_6 B_1 + \overline{Z_6} C_4 \tag{7}$$

4.2. Interlock and failure analysis

Interlock analysis: it can be seen from Figure 2 that C2 and C3 cannot be combined at the same time, because they cannot be connected to the same planetary row component P2-R3 at different speeds. C2 (or C3) and B2 cannot be combined at the same time, because component P2-R3 cannot input and brake at the same time. C4 and B1 cannot be combined at the same time, because components R2 cannot input and brake at the same time. From the control logic of shift valves H1 to H6 from Figure 3 to Figure 8, the input oil circuit of one latter shift valve is on the basis of the power of the former one, so the six shift valves are interlocked. The combination and separation of the shift elements mentioned above will not interfere with each other and can meet the requirements of interlock.

Failure analysis: gear R does not pass through the shift valve, so there is no possibility of shift valve failure. For the lever position D, if all shift valves fail to work, C1 and B2 will be combined, and the transmission will operate in gear D1. The premise of normal operation of H6 is that H1 is powered on. If any one or several shift valves from H1 to H5 fail to work, the subsequent shift valves will fail successively. For example, H1, H2 and H3 work normally while H4 does not, then there is no hydraulic oil passing through H5, and the transmission will operate in gear D3.

Through the above analysis, the designed power shift electro-hydraulic-controlled system is shown in Figure 9.

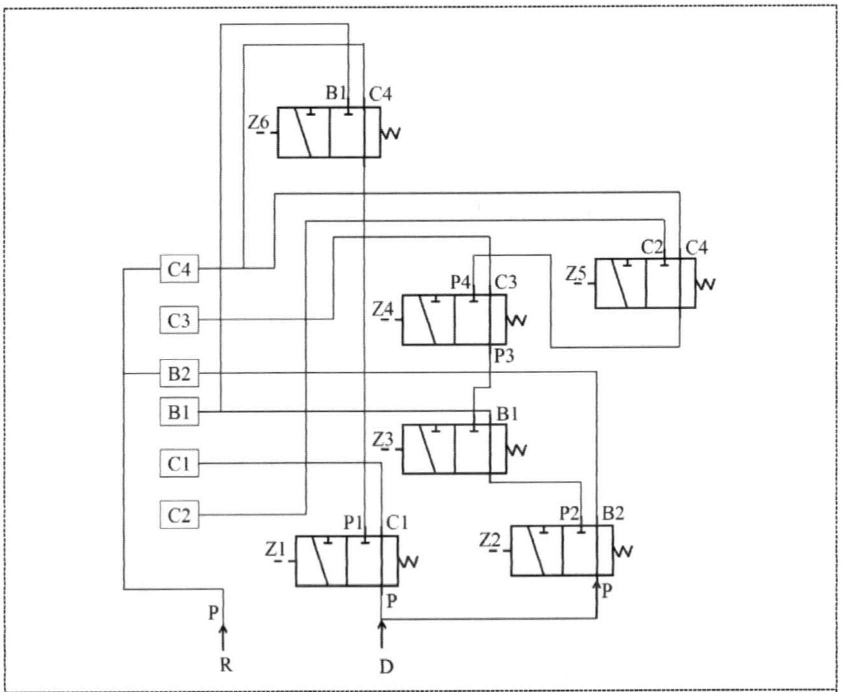

Figure 9. Diagram of the power shift electro-hydraulic-controlled system.

4.3. Verification analysis

According to the analysis in Section 4.1, the corresponding shift valve controlling status is shown in Table 2.

Table 2. Shift valve controlling status.

Gear lever positions	Gear	H1	H2	H3	H4	H5	H6
	1	off	off				
	2	off	on	off			
	3	off	on	on	off		
D	4	off	on	on	on	off	
	5	off	on	on	on	on	
	6	on	on	on	on	on	off
	7	on	on	on	on	on	on
R	R						

According to Table 2 and equation (2) to (7), it is analyzed as follows. Position R is directly combined with the shift element controlled by the manual valve. Position D is powered on and powered off by the corresponding shift valve H1 to H6 in Table 1 to achieve the corresponding gears, specifically as follows:

In gear D1, H1 and H2 are powered off. From logical equations (2) and (3), $P=\overline{Z_1}C_1$ and $P=\overline{Z_2}B_2$ can be calculated, and C1 and B2 are combined.

In gear D2, H1 and H3 are powered off, and H2 is powered on. From logical equations (2) to (4), $P=\overline{Z_1}C_1$, $P=Z_2P_2$ and $P_2=\overline{Z_3}B_1$ can be calculated, and C1 and B1 are combined.

In gear D3, H1 and H4 are powered off, and H2 and H3 are powered on. From logical equations (2) to (5), $P=\overline{Z_1}C_1$, $P=Z_2P_2$, $P_2=Z_3P_3$ and $P_3=\overline{Z_4}C_3$ can be calculated, and C1 and C3 are combined.

In gear D4, H2, H3 and H4 are powered on, and H1 and H5 are powered off. From logical equations (2) to (6), $P=\overline{Z_1}C_1$, $P=Z_2P_2$, $P_2=Z_3P_3$, $P_3=Z_4P_4$ and $P_4=\overline{Z_5}C_4$ can be calculated, and C1 and C4 are combined.

In gear D5, H1 is powered off, and H2, H3 H4 and H5 are powered on. From logical equations (2) to (6), $P=\overline{Z_1}C_1$, $P=Z_2P_2$, $P_2=Z_3P_3$, $P_3=Z_4P_4$ and $P_4=Z_5C_2$ can be calculated, and C1 and C2 are combined.

In gear D6, H1, H2, H3, H4 and H5 are powered on, and H6 is powered off. From logical equations (2) to (7), $P=Z_1P_1$, $P=Z_2P_2$, $P_2=Z_3P_3$, $P_3=Z_4P_4$, $P_4=Z_5C_2$ and $P_1=\overline{Z_6}C_4$ can be calculated, and C2 and C4 are combined.

In gear D7, H1, H2, H3, H4, H5 and H6 are all powered on. From logical equations (2) to (7), $P=Z_1P_1$, $P=Z_2P_2$, $P_2=Z_3P_3$, $P_3=Z_4P_4$, $P_4=Z_5C_2$ and $P_1=Z_6B_1$ can be calculated, and C2 and B1 are combined.

The result of logical equations is completely consistent with the working state of each shift element in Table 1, which verifies the correctness of the design of electro-hydraulic-controlled system by logic method.

5. Conclusion

In view of the self-designed seven-speed planetary gear automatic transmission, the power shift electro-hydraulic-controlled system is designed by logic method. Based on the analysis of transmission structure and shift elements, the shift valves and logical equations are designed for different gears in turn, and the interlock and failure analysis are carried out. The final control system is a logical combination containing six two-position three-way electro-hydraulic-controlled shift valves and one manual control valve. At the same time, the logical equations are used to verify the system function, so as to achieve the purpose of self-test design. The whole design process goes through design, analysis, verification and other links, which belongs to a "closed-loop" design process. In the case that the transmission structure diagrams and the working status of each shift element are known, the design of the power shift electro-hydraulic-controlled system can be carried out clearly, efficiently and accurately.

Acknowledgments

This work was financially supported by the Professor Training Project of Nanning University (2019JSGC03).

References

[1] Suzuki T, Sugiura H, Niinomi A, et al. (2017) New RWD 10 speed automatic transmission for passenger vehicles. SAE International Journal of Engines, 10(2): 695-700.

[2] Ebner W, Fuchs J, Brasseur G. (2017) A model-based design approach for an optimal electro-hydraulic system within automatic transmissions. IFAC-PapersOnLine, 50(1): 7606-7612.

[3] Mishra K D, Cardwell G, Srinivasan K. (2019) Modeling, Control, and Adaptation for Shift Quality Control of Automatic Transmissions. No. 2019-01-1129, SAE Technical Paper.
[4] Jiang J N, Wang S H, Sun W B. (2017) Simulation and optimization of automatic transmission hydraulic control module//2017 2nd International Conference on Robotics and Automation Engineering (ICRAE). IEEE, 243-247.
[5] Oh J, Park J, Cho J, et al. (2017) Influence of a clutch control current profile to improve shift quality for a wheel loader automatic transmission. International Journal of Precision Engineering and Manufacturing, 18(2): 211-219.
[6] Fan Y. (2015) The Transmission Design of the Seven-speed Planetary-gear AT Based on the Lever Method. Applied Mechanics and Materials, 727: 449-452.

Evaluation and selection of auto spare parts suppliers in supply chain

Rong Zeng

Mechanical Engineering Department, JIANGHAN University, Wuhan, Hubei, 430056, China

Abstract: In order to reduce the cost of automobile manufacturing, it combines with the improved multi-objective analytic hierarchy process (AHP), establishes a supplier evaluation system to make a choice of suppliers. Taking the procurement status of a specific company as an example, the rationality of the choice of parts suppliers is analyzed.

1. Introduction

Nearly 70% of the total cost of automobile manufacturing comes from purchased auto parts. If the procurement cost is reduced by 1%, the profit will be increased by 5% to 10%, which shows that the evaluation and selection of auto parts suppliers in the supply chain sector is of great significance to automobile manufacturers. Therefore, a scientific supplier evaluation system should be established, and then the supplier should be evaluated by reasonable methods[1-2].

2. Selection and Evaluation Supplier System

2.1 improved multi-objective AHP

The traditional analysis hierarchical method uses nine scales to compare the indicators of the same level, and the actual operation is more difficult[3]. The improved analysis hierarchical method uses three scales, the two comparisons only judge whether it is important or not, or equally important. Here are the steps.

(1) Confirm the target hierarchy

The target to be evaluated is layered, divided into the highest, middle and bottom ground according to the target influencing factors.

(2) Constructing a judgment matrix

After the establishment of the multi-hierarchy model, the two factors of the same level are compared with each other, and the relative scale is used to obtain the comparison matrix *C*.

$$C=\begin{bmatrix} a_{11} & a_{12} & \cdots & a_{1n} \\ a_{21} & a_{22} & \cdots & a_{2n} \\ \vdots & \vdots & \vdots & \vdots \\ a_{n1} & a_{n2} & \cdots & a_{nn} \end{bmatrix}$$

Scale scales are based on the following comparison matrix:

$$C_{ij}=\begin{cases} 2 & i \text{ is more important than } j \\ 1 & i \text{ is as important as } j \\ 0 & i \text{ is not important than } j \end{cases}$$

$$(i,j=1,2,3\ldots\ldots n)$$

Sort the importance of individual elements with the following formula:

$$F_i = \sum_{j=1}^{n} c_{ij} (i = 1,2,\ldots n) \quad (1)$$

f_{max} is the max Sorting Index, f_{min} is the minimum Sorting Index, C_{max} is the max factor in the ranking index, C_{min} is the minimum factor in the ranking index, and the two factors as the basic comparative factors, after the managers compare, assess their relative importance b_m (>1) , and then use the following conversion formula to find the relative importance between the elements:

$$f_{ij} = \begin{cases} \frac{f_i - f_j}{f_{max} - f_{min}}(b_m - 1) + 1 (f_i - f_j \geq 0) \\ 1 \Big/ \frac{f_j - f_i}{f_{max} - f_{min}}(b_m - 1) + 1 (f_j - f_i \geq 0) \end{cases} \quad (2)$$

The corresponding judgment matrix:

$$F = \begin{bmatrix} f_{11} & f_{12} & \cdots & f_{1n} \\ f_{21} & f_{22} & \cdots & f_{2n} \\ \vdots & \vdots & \vdots & \vdots \\ f_{n1} & f_{n2} & \cdots & f_{nn} \end{bmatrix}$$

Solve feature vector w

First calculate the product of each row of elements of the judgment matrix F, then calculate the three square root $\overline{W}_i$t of M_i, and then normalize the vector (so that the sum of the elements in the vector is equal to 1) $\overline{W}_i$= ($\overline{W}_1$, $\overline{W}_2$......$\overline{W}_n$) .

The naturalization formula is: $$w_i = \overline{w_i} \Big/ \sum_{i=1}^{n} \overline{w_i} \quad (3)$$

This gives you the individual proportion of each metric in each layer, as well as the weight between the layers.

Consistency test

First of all, the maximum characteristic value of the judging matrix is determined λ_{max}:

$$\lambda_{max} = \frac{1}{n} \sum_{i=1}^{n} \frac{(Fw)_i}{w_i} \quad (4)$$

($F \cdot w$ means the product of the judgment matrix and the characteristic vector.); w_i means the ith element of the feature vector w.

Then the consistency indicator can be calculated: $CI = \frac{\lambda_{max} - n}{n-1}$ (5)

(CI represents consistency indicator, CI equals 0 is completely consistent, CI close to 0 means to have satisfactory consistency, the greater the CI inconsistency is more serious)

Then calculate the test coefficient: $CR = \frac{CI}{RI}$ (6)

(RI represents a random consistency indicator, CR represents a test coefficient, and when CR 0.1, the result of the judgment matrix passes the consistency test; if the CR is 0.1, then the judgment matrix does not meet the consistency test, and the decision matrix is invalid.)

2.2 The establishment of the supplier evaluation system

In the selection of suppliers[4], the establishment of the first-level indicators to be able to truly reflect the actual situation of suppliers, the paper intends to choose the quality(mark A), delivery capacity(mark B), price(mark C), technical level(mark D) and service level (mark E)of these five as a first-level indicators, and the second-level are Product pass rate(mark A_1), Quality inspection capability(mark A_2), Quality certification(mark A_3); Timely delivery(mark B_1), Accurate delivery(markB_2), Delivery efficiency(mark B_3);Product prices(mark C_1) , Price stability(mark C_2), Terms of payment (mark C_3); Self-design (mark D_1), Degree of automation(mark D_2), Infrastructure(mark D_3), Technical improvement capabilities(mark D_4), Service response time(mark E_1), Level of after-sales service(mark E_2), Service attitude(mark E_3).

After understanding the characteristics of the supply chain, the next step is to determine the weight of the indicator[5]. First of all, the above evaluation indicators, combined with the three scale method to construct the comparison matrix, to obtain the following table 1 as shown:

Table 1. First-level indicator comparison matrix

	A	B	C	D	E	f_i
A	1	2	2	2	2	9
B	0	1	2	2	2	7
C	0	0	1	2	2	5
D	0	0	0	1	2	3
E	0	0	0	0	1	1

Using formula (2) to obtain the judgment matrix F of the first-level indicator

$$F=\begin{bmatrix}1 & 1.25 & 1.5 & 1.75 & 2\\ 0.8 & 1 & 1.25 & 1.5 & 1.75\\ 0.67 & 0.8 & 1 & 1.25 & 1.5\\ 0.57 & 0.67 & 0.8 & 1 & 1.25\\ 0.5 & 0.57 & 0.67 & 0.8 & 1\end{bmatrix}$$

Using formula (3), the product of each row of the matrix is calculated, and the result is M_1=6.5625 M_2=2.625 M_3=1.005 M_4=0.3819 M_5=0.15276

Add M_1 to M_5 and open the root three times to get $\sum_{i=1}^{n}\varpi_i$ =5.5134

Then ω_1=0.3395 ω_2=0.2503 ω_3=0.1817 ω_4=0.1316 ω_5=0.097

So:

$$F\cdot\omega=\begin{bmatrix}1.3492\\ 1.1016\\ 0.9072\\ 0.7490\\ 0.6498\end{bmatrix}$$

The next step is to use formula (4) to calculate $\lambda_{\max}$=5.1855

use formula (5) to calculate CI=0.0464

use formula (6) to calculate CR=0.0414

Sort out the judging matrix and weights shown in Table 2 below:

Table 2. First-level indicator judgment matrix and weight

	A	B	C	D	E	weight
A	1	2.25	3.5	4.75	6	0.3395
B	0.44	1	2.25	3.5	4.75	0.2503
C	0.29	0.44	1	2.25	3.5	0.1817
D	0.21	0.29	0.44	1	2.25	0.1316
E	0.17	0.21	0.29	0.44	1	0.0970

b_m=5 $\lambda_{\max}$=5.1855 *CI*=0.0464 *CR*=0.0414

As a result of 1CR=0.0414<0.1, the constructed first-level indicator judgment matrix meets the consistency requirements.

Using the same methods and steps, the indicator weights of the secondary indicators are calculated.

After the weight of the evaluation index is established, the evaluation matrix Q is constructed. Then, combined with the evaluation index comprehensive weight and evaluation matrix Q, the potential suppliers of their own scores.

The final score of the potential supplier is: $R=\omega\times Q$ (7)

3. Instance analysis

Taking the procurement management of gears of a car manufacturer as an example, it is reasonable to re-judge the supplier's choice. After selecting elimination at the procurement level, there are three potential suppliers left. A comprehensive score of 16 evaluation indicators for the three supplier spare parts factories is shown in the 3-1 evaluation matrix in table 3.

Table 3. Evaluation matrix

	Manufacturers 1	Manufacturers 2	Manufacturers 3
A_1	6	4	5
A_2	4	5	6
A_3	6	5	5
B_1	6	6	7
B_2	7	5	6
B_3	8	7	5
C_1	8	6	7
C_2	7	5	4
C_3	8	7	7
D_1	6	6	7
D_2	7	5	8
D_3	7	6	6
D_4	9	5	7
E_1	8	6	9
E_2	5	9	6
E_3	6	5	7

Using formula (7), the comprehensive score R of the three suppliers was derived

R$=\omega\times Q=(6.4291 \quad 5.4647 \quad 6.2402)$, Thus, Supplier 1 has the highest score, and the car manufacturer should select Supplier 1 as its best supplier of gears. In fact, the company's previous lying supplier was also this one, indicating that the evaluation system was reasonable.

4. Conclusions

Using the improved hierarchical analysis method, this paper re-establishes the five first-level indicators of quality, delivery capacity, price level, technical capability, service level, etc., quality inspection system, quality certification, delivery timeliness, delivery accuracy, delivery efficiency, product price, price stability, payment terms, independent design, The respective weights of 16 secondary indicators, such as automation degree, infrastructure scale, technology improvement capability, service response time, after-sales service level, service attitude, etc., and the evaluation matrix Q and scoring system R are constructed for evaluation and selection of potential suppliers. It can make the subjective factors of decision makers be greatly reduced, the evaluation results can be more scientific and objective, the credibility can be improved, and the accuracy of the system is verified by example, which provides a reference for the automobile manufacturers' purchasing management of auto parts.

Acknowledgments

This paper is funded by the university's high-level talent research project, the project number is 100606650001, and the name is “auto parts supply chain resource optimization research”.

References

[1] ZHOU Liang, LI Qing-min, PENG Ying-Wu. Optimization model for multi-level and multi-echelon spare part allocation in dynamic support structure[J]. Hang kong Xue bao/acta Aeronautica Et Astronautica Sinica, 2017.
[2] Sarbijan M S, Markabi M S. Using Grey Data Envelopment Analysis for Evaluation and Selection of Efficient Suppliers Based on Uncertainty[J]. 2015, 4(2):325-335.
[3] Costantino N , Dotoli M . A cross efficiency fuzzy Data Envelopment Analysis technique for supplier evaluation under uncertainty[C]// Emerging Technologies & Factory Automation. IEEE, 2013.
[4] Chuan-Jun Su, Yin-An Chen. Risk assessment for global supplier selection using text mining[J]. Computers & Electrical Engineering, 2018, 68:140-155.
[5] Deepika M , Kannan A S K . Global supplier selection using intuitionistic fuzzy Analytic Hierarchy Process[C]// 2016 International Conference on Electrical, Electronics, and Optimization Techniques (ICEEOT). IEEE, 2016.

IOP Conf. Series: Materials Science and Engineering **787** (2020) 012021 doi:10.1088/1757-899X/787/1/012021

Study On Train Routing Plan Of Y-Line Of Urban Rail Transit Considering Passenger Travel Purpose

Yu Liu[1,2*], Baoming Han[3]

[1] SinoRail Network Technology Research Institute, Haidian District, Beijing , 100044, People's Republic of China

[2] China Railway Information Technology Co., Ltd. Haidian District, Beijing , 100044, People's Republic of China

[3]School of traffic and transportation, Beijing Jiaotong University, Haidian District, Beijing, 100044, People's Republic of China

[*]Corresponding author's e-mail: liuyu_0618@sina.com

Abstract. Through the questionnaire survey on the passengers of the case line, the travel purpose of the passengers is divided into three categories: leisure, commuting and business. Based on the analysis of the rules of passenger's choice of route, the model of passenger's choice of route and the model of route planning are established, and the parameters of logic model are calibrated according to the results of questionnaire survey, and the generalized cost of passenger's travel is obtained by considering the degree of train congestion. The validity of passenger flow distribution model and routing planning model is verified by solving the case line model.

1. Introduction

With the development of urban rail transit in China, some new lines end at the middle station of the existing rail transit line, or the extension line from the middle station of the existing rail transit line. Thus, the existing rail transit line and the new line or extension line form a "Y" shape line, which is called a "Y" shape line. How to analyze the passenger flow of Y-line, understand the demand of passenger flow and reasonably organize the train routing of Y-line, make the train operation and passenger flow demand more match, and provide efficient and high-quality transportation service is an important content that needs to be studied at present.

Zhang, Liu, etc. considered the heterogeneity of passengers in their study[1-2], Zhao studied the factors affecting the choice of passenger routes[3], Yan divided passengers into commuter and non commuter passenger flows[4], and Liu considered the direct preference of passengers in the study[5].

There are few researches on passenger classification in the existing researches, most of them are based on single line. In the existing studies, passenger travel choice behavior, passenger flow distribution and train routing are considered separately, and passenger flow distribution is only carried out under the condition of established routing, which fails to consider passenger flow choice behavior and train routing formulation in a unified way. Therefore, this paper unifies passenger classification, route selection preference and train routing to get the optimal train routing plan under specific classification.

2. Problem description

2.1. Train routing plan of Y-line

The train routing plan is a plan that specifies the train operation interval and departure pairs. A reasonable train routing plan can meet the needs of passengers, reduce the operating costs of the operating enterprises, and achieve a win-win situation between the operating enterprises and passengers. To make a reasonable train routing plan, first of all, it is necessary to sort out the passenger demand, secondly, it is necessary to understand the basic situation of the line, such as the location of turn back station, station type, etc., finally, it is necessary to formulate the corresponding solution model and get the optimal routing results.

The train routing form of Y-type line is the combination and expansion of the basic train routing form of urban rail transit. According to the operation section of train routing and the main branch line of Y-type line, the train routing of Y-type line can be divided into three categories: independent operation, direct operation and mixed operation. As shown in Figure 1.

Table 1. Typical train route plan of Y type line

Independent operating	Direct operating routes		Mixed routing mode
	all direct operating	partial direct operating	
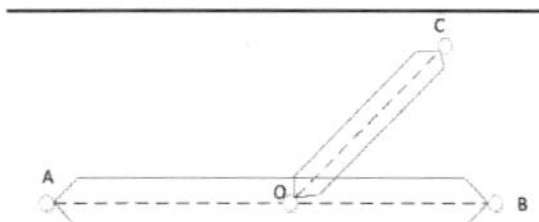	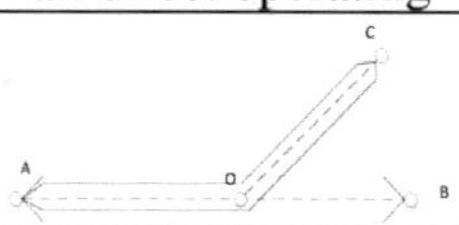	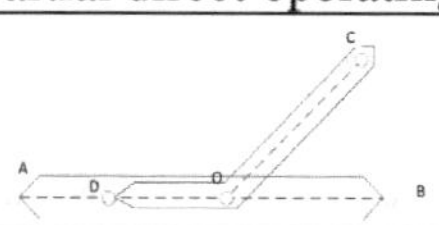	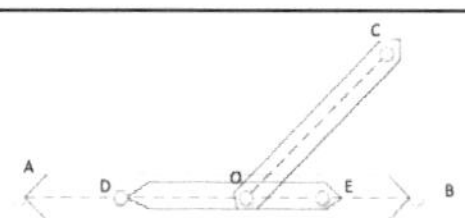

2.2. Passenger's demand for route selection

Urban rail transit should not only meet the needs of passengers, but also ensure the service quality and efficiency. Generally speaking, there are at least two train routes on the Y-type line, and there may be multiple routes between the OD of passengers. Different passengers will have different preferences when choosing the route. Different passengers have different preferences for waiting time when they take subway, and passengers who need short travel time will be sensitive to waiting time. The degree of congestion in the car is one of the most direct physiological manifestations of passengers' travel comfort. The degree of congestion in the car is expressed as the full load rate of the section, for the train, it is expressed as the full load rate of the train or the density of passengers, and the degree of congestion directly affects passengers' feeling. For Y-type line, if the line adopts the traditional independent operation routing, all passengers from the main line to the branch line need to transfer, that would greatly increase the travel time cost of passengers.

When the conditions of facilities and equipment are determined, in order to make the train routing serve the passengers better and save the cost of operation enterprises, it is necessary to deeply understand the characteristics of line passenger flow and the choice of passengers when they travel, and establish the Logit model of passenger flow distribution with the actual passenger selection characteristics of the route. and lay the foundation for the development of the train routing. This paper will elaborate on this problem in the next chapter.

3. Path selection

In order to explore the passenger's demand and path selection preferences, the passenger route selection Logit model of Y-line is established and parameter-calibrated. This paper investigates the passengers of Guangzhou Metro Line 3.

3.1. Passenger flow survey

In order to understand the passenger flow structure of Guangzhou Metro Line 3, a total of 300 questionnaires were issued and 271 valid questionnaires were collected.

IOP Conf. Series: Materials Science and Engineering **787** (2020) 012021 doi:10.1088/1757-899X/787/1/012021

3.1.1. Passenger property

According to the results of the questionnaire, the gender, age and income ratio of passengers are as follows.

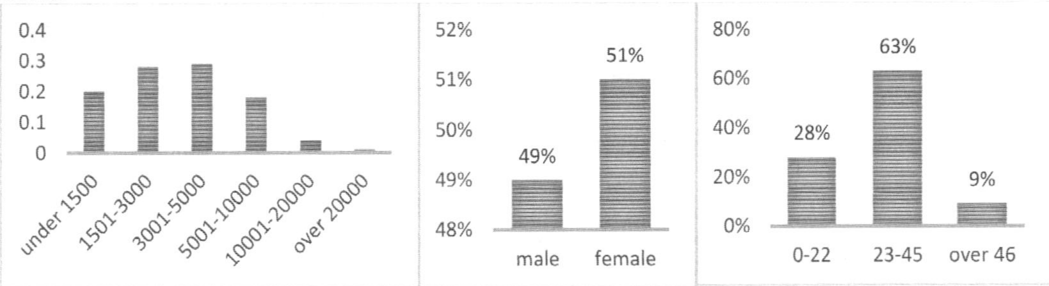

Figure 1. Passenger flow structure

3.1.2. Path selection

In order to explore the relationship between personal attributes of passengers and travel path selection, the following scenarios and questions are set in the questionnaire.

"If you travel during the morning rush hour and the basic travel time to your destination is 28 minutes, which route will you choose?

(1) Waiting for 5 minutes, no transfer, total travel time 36 minutes.(2) Wait 3 minutes, transfer 2 minutes, total travel time 33 minutes.(3) Transfer 0-2 times, wait 1 minute, transfer 0-8 minutes, total travel time 29-38 minutes."

The results are as follows.

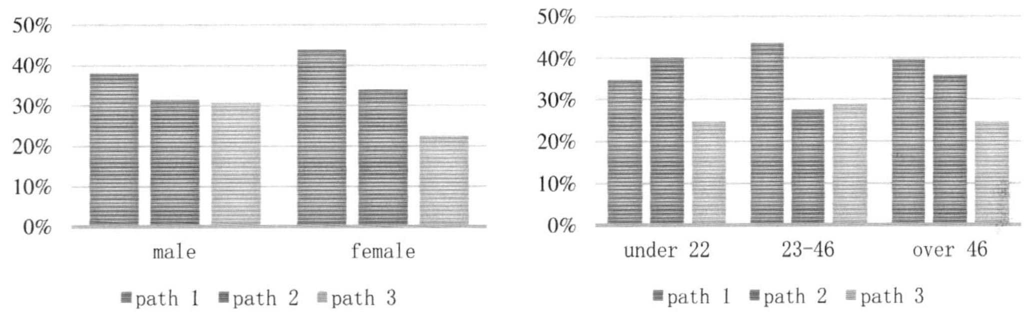

a) genders b) ages

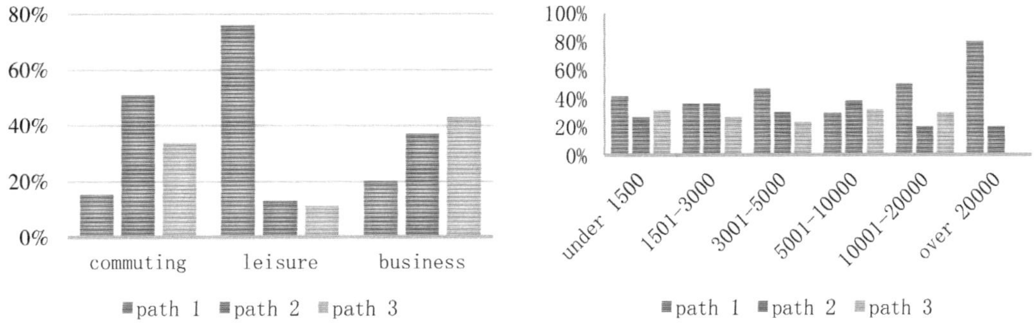

c) travel purpose d) income

Figure 2. Path selection for passengers with different attributes

3.2. Parameter calibration

3.2.1. Chi-square test

In order to verify whether there is correlation between passenger attributes and path selection, chi square test is carried out on the result of passenger selection.

Table 2. Chi-square test results

Properties	Parameters	Value	df	Sig.
Gender	Pearson Chi-Square	.106	1	.744
	LR	.106	1	.745
Age	Pearson Chi-Square	4.087a	4	.394
	LR	4.064	4	.397
Income	Pearson Chi-Square	14.264	3	.003
	LR	13.320	3	.004
Travel purpose	Pearson Chi-Square	55.233	5	.043
	LR	59.168	5	.041
	N of Valid Case	271		

Table 2 shows that the passenger's travel purpose and income are highly correlated with the passenger's route selection behavior. Since the number of passenger samples at some income levels is small in this survey, in this paper, the passengers are classified according to the travel purpose with a relatively significant relationship with the passenger travel path.

3.2.2. Parameter calibration

Compared with the road network, the common line operation section of each route has less independent influence on different paths, so different travel paths can be regarded as mutually independent.

Logit model has many forms of utility function, but the most classical and widely used is the linear form of utility function. The formula is as follows:

$$V_{lm}^{t,k} = \sum_{i}^{N} \theta_i^t X_i \tag{1}$$

X_i refers to the i-th factor that affects the passenger path selection. Each factor is independent and distributed. θ_i^t is the coefficient of relevant factors. The specific value is calibrated according to the questionnaire survey. According to the above survey results, this paper constructs the utility function of passenger travel path selection as follows:

$$V_{lm}^{t,k} = \theta_1^t t_{lm,c}^k + \theta_2^t t_{lm,jd}^k + \theta_3^t t_{lm,hd}^k + \varepsilon_{lm}^k \tag{2}$$

$V_{lm}^{t,k}$ ——The utility value of the t-th passenger to the k-th path between lm ,

$t_{lm,c}^k$ ——The travel time of the k-th path, including the running time and stop time of the train;.

$t_{lm,jd}^k$ ——Arrival waiting time of the k-th path.

$t_{lm,hd}^k$ ——Waiting time for transfer of the k-th path.

θ_1^t 、θ_2^t 、θ_3^t ——The coefficients of the above three influencing factors are respectively calibrated according to the questionnaire data.

$t = 1$、2、3——represent the purpose of passenger travel for commuting, leisure and business.

ε_{lm}^k ——The random utility term of the k-th path obeys Gumble distribution.

This paper introduces another concept in economic theory marginal utility. In this paper, the marginal utility of passenger transfer time and waiting time is calculated based on the travel time of passengers. The waiting time and transfer time of passengers in travel are all transformed into the travel time of passengers. Thus, the generalized travel time cost of passengers is obtained. The calculation formula of generalized travel time is as follows.

$$C_{lm}^{t,k} = t_{lm,c}^{k} + MRS_{jd}^{i,t} t_{lm,jd}^{k} + MRS_{hd}^{i,t} t_{lm,hd}^{k} \quad (3)$$

$$MRS_{hd}^{i,t} = \theta_1^i / \theta_1^t \quad (4)$$

In this paper, SPSS software is used for regression analysis of passenger questionnaire data, and the results of model estimation and parameter calibration are as follows.

Table 3. Result Model fitting

Travel purpose	-2Likelihood	Sig.	Cox and Snell R^2	Nagelkerke R^2
Commuting	73.453	0.021	0.467	0.54
Entertainment	80.274	0.027	0.408	0.489
Business	75.895	0.038	0.386	0.447

From table 3, it can be seen that the significance level of model fitting is 0.038, which is less than the hypothesis level of 0.05, so it can be seen that the parameters of the model are not all zero, the model hypothesis factors and dependent variables have strong significance, and the overall fitting results are credible.

Table 4. Estimation of parameters

	B	*Sig.*		*B*	*Sig.*		*B*	*Sig.*
θ_1^1	-1.805	0.013	θ_1^2	-2.017	0.008	θ_1^2	-3.327	0.019
θ_2^1	-2.742	0.021	θ_2^2	-2.042	0.043	θ_2^2	-3.531	0.027
θ_3^1	-2.243	0.042	θ_3^2	-2.847	0.021	θ_3^2	-3.494	0.009

According to the formula (3), the marginal substitution rate of each variable for travel time is as follows.

Table 5. Marginal rate of substitution

marginal substitution rate	$MRS_{jd}^{2,1}$	$MRS_{hd}^{3,1}$	$MRS_{jd}^{2,2}$	$MRS_{hd}^{3,2}$	$MRS_{jd}^{2,3}$	$MRS_{hd}^{3,3}$
value	1.52	1.24	1.01	1.41	1.13	1.12

4. Routing model

4.1. Basic assumptions

1) Taking hours as the unit, this paper mainly considers the train routing plan in peak hours;

2) The train stops at each station, and the train formation in each routing is the same;

3) Assuming that the parking capacity of the depot and the station storage line is sufficient, the number of trains that can be used is unlimited;

4) the routing plan of up and down trains is the same, and the trains depart in pairs;

5) passengers are evenly distributed on the train;

6) only the internal passenger flow of a single line is considered, and the transfer between other lines and the research line is not considered.

4.2. Basic assumptions

In order to establish the train routing model of Y-line, the parameters set in this paper are shown in table 6.

Table 6. Parameters

Symbol	Meaning
V	Station set, v_i means different stations, n is the total number of stations on the line
E	Interval set, e_i represents the i-th interval, m is the total number of intervals
R	Set of all routes, r_i represents the i-th route, N_r is the total number of routes
E_{ri}	Set of intervals included in the i-th train routing
V_{ri}	Set of stations included in the i-th train routing
E_{lm}^k	Set of intervals contained in the k-th path between l and m
V_{lm}^k	Set of stations contained in the k-th path between l and m
T	Time frame for preparing train routing plan, in hours
α	Unit purchase cost of train, unit: yuan / (train × minute)
β	Unit running cost of train, yuan / (train × km)
l_{e_i}	Mileage of interval e_i, KM
l_{r_i}	Mileage of route r_i, KM
$t_{折}$	Reentry time of turn back station, minutes
t_{e_i}	Running time of interval e_i, minutes
t_{vi}	Stop time at station i, minutes
a_{jd}	Arrival waiting time coefficient
a_{hd}	Transfer time coefficient
λ	Unit time cost of passengers, yuan / minute
q_{lm}	Passenger flow between l and m, in units of people
$p_{lm}^{t,k}$	The probability of selecting the k-th route for the T-class passengers between l and m
p_{lm}^t	Passenger flow proportion of class T passengers between l and m
Q_{zi}	Rated seats of train
Q_{bi}	Standard passenger capacity of train
$\delta_{r_i}^{e_j}$	0-1 variable, 1 when x contains V, otherwise 0
Nr_{min}	Minimum total number of train routing on the line
Nr_{max}	Maximum total number of train routing on the line
n_{min}	The lower limit of the number of routes containing interval e_i
n_{max}	The upper limit of the number of routes containing interval e_i
$h_{ej,min}$	Lower limit of train departure interval in section e_i, min
$h_{ej,max}$	Upper limit of train departure interval in section e_i, min
η	Capacity rate
ξ	Weight of enterprise operation cost
A，B ρ γ	Penalty coefficient due to congestion
x_i	Passenger flow of section e_i

$\varphi_{lm}^{k}(r_i,e_j)$	0-1 variable, $\varphi_{lm}^{k}(r_i,e_j)=1$ when the e_i interval of r_i routing plan is on the k-th path between l and m
$\delta_{r_i}^{v_j}$	0-1 variable, $\delta_{r_i}^{v_j}=1$ when r_i contains v_j
$t_{lm,c}^{k}$	Running time of the k-th path between l and m, min
$t_{lm,jd}^{k}$	Arrival waiting time of the k-th path between l and m, min
$t_{lm,hd}^{gk}$	Transfer waiting time of the k-th path between l and m, min
s_{lm}^{k}	0-1 variable, $s_{lm}^{k}=0$ when the k-th path between l and m contains only 1 train routing
x_i	Passenger capacity of train in section e_i
n_{lm}^{k}	The number of routes included in the k-th path between l and m
δ_i	0-1 variable, $\delta_i=1$ if routing I is selected by the current solution, otherwise 0
f_{r_i}	Number of departures per hour for route r_i

4.3. *Objective function*
Objective function includes operating enterprise cost and passenger cost.

4.3.1. *Operating enterprise cost*
Operating enterprise costs include train purchase cost Z_1 and train travel kilometers fee Z_2 ,as follows.

$$min\ Z_1 = 1.2\alpha\sum_{i=1}^{N_r}\left(\sum_{ei\in E_{ri}}\sum_{v_i\in V_{ri}}(t_{ei}+t_{vi})+2t_{折}\right)\cdot\delta_i f_{r_i} \tag{5}$$

$$min\ Z_2 = 2\beta\sum_{i=1}^{N_r}\delta_i f_{r_i} l_{r_i} \tag{6}$$

4.3.2. *Passenger cost*
The cost of passengers includes three parts: the time when passengers are on the train, the time when passengers are waiting for the train at the station and the time when they are waiting for transfer, which are indicated by $t_{lm,c}^{k}$, $t_{lm,jd}^{k}$, and $t_{lm,hd}^{k}$, respectively.

$$t_{lm,c}^{k} = \sum_{e_i\in E_{lm}^{k}} t_{e_i} + \sum_{v_s\in V_{lm}^{k}} t_{v_i} \tag{7}$$

$$t_{lm,jd}^{k} = \frac{a_{jd}T}{f_{r_i}} \qquad \forall r_i \in y_k \tag{8}$$

$$t_{lm,hd}^{k} = s_{lm}^{k}\sum_{r_m}(a_{hd}T / f_{r_m}) \tag{9}$$

$$s_{lm}^{k} = \begin{cases} 0 & n_{lm}^{k} = 1 \\ 1 & n_{lm}^{k} \geq 1 \end{cases} \tag{10}$$

When the number of passengers in the train is less than the number of rated seats, passenger feels most comfortable. When the number of passengers on the train is between the number of rated seats and standard capacity, passenger feels less comfortable. When the number of passengers on the train is between standard capacity and the maximum passengers number, passenger feels worst. Penalty coefficient is exponentially related to congestion.

$$U_{e_i}=\begin{cases}0 & x_i \le Q_{zi} \\ (\frac{x_i - Q_{zi}}{Q_{zi}})^{\rho} \bullet A & Q_{zi} \le x_i < Q_{bi} \\ (\frac{x_i - Q_{zi}}{Q_{zi}})^{\rho} \bullet A + (\frac{x_i - Q_{bi}}{Q_{bi}})^{\gamma} \bullet B & Q_{bi} \le x_i\end{cases} \tag{11}$$

Therefore, passenger travel costs is as follows.

$$\min Z_3=\lambda \sum_{l=1,m=2}^{n} \sum_{k=1}^{K} \sum_{t=1}^{3} q_{lm} \cdot p_{lm}^{t} \cdot P_{lm}^{t,k} \cdot [(1+U_{e_i})t_{lm,c}^{k} + \frac{a_{jd}T}{f_{r_i}} + s_{lm}^{k} \sum_{r_m} (\frac{a_{hd}T}{f_{r_m}})] \tag{12}$$

4.4. Restrictions

(1) There is a limit to the number of routes each station and intervals belongs to.

$$n_{min} \le \sum_{i=1}^{N_r} \delta_i \delta_{r_i}^{e_j} \le n_{max}, \forall e_j \in E \tag{13}$$

(2) Total number of routes.

$$Nr_{min} \le \sum_{i=1}^{N_r} \delta_i \le Nr_{max} \tag{14}$$

(3) Routing length.

$$\sum_{j=1}^{n} \delta_{r_i}^{v_j} \ge w, \forall r_i \in R \tag{15}$$

(4) Section passenger flow and section capacity determine service frequency.

$$f_{r,min} \le f_{r_i} \le f_{r,max}, r_i \in R \tag{16}$$

(5) Passengers must be transported by train.

$$\sum_{k \in K} q_{lm}^{k} = q_{lm} \tag{17}$$

(6) Interval between trains

$$h_{e_j,min} \le T / \sum_{i=1}^{N_r} \delta_{r_i}^{e_j} \cdot f_{r_i} \le h_{e_j,max}, \forall e_j \in E \tag{18}$$

(7) The number of passengers on the train shall not exceed the full load rate of the interval.

$$\frac{\sum_{v_l,v_m \in V} \sum_{k \in K} \sum_{i=1}^{N_r} q_{lm}^{k} \varphi_{lm}^{k} (r_i, e_j)}{Q \sum_{i=1}^{N_r} f_{r_i} \delta_{r_i}^{e_j}} \le \eta, \forall e_j \in E \tag{19}$$

4.5. Multi objective function to single objective function

In this paper, ξ is defined as the weight of enterprise operation cost, and $1-\xi$ as the weight of passenger travel cost.

$$\min Z_4 = \xi(Z_1 + Z_2) + (1-\xi)Z_3 \tag{20}$$

5. Solving algorithm

Enumeration is a classical algorithm to solve the problem under certain conditions. By enumerating and calculating the target value of all the cases that meet the set limit conditions, the optimal solution can be obtained. In this paper, enumeration method is used to solve the model.

6. Case analysis

6.1. Case description and parameter value

The following figure shows the urban rail transit lines of a city, which has totally 29 sections.

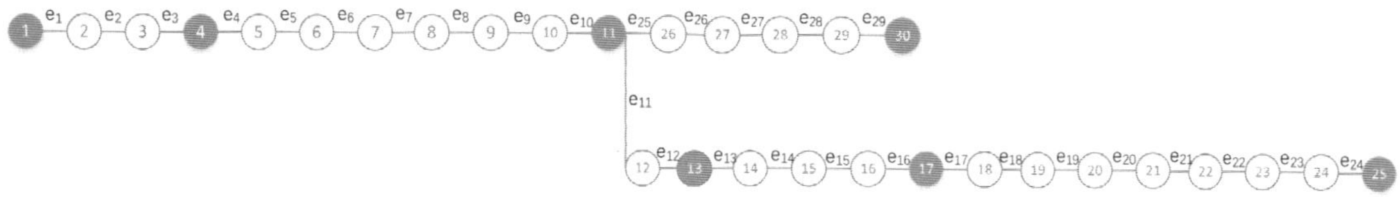

Figure 3. Case line

Table 7. Station spacing and interval running time

No.	Stopping time	Running time(s)	Length (km)	No.	Stopping time	Running time(s)	Length (km)
1	45	135	3.9	16	25	95	1.4
2	25	155	6.2	17	45	135	3.4
3	45	135	3.0	18	30	150	0.833
4	30	150	2.2	19	30	150	2.3
5	30	90	1.9	20	35	205	6.3
6	35	85	2.6	21	40	260	5.5
7	40	140	2.1	22	30	150	2.5
8	30	90	1.2	23	30	150	3.6
9	30	90	1.5	24	45	75	1.2
10	45	195	1.3	25	45		
11	45	75	1.2	26	35	145	1.3
12	35	85	0.972	27	30	90	0.843
13	30	150	1.2	28	35	85	0.89
14	35	85	1.9	29	40	80	1.6
15	40	80	1.3	30	45	145	2.4

Table 8. Model parameter setting

Parameters	Value	Parameters	Value
$z_{i,o}$, $z_{i,d}$	135s	ξ	0.2
α	4 yuan/train·min	η	120%
β	60 yuan/train·min	Nr_{min}	2
T	1 hour	Nr_{max}	3
a_{jd} , a_{hd}	0.5	$h_{ej,min}$	105 s
λ	0.8yuan/min	$h_{ej,max}$	15 mins
Q_{zi}	192 per train	Q_{bi}	1350 per train
ρ	0.01	γ	0.02
A	1	B	2

6.2. Results

The algorithm in Chapter 5 is used to get the optimal train routing plan, the result shows that the best routing plan that contains two routings.

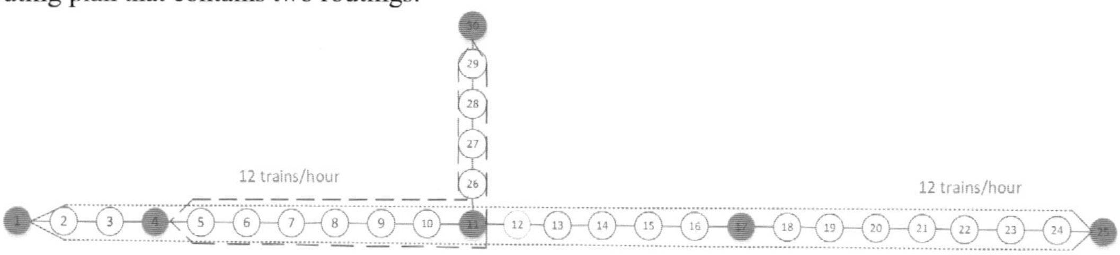

Figure 4. Model result

6.3. Comparison

6.3.1. Compared with actual routing plan

According to the model solution result, we compared the results with actual routing plan as follows.

Table 9. Compared with actual routing plan

	Routing (¥)	Frequency (¥)	Purchase (¥)	Operating (¥)	Waiting (¥)	Transfer (¥)	In vehicle cost(¥)	Total cost(¥)
Actual Routing	1-30	14	3 745	58 620	113 345	18 278	605 261	197 268
	4-17	7						
	11-25	7						
Solution result	1-25	12	3 580	57 843	116 620	13 451	592 224	190 618
	4-30	12						
Fluctuation range		8	-165	-777	3 275	-4 827	-13 037	-6 650
Increase proportion		-14.3%	-4.41%	-1.33%	2.89%	-26.41%	-2.15%	-3.37%

The results of the above table show that the model in this paper can reduce the number of departure pairs, train purchase cost and running kilometres cost to meet the needs of passengers, so as to reduce the operation cost of enterprises. In addition, the solution of the model also reduces the travel cost of passengers, thus reducing the total cost by 3.37%.

6.3.2. Sensitivity analysis of passenger flow structure

The chapter 3 of this paper shows that passengers with different travel purposes have different choices of travel routes. This section analyzes the results of the proportion of passengers with different travel purposes on the train routing.

The proportion of commuting, leisure and business office is 0.38, 0.35 and 0.27 respectively in the case line. But in this section, we study how the routing plan changes when passenger flow construction changes. We let passenger flow with commuting, leisure and business purpose as the main passenger flow separately, assuming that the proportion of passenger flow in the three cases is 6:2:2, 2:6:2 and 2:2:6 respectively, and get the optimal traffic results as follows.

Table 10. Optimal Routes with Different Passenger Flow Structures

Main Passenger	Routing	Frequency /hour	Purchase	Operating	In vehicle	Waiting	Transfer	Total cost
Commuting	1-30 4-17 11-25	16 8 8	4 045	60 620	575 261	98 278	113 345	201 109
Leisure	1-30 4-25	20 10	3 830	59 562	591 508	94 537	13 424	190 607
Business	1-25 4-30	12 12	3 580	57 843	592 224	116 620	67 452	193 598

It can be seen that when the main passenger travel purpose of the line is commuting, the optimal combination of routes is three routes, at which time the departure frequency is the largest and the passenger's travel time is the shortest; when the line is dominated by leisure passengers, the optimal route is two direct routes, at which time the line's travel time is the shortest; when the line is dominated by business passengers, the total departure pairs and the enterprise cost is the least.

7. Conclusion

In this paper, the optimal routing plan obtained by the model is compared with the actual routing plan. It is found that both the optimal routing plan and the actual routing plan of the model have direct routing, which is determined by the passenger flow characteristics of the line. At the same time, the total cost of the model results is 3.37% lower than the total cost of the actual routing plan, which verifies the effectiveness of the model results. In addition, this paper obtains the optimal results by changing the proportion of passengers with different travel purposes, and finds that the optimal results are also different, which verifies the effectiveness of the passenger flow distribution model.

Acknowledgments

The study is sponsored by National Engineering Laboratory project for the Safety Technology of Urban Rail Transit System (Development and Reform Office High Technology [2016] No. 583); 2018 China Railway Information Technology Co., Ltd. Science and Technology Research and Development Program Major Project JGZG-CKY-2018013 (2018A01).

References

[1] Liu S, Yao E, Zhang Y. (2014) Personalized Route Planning Algorithm for Urban Rail Transit Passengers. Journal of Transportation Systems Engineering and Information Technology, 14(05):100-104+132.

[2] Zhang Y, Yao E, Liu S, Cai C. (2018) Metro Passengers' Semi-compensatory Route Choice Modeling and Application. Journal of The China Railway Society. 2018,40(02):1-7.

[3] Zhao K, Li H. (2018) Metro Passengers' Semi-compensatory Route Choice Modeling. Railway Transport and Economy. 40(11):110-115.

[4] Yan H. (2015) Personalized travel path recommendation algorithm for rail transit passengers. Computer Programming Skills & Maintenance. (22):87-88+91.

[5] Li D, Liu Z, Wang X, Lan Z, Liu Y. (2018) Train Routing Planning of Y-type Urban Rail Transit Considering Passenger Choice Behavior. China Railway Science. 39(04):114-122.

IOP Conf. Series: Materials Science and Engineering **787** (2020) 012022 doi:10.1088/1757-899X/787/1/012022

The Development of the Assistant Driving and Monitoring and Assessment System for Missile Weapon Special Vehicles

Hong-liang Cui

High-tech Institute, Fan Gong-ting South Street on the 12th, QingZhou 262500, China.

Abstract. Based on the analysis and study of the characteristics of missile weapon special vehicles, combined with the problems existing in the driving operation training process of the military drivers, an assistant driving and monitoring and assessment system for missile weapon special vehicles is developed. The design ideas of the hardware modules of the system and the realization methods of the software design of each subsystem are introduced in detail. Finally, the application effect of the system in the training of equipment formations of National Day parade is illustrated.

1. Introduction

The missile weapon special vehicles of the Rocket Force are nearly ten meters longer than the ordinary transport vehicles. There are problems in the launching position, such as long positioning time, low positioning accuracy, and difficult positioning at night. In the process of military parade training, there are some problems such as the inability to accurately locate due to the existence of multiple vehicles and the lack of training and assessment methods, resulting in low efficiency of the military parade training and the inability to record the operation actions in the driving process. In view of the problems existing in the missile weapon special vehicles, using advanced technologies such as the laser ranging, satellite positioning, panoramic imaging and electronic assessment, adopting mature I^2C bus technology and ARM embedded technology, the assistant driving and monitoring and assessment system for missile weapon special vehicles is developed to solve the problems of long positioning time and low positioning accuracy of vehicles in the launching position, which provides an effective means to ensure the safe driving of the missile weapon special vehicle drivers, improves the training efficiency of the military parade formations of the missile weapon special vehicles, and provides methods for the driving skill assessment of the missile weapon special vehicle drivers.

2. The Overall Structure of the System

The assistant driving and monitoring and assessment system of missile weapon special vehicles is composed of four parts: the satellite positioning module, the laser ranging module, the panoramic imaging module, and the assessment and evaluation module, as shown in Fig.1.

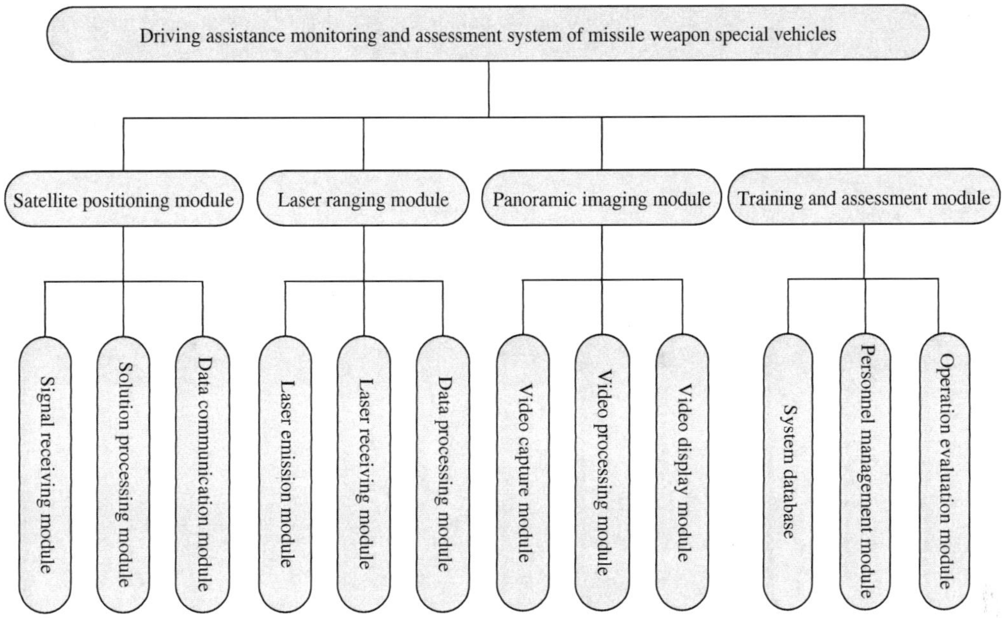

Figure 1. The composition diagram of assistant driving and monitoring and assessment system for missile weapon special vehicles

The satellite positioning module consists of the signal receiving module, the solution processing module, the data communication module, etc. It mainly completes the functions of vehicle positioning, ranging, etc., which can provide the relative position information between vehicle teams for vehicle array commanders in the process of military parade training, guide the launching vehicle drivers to locate the launching vehicles quickly and accurately, and reduce the missile launching preparation time.

The laser ranging module consists of the laser launching module, the laser receiving module and the data processing module, which can accurately judge the distance between the vehicle body and the surrounding objects, guarantee the constant speed, equal distance and equal interval travel of the vehicle square formations of the military parade, and can guarantee the missile force to avoid the vehicle scratch accident in the night synthetic training.

The panoramic imaging module is composed of the video capture module, the video processing module and the video display module, which can complete video capture, real-time video processing, data storage, video panoramic display and other functions around the vehicle, eliminate the blind area of the driver's vision, and ensure the driver's driving safety.

The training and assessment module is composed of the system database, the personnel management module and the operation evaluation module, which completes the training and assessment of the driving skills of the missile weapon special vehicle drivers and completes the comprehensive training and scoring of the vehicle square formations in the process of military parade training, providing a scientific and effective platform for quantitative evaluation and auxiliary training for driver training.

The assistant driving and monitoring and assessment system of missile weapon special vehicles integrates positioning, camera, ranging, assessment, display and terminal processing. The driver can adjust timely according to the transmitted data and image, and realize all-round real-time monitoring of the missile weapon special vehicle, the safe distance alarm, and the scientific and effective assessment of driver's driving skills.

3. The Design and Implementation of System Hardware

The assistant driving and monitoring and assessment system for missile weapon special vehicles combines skill training, collaborative drilling, assistant driving and assessment organically. Based on the embedded ARM, it adopts advanced technologies such as the laser ranging, the satellite positioning, the panoramic imaging, and the electronic assessment, combines bus control technology, and adopts modular design idea. The main equipment selection and connection diagram of the system is shown in Fig.2:

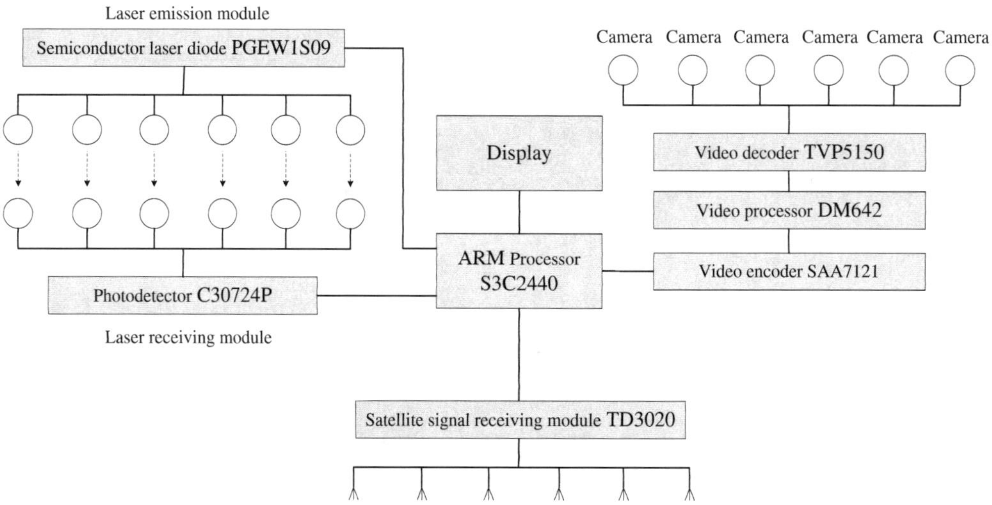

Figure 2. Main equipment selection and connection diagram of the system

Hardware equipment mainly includes: embedded ARM system S3C2440, video decoder TVP5150, video processor DM642, video encoder SAA7121, satellite signal receiving device TD3020, laser transmitting device PGEW1S09, laser receiving device C30724P, display device and signal conditioning board.

The system adopts the architecture of ARM+FPGA, and uses the powerful processing ability and rich interface of embedded ARM to ensure the real-time performance of the system. FPGA is a programmable logic device with flexible configuration, which can be developed flexibly and conveniently to improve the working efficiency of the system.

3.1 The Design of the Satellite Positioning Subsystem

The satellite positioning subsystem mainly completes the reception and analysis of navigation satellite signals, and displays the position information on the display.

The satellite receiving module selected by the system is TD3020 dual system high-performance satellite receiving module, which supports dual-mode navigation positioning information of BD2 and GPS frequency points at the same time. By setting the working mode, three working modes can be used: single BD2 positioning, single GPS positioning and BD2/GPS hybrid positioning.

When the satellite receiving chip works normally, it outputs the standard NMEA-0183 format positioning data and 1PPS timing signal through the serial port, which provides BD2/GPS dual-mode solution with high precision, low power consumption and low cost for the missile weapon special vehicles.

3.2 The Design of the Laser Ranging Subsystem

The laser ranging subsystem uses the principle of laser ranging to realize the function of obstacle distance detection. According to the distance of the obstacle, it will automatically alarm, so as to achieve the purpose of all-round anti-collision of missile weapon special vehicles. Through the laser ranging system, the vehicle square formations participating in the military parade can travel in formation according to the requirements of horizontal alignment and vertical equidistance in the process of military parade, reducing the excessive dependence on the operation skills of special vehicle drivers. The design schematic diagram of the laser ranging subsystem is shown in Fig.3.

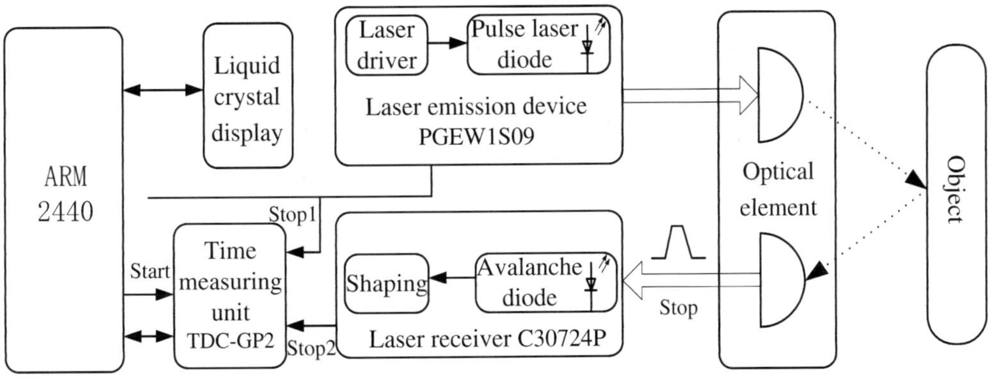

Figure 3. Schematic diagram of the design of the laser ranging subsystem

PGEW1S09 is selected as the laser emission device, the integrated chip C30724P is selected as the laser receiving device and the TDC-GP2 chip with higher accuracy is selected as the time measuring device. The embedded ARM controls the TDC to time, controls the laser emission device to send strong and narrow pulses, controls the laser receiving device to receive the feedback signal in real time, carries out the distance calculation in real time, and drives the LED display module to display the distance information. If the distance is within the dangerous range, it controls the alarm circuit to send out the sound and light alarm signal.

3.3 The Design of the Panoramic Imaging Subsystem

The system covers all areas around the vehicle body through six wide-angle cameras on the vehicle. Through the high-speed operation of ARM, the parameter information of six camera splicing relationship is obtained. According to the calculated parameter information, the panoramic top view image is generated. The core processing unit is DM642, a multimedia processor with high cost performance from TI company.

After the DM642 captures 6 source images, the video decoder TVP5150 decodes them. Then the image parameter information is synthesized by ARM, and the processed panoramic image information is output through video port and displayed on the display. The video display process is the opposite of video capture. The digital video processed by DM642 is encoded by SAA7121 and output in real time. The driver can control the missile weapon special vehicle conveniently, ensure the driving safety of the vehicle, record the driving track of the vehicle in real time, and provide the basic basis for the assessment system.

3.4 The Design of the Training and Assessment Subsystem

The training and assessment system is mainly aimed at the training and assessment of the driving skills of drivers of the missile weapon special vehicles. It trains and inspects the drivers' familiarity with vehicles, operating accuracy, and the disposal of special conditions of vehicles, and gives a score for drivers' operation.

During the training, it gives the user tips on how to operate, the key points of operation, the consequences caused by misoperation and how to deal with special situations. In the assessment mode, the program will not show any prompts, and only record all the information of the driver's operation actions in detail in the background, including operation time, operation actions, handling of special situations, etc., for the assessment of the assessment results.

4. The Design and Implementation of System Software

According to the characteristics of the missile weapon special vehicles of the Rocket Force, combined with the design of system hardware, Microsoft C#.NET is selected as the development tool, and modular programming method is adopted. The software system is divided into four functional modules: system software, main control software, communication software and management software, as shown

IOP Conf. Series: Materials Science and Engineering **787** (2020) 012022 doi:10.1088/1757-899X/787/1/012022

in Fig.4.

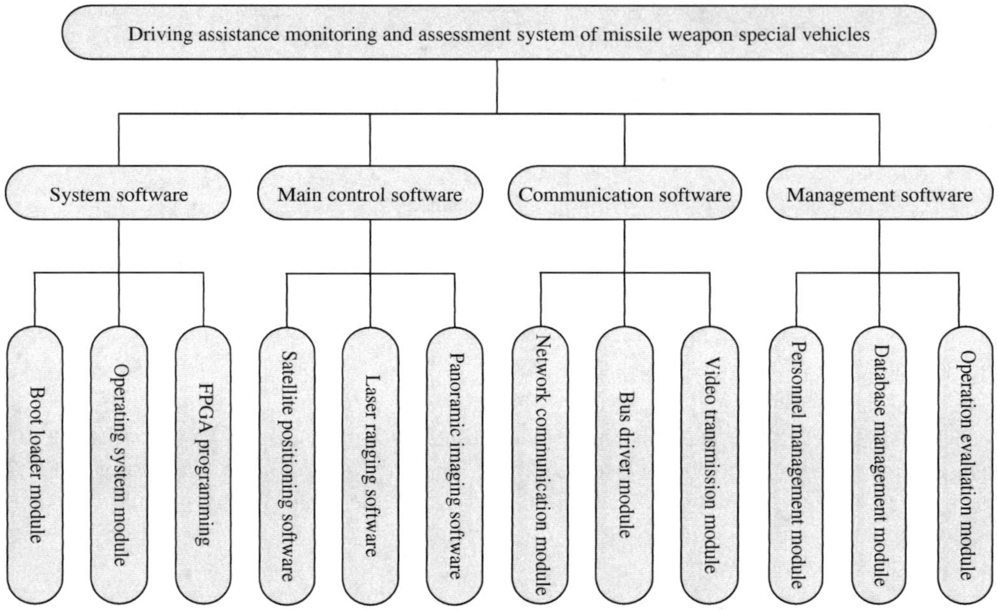

Figure 4. The software system structure diagram

The system software is composed of boot loader, Linux operating system and FPGA programming, which can control, monitor and manage other devices. The boot loader exits and disappears after the boot function is completed.

The main control software consists of satellite positioning software, laser ranging software and panoramic imaging software. The satellite positioning software completes the processing of the positioning information data collected by the satellite positioning module, and obtains the positioning information. The laser ranging software carries out the high-precision time interval measurement through the propagation delay control of the internal gate circuit, and calculates the accurate distance information. The panoramic imaging software is to complete the collection, processing and output of video data and other functions.

The communication software is composed of three parts: the network communication module, the bus driver module and the video transmission module, which realizes the communication between the lower computer and the upper computer as well as the network communication function, bus driver loading, video transmission and other functions.

The management software consists of three parts: the personnel management module, the database management module and the operation evaluation module, which can store driver information, update database and make operation evaluation rules.

5. Conclusions

The assistant driving and monitoring and assessment system for missile weapon special vehicles developed in this paper has novel design, high efficiency cost ratio and wide application object. It has been applied in a special equipment driver training base of the Rocket Force and the equipment formation training of National Day parade, and successfully completed the training assessment task of relevant special vehicle drivers and the training task of the equipment formation of National Day parade, as shown in Fig.5. The application results show that the system runs stably and is based on the support of assistant driving and assessment and training of special vehicles. With lower cost, the system can complete multi-channel real-time timing, positioning, monitoring and assessment functions of multiple vehicles, and accurately measure the moving parameters of equipment formations such as uniform speed, alignment and equidistance. The system builds a hardware platform for the training and assessment of the military missile special vehicle drivers and the military parade training of the equipment formations, reduces the cost of the special vehicle driving training, improves the driving training effect of the special vehicles, and has great military significance and greater economic benefits.

Figure 5. The practical application diagram of the system in training of equipment formations of military parade

References

[1] Hong-liang Cui.Development of Crash-proof Ranging Alarm for Missile Launching Vehicle Based on Single Chip Microcomputer[J].Electronic products world,2012(12)
[2] Ming Chang.Overview of Training and Assessment System for Vehicle Parade[J].Automobile Application,2009(10)
[3] Geng Xiao.Real-time Vehicle Detection System Based on Video[J].Computer Simulation,2015(09)
[4] Chaoyang Zou.Research on Real-time Video Monitoring Technology Based on Multi-camera Panoramic Image Mosaic[D].South China University of Technology,2014
[5] Wenxue Wang.Real-time Video MosaicTechnology Research[D].Beijing University of Technology,2014
[6] Yong Bai.Research on the Panoramic Parking System[J].Auto Parts,2014(08)

RBF Neural Network Based Correction Iterative Learning Control for Direct-drive Pump-controlled Clutch Actuator

Jieyu Wang[1], Bingzhao Gao[1] and Hong Chen[2,1*]

[1] State Key of Automotive Simulation and Control, Jilin University, Changchun, Jilin, 130000, China

[2] Clean Energy Automotive Engineering Center, Tongji University, Shanghai, 201804, China

[*]Corresponding author's e-mail: achenh@jlu.edu.cn

Abstract. This paper mainly focuses on the motion control of a novel direct-drive pump-controlled clutch-actuator system. The RBF neural network and iterative learning control method are combined to perform fast and accurate position tracking for dynamic nonlinear systems with high uncertainty. And the simulink and AMEsim co-simulation is carried out under different conditions. The simulation results confirm that the proposed RBF neural network based correction ILC controller has sufficient practicability and is suitable for the direct-drive pump-controlled actuator motion control conditions.

1. Introduction

Electro-hydraulic actuators are widely used due to the advantages of large capacity, fast response and space-saving arrangement [1]. And the actuator control strategy affects the performance of the clutch and the ride comfort of the vehicle [2]. Because of some uncertainties and nonlinearities, such as time-varying leakage, friction and load force, it is a difficult task to model nonlinear effects explicitly [3].

Iterative learning control (ILC) can generates excellent control effect through practice in the iteration domain, which is highly robust to system [4]. Many PID type ILC schemes are used on deterministic nonlinear systems to effectively control the system to track the given trajectory [5]. ILC essentially achieves ideal control effects by storing and memorizing dynamic errors over multiple iterations [6]. For real clutch actuator, the controller is required to be able to deal with model inaccuracy or model change between iterations. So a controller with the ability to approximate nonlinear systems over a cycle used in combination with the ILC is the best approach to improve tracking accuracy and learning ability [7]. Radial basis function neural network (RBF NN) has been successfully applied to complex time-varying uncertain nonlinear plants [8]. For example, adaptive neural control is formed for strict-feedback stochastic nonlinear system and for unknown dead zones of nonlinear systems [9].

This paper is related to the motion control of a novel direct-drive pump-controlled clutch actuator. The actuator system eliminates control valves, and dc motor and pump are directly connected to provide hydraulic pressure that drives the clutch to disengagement or engagement [10]. In order to generate the desired control effect, RBF NN based correction ILC controller is proposed for the plants. The rest of this paper is organized as follows: In section 2, the novel actuator structure is expressed and the physical model is described in detail. Section 3 proposes the RBF NN based correction ILC controller. In section 4, some simulations under various controllers are reported and analysed. And, section 5 presents conclusions about this paper and provides suggestions for future work.

2. Actuator system modeling

The direct-drive pump-controlled clutch actuator system is shown in Figure 1. DC motor transmits the torque to the external gear pump to form the hydraulic pressure. The release bearing (hydraulic cylinder) is moved forward by the pressure and pushes the diaphragm spring, thus completing the disengagement of the clutch. As the motor rotates in the reverse direction, a vacuum is formed in the oil passage, and the diaphragm returns to its original position under the elastic force of the diaphragm spring.

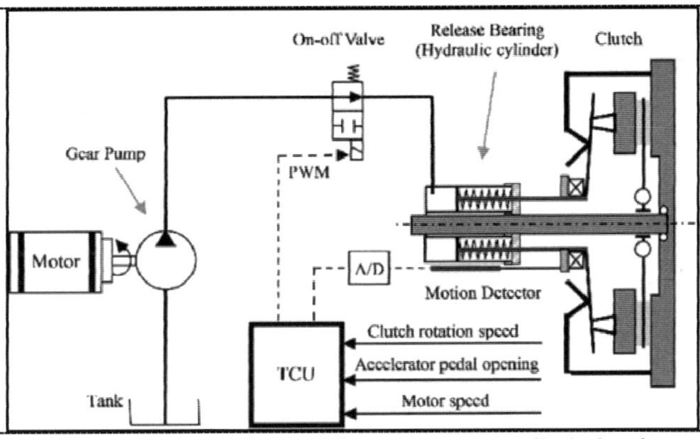

Figure 1. Schematic of direct-drive pump-controlled clutch actuator

2.1. System modeling

2.1.1. DC motors. The brush DC motor driven by an H-bridge circuit can be described as

$$L_a \dot{I}_a = V_{bat} u - I_a R_a - k_v \omega \tag{1}$$

$$T_m = k_t I_a - c_m \omega - J_m \dot{\omega} \tag{2}$$

where L_a is the armature inductance, I_a is the armature current, V_{bat} is the battery voltage, u is the PWM duty ratio, R_a is the armature resistance, k_v is the back electromotive force coefficient, T_m is the torque, c_m is the motor rotation damping coefficient, k_t is the torque coefficient, and J_m is the inertia.

2.1.2. External gear pump. The motor and the external gear pump are directly connected. The rotation of the motor affects the flow of hydraulic oil between the hydraulic cylinder and the tank. The instantaneous pressure change of the pump and output torque can be obtained as follows:

$$\dot{p}_p = \frac{\beta}{V_1}(V_{th}\omega\eta_{vol} - Q_{clt} - Q_{leak}) \tag{3}$$

$$T_a = T_m = \frac{T_{th}}{\eta_m} = \frac{T_{th}\omega}{\eta_m \omega} = \frac{Power}{\eta_m \omega} = \frac{V_{th} p_p \omega}{\eta_m \omega} \tag{4}$$

where V_{th} is the theoretical pump displacement, ω is the pump speed (equal to motor speed), p_p is the export pressure, β is the fluid bulk modulus, η_{vol} is the volumetric efficiency, Q_{clt} is the flow loss in the pipeline, Q_{leak} is the leakage flow at the port, *Power* is the pump power (equal to motor power), T_a is the actual output torque, T_{th} is the theoretical torque, η_m is the mechanical efficiency.

2.1.3. Driven component. The hydraulic pressure received by the release bearing is transmitted directly to the diaphragm spring. The model can be described as follows:

$$m_{clt}\ddot{x}_{clt} = p_{clt}A_{clt} - F_{clt}\left(x_{clt}\right) - F_{damp}\left(\dot{x}_{clt}\right) \tag{5}$$

where A_{clt} is the piston area of hydraulic cylinder, p_{clt} is the pressure difference, m_{clt} is the piston mass, and F_{clt} is load force of the diaphragm spring, F_{damp} is the combination of damping force and friction.

2.2. Model reduction

Based on the structural mentioned above, the dynamic model is reasonably simplified as follows:

- The influence of motor inductance can be ignored, since the value (La/Ra is about 1.3×10^{-6}) corresponding to the time constant is relatively small;
- the hydraulic flow is laminar flow, considering the consideration of the tube diameter size and hydraulic oil (DOT4) performance;
- the flow loss in the pipeline is zero, Q_{clt}=0, $p_{clt}=p_p$ (since the pipeline hydraulic pressure is $1.66Mpa$, and the maximum pressure loss along the path is: $1.18\times10^{-6}Mpa$).
- the forces can be simplified near the secondary level to establish a linear system:

$$F_{clt}\left(x_{clt}\right)=k_{clt}\cdot x_{clt}, k=1.8\times10^{5}N/m \tag{6}$$

$$F_{damp}\left(\dot{x}_{clt}\right)=\sigma\cdot\dot{x}_{clt}\ , \sigma=60N/(m/s) \tag{7}$$

Thus, we have the system dynamics of the actuator system

$$K_2\ddot{x}_{clt}+K_1\dot{x}_{clt}+K_0x_{clt}-K_Uu=0 \tag{8}$$

where

$$K_2=\frac{J_mA_{clt}}{V_{th}\eta_{vol}+V_{th}\omega\left(\partial\eta_{vol}/\partial\omega\right)}+\frac{V_{th}m_{\text{clt}}}{A_{\text{clt}}\eta_m} \tag{9a}$$

$$K_1=\frac{\left(k_tk_v+R_ac_m\right)A_{clt}}{R_aV_{th}\eta_{vol}}+\frac{V_{th}\sigma}{A_{\text{clt}}\eta_m} \tag{9b}$$

$$K_0=\frac{V_{th}k_{\text{clt}}}{A_{\text{clt}}\eta_m} \tag{9c}$$

$$K_U=\frac{k_tV_{\text{bat}}}{R_a} \tag{9d}$$

The parameters involved in the dynamic system are given in Table 1.

Table 1. Parameters used in the model

Parameter	Value(unit)
Battery voltage (V_{bat})	12V
Armature inductance (L_a)	0.0003H
Torque coefficient (k_t)	0.0405N m/A
Back electromotive force coefficient (k_v)	0.0405V/(rad/s)
Resistance of the armature circuit (R_a)	0.2 Ω
Rotating damping coefficient of the motor (c_m)	0.001N ms/rad
Inertia (J_m)	0.0001 kg m^2
Piston area of hydraulic cylinder (A_{clt})	578.25mm^2
Piston mass (m_{clt})	0.312 kg
Theoretical volumetric displacement (V_{th})	0.8 ml/r

The time-varying nonlinear dynamic effects are reasonably simplified. The remnant tracking error incurred in each iteration period of the standard linear system should, thus, be considered. And an

IOP Conf. Series: Materials Science and Engineering **787** (2020) 012023 doi:10.1088/1757-899X/787/1/012023

adequate modeling tool for nonlinear functions should be used to solve the challenging control problem.

3. Controller design

The main work of this part is the design of RBF NN based correction ILC controller for the novel actuator, the control algorithm can be seen in Figure 2. The RBF NN model makes full use of the information of the real plant, which is stored in memory, and provides correction for the next cycle.

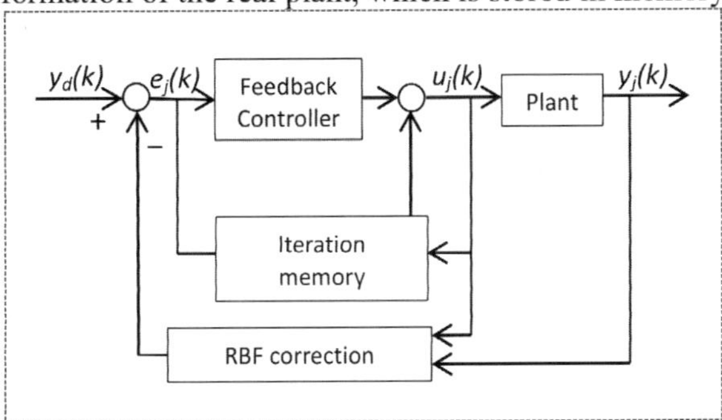

Figure 2. Schematic of actuator controller

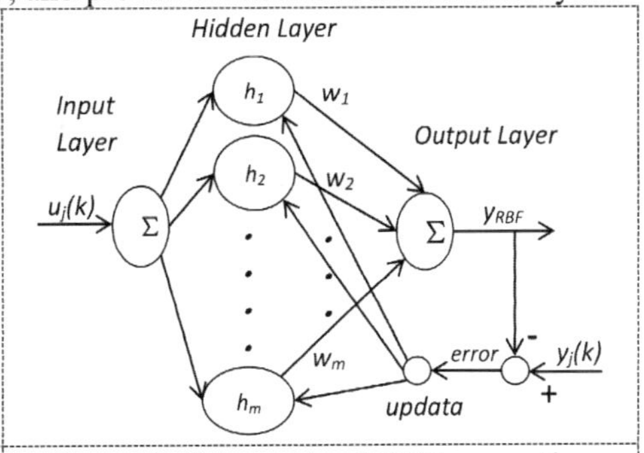

Figure 3. Schematic of RBF correction

3.1. ILC controller scheme

ILC is a memory-based control scheme that needs to store the tracking errors and control efforts of previous iterations so that the present control can be constructed to achieve satisfactory performance. The PD-type learning function used in nonlinear systems can be written as

$$u_{j+1}(i) = u_j(i) + k_p e_j(i+1) + k_d\left[e_j(i+1) - e_j(i)\right] \quad i = 0,1,...,M;\, j = 0,1,... \tag{10}$$

where j denotes the number of iterations, i denotes the number of times in the time domain, $i \in [0,M]$, $M=t/ts$ (t is the run duration, ts is the sampling time). k_p and k_d are the proportional and derivative gain.

For the above actuator system, the discrete state equations are as follows:

$$\begin{cases} x_j(i+1) = A_d x_j(i) + B_d u_j(i) \\ y_j(i) = C_d x_j(i) \end{cases} \tag{11}$$

The input, output and tracking error vectors have the following relationships in iteration j:

$$Y(j) = \left[y_j(1) \quad y_j(2) \quad \cdots \quad y_j(M)\right] \tag{12a}$$

$$U(j) = \left[u_j(0) \quad u_j(1) \quad \cdots \quad u_j(M-1)\right] \tag{12b}$$

$$E(j) = \left[yd_j(1) - y_j(1) \quad \cdots \quad yd_j(M) - y_j(M)\right]^T \tag{12c}$$

Hence, Then, the tracking error for the next iteration is

$$E(j+1) = H_c \cdot E(j) \tag{13}$$

where

$$H_c = I - \left(k_p + k_d\right)H + k_d HF \tag{14a}$$

$$H = \begin{bmatrix} C_d B_d & 0 & \cdots & 0 \\ C_d A_d B_d & C_d B_d & \cdots & 0 \\ \vdots & \vdots & \ddots & \vdots \\ C_d A_d^{M-1} B_d & C_d A_d^{M-2} Bd & \cdots & C_d B_d \end{bmatrix} \quad F = \begin{bmatrix} 0 & 0 & 0 & \cdots & 0 & 0 \\ 1 & 0 & 0 & \cdots & 0 & 0 \\ 0 & 1 & 0 & \cdots & 0 & 0 \\ \vdots & \vdots & \vdots & \ddots & \vdots & \vdots \\ 0 & 0 & 0 & \cdots & 1 & 0 \end{bmatrix}_{M\times M} \tag{14b}$$

In order to converge the system, we need to choose appropriate selection of the gains kp and kd so that all eigenvalues of Hc lie into the unit circle. That is

$$\left|1 - C_d B_d (k_p + k_d)\right| < 1 \tag{15}$$

3.2. RBF NN based correction

As shown in Figure 3, the RBF NN can correct for uncertain nonlinear disturbances in a cycle. The input of RBF NN is the control input u for each cycle. The output and neurons of the hidden layer $h=[h_1,h_2,\dots,h_m]^T$ are activated by Gaussian radial base function as:

$$y_{RBF}(k) = h_1 w_1 + \cdots h_j w_j \cdots + h_m w_m \tag{16}$$

$$h_j = \exp\left(-\frac{\left\|u_k - c_j\right\|^2}{2b^2}\right) \tag{17}$$

where $j=1,2,\dots,m$, m is the number of hidden layer; c_j and b are the central vector and the base width vector of the jth neuron, $c_j=[c_{11},c_{12},\dots,c_{1m}]$, w_j is the weight vector, $w_j=[w_1,w_2,\dots,w_m]^T$, the weights adaptation rule derived from the steepest descent method can be expressed as

$$\Delta w_j(k) = -\eta\left(\partial E(k)/\partial w_j\right) = \eta e_j(k) h_j$$
$$w_j(k) = w_j(k-1) + \eta \Delta w_j(k) + \alpha(w_j(k-1) - w_j(k-2)) \tag{18}$$

where η is learning rate, $\eta \in [0,1]$; α is momentum factor, α and $\alpha \in [0,1]$.

The criterion on function used performance index of the RBF NN correction can, thus, be defined as:

$$E(k) = \frac{1}{2}\left[y_{plant}(k) - y_{RBF}(k)\right]^2 \tag{19}$$

4. Simulation result and analysis

In this section, the effectiveness of the controller is verified by Simulink and AMEsim co-simulation.

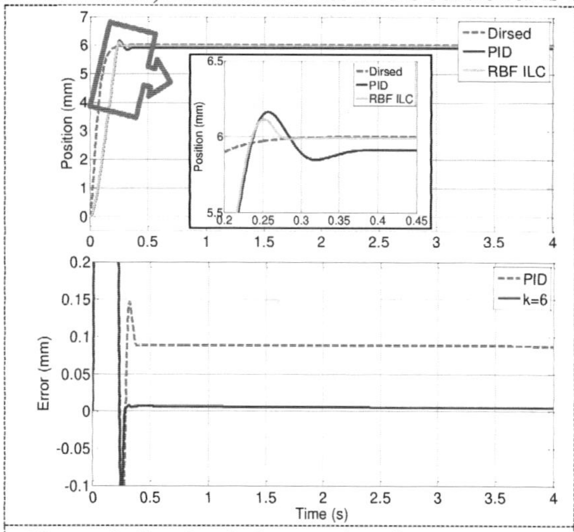

Figure 4. Co-simulation results of step response

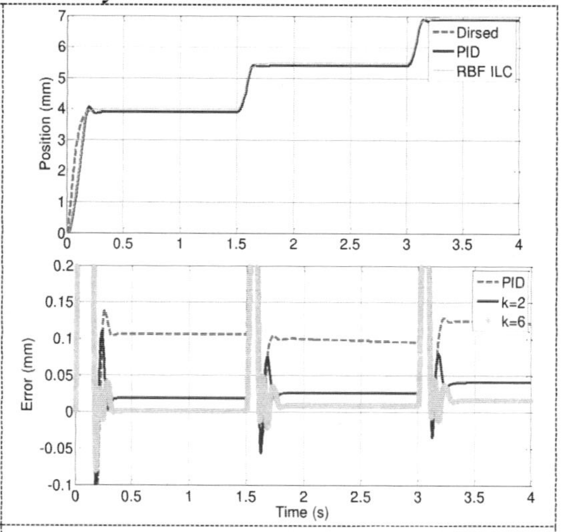

Figure 5. Results of multi-step response

In order to cover more operating conditions of the clutch actuator, step signal (*6mm*) tracking and multi-step step (*4mm 5.5mm 6mm*) signal tracking are simulated, and the results are shown in Figure 4 and Figure 5 (In the figure, k represents the number of iterations). It can be seen that the steady-state tracking error of the proposed controller is close to 0mm after 6 iterations, which is much smaller than the PID (k=1) error of 0.1mm. Specifically, the proposed controller has reduced the overshoot.

In addition, set the time-varying sinusoidal signal as a tracking reference to test the proposed controller's transient performance, as actual driving state conditions may change at any time. The sinusoidal response is shown in Figure 6. Obviously, the control effect is still very impressive, that is, the tracking error is within 0.03mm. And the norm of the error under the sinusoidal signal response is shown in Table 2. The tracking error decreases rapidly with the number of iterations. So, the scheme is extremely critical for the combination of ILC and RBF to be applied to the novel clutch actuator.

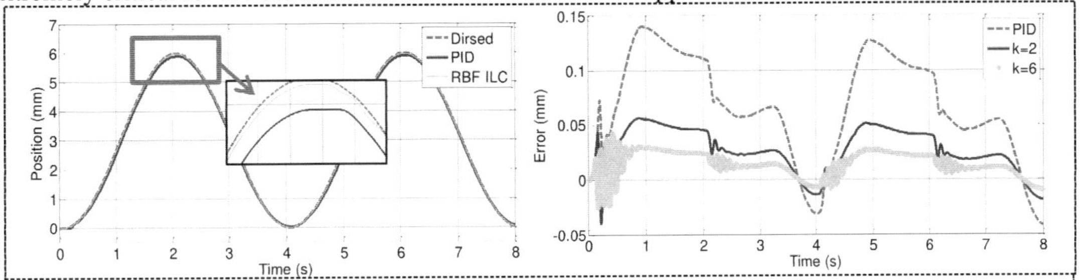

Figure 6. Schematic of direct-drive pump-controlled clutch actuator

Table 2. the norm of the error under the sinusoidal response

	k=1	k=2	k=3	k=4	k=5	k=6
Norm	246.605	100.355	75.328	63.193	57.339	54.673

5. Conclusion

This paper mainly focuses on the motion control problem of a novel direct-drive pump-controlled clutch actuator. The control challenge is caused by the fluid compressibility, friction and leakage. The RBF NN based correction ILC method is ued to implement accurate position tracking for nonlinear system. And it is solved by learning the dynamic characteristics through ILC and neural network. The Simulink and AMEsim co-simulation results demonstrate that the proposed controller is enough sustainable and applicable for the actuator motion control conditions. In future research, the actuator should be controlled to operate under more complex driving conditions, which requires a more stable algorithm. Moreover, clutch wear and changes in hydraulic oil temperature will be discussed in details.

References

[1] Du, C., Plummer, A., & Johnston, D. (2017). Performance analysis of a new energy-efficient variable supply pressure electro-hydraulic motion control method. Control Engineering Practice, 60, 87–98.

[2] Gao B, Chen H, Liu Q, et al. Position control of electric clutch actuator using a triple-step nonlinear method[J]. IEEE Transactions on Industrial Electronics, 2014, 61(12): 6995-7003.

[3] Borghi, M., Zardin, B., & Specchia, E. (2009). External gear pump volumetric efficiency: Numerical and experimental analysis. In SAE 2009 Commercial Vehicle Engineering Congress and Exhibition.SAE Technical Paper No. 2009-01-2844.

[4] Bristow D A, Tharayil M, Alleyne A G. A survey of iterative learning control[J]. IEEE control systems magazine, 2006, 26(3): 96-114.

[5] Madady A. PID type iterative learning control with optimal gains[J]. International Journal of Control, Automation, and Systems, 2008, 6(2): 194-203.

[6] Ahn H S, Moore K L, Chen Y Q. Iterative learning control: robustness and monotonic convergence for interval systems[M]. Springer Science & Business Media, 2007.

[7] Yin C, Xu J X, Hou Z. A high-order internal model based iterative learning control scheme for nonlinear systems with time-iteration-varying parameters[J]. IEEE Transactions on Automatic Control, 2010, 55(11): 2665-2670.

[8] Lee C M, Ko C N. Time series prediction using RBF neural networks with a nonlinear time-varying evolution PSO algorithm[J]. Neurocomputing, 2009, 73(1-3): 449-460.
[9] Wang, H., Chen, B., Lin, C.: Direct adaptive neural controlfor strict-feedback stochastic nonlinear systems. NonlinearDyn. 67(4), 2703–2718 (2012)
[10] Gao B, Li X, Zeng X, et al. Nonlinear control of direct-drive pump-controlled clutch actuator in consideration of pump efficiency map [J]. Control Engineering Practice, 2019, 91: 104110.

Vehicle Mass Identification Based on Two-axle and Two-mass Vibration Model

Zhenyu Wang[1*], Ming Li[2] and Haoyu Wang[3]

[1-3] School of automotive engineering, Wuhan University of Technology, Wuhan, Hubei, 430070, China

[*]Corresponding author's e-mail: 2640783420@qq.com

Abstract. The mass of the vehicle is an important parameter, through which many other performance indicators of the vehicle can be measured. The traditional method of vehicle mass identification is mainly through fixed platform detection station and weighing in motion. In this paper, a two-axle and two-mass vibration model is established to identify the vehicle mass, which takes the free vibration frequency of suspension as input and outputs the vehicle mass. Firstly, when the vehicle passes a specific excitation, the two-mass model composed of suspension and wheels generates free vibration, and then its main frequency is obtained. And then the mathematic relation between the vehicle mass and the main frequency is established through the two-axle vibration model, and the vehicle mass is obtained. Simulation in TruckSim shows that the average relative error between the identified mass and the actual mass is 6.27%. In practical application, the free vibration frequency of the suspension can be obtained by image recognition or displacement sensor, so as to identify the vehicle mass.

1. Introduction

Vehicle mass is a significant parameter of vehicles. It can be used for vehicle overload identification and intelligent transportation system [1]. The current methods to identify mass mainly cover the embedded method and the non-embedded method [2]. Embedded means to add weight measurement equipment inside the vehicle, while non-embedded means to carry out contact or non-contact weight measurement outside the car.

In the research of embedded method, Phong x. Nguyen has proposed a method based on the relationship between vehicle's vertical acceleration and mass to identify the vehicle mass through the accelerometer on the phone [2]. And Rongjun Zhang has used on-board sensors to detect the deformation in a certain area of the axle, and converted the strain into the electrical signal representing the axle mass of the vehicles [3]. Furthermore, LI Hui-min has designed a resistance displacement sensor to measure the distance between the vehicle body and the axle, converted the displacement into vehicle mass signal by Hooke Law, and modified the results with inclination Angle, speed and acceleration sensors to apply to different conditions [4].

The non-embedded method is mainly to identify the vehicle mass statically by placing loadometer on the road surface at a fixed position, which is also the principle method of vehicle mass recognition at present. However, this method will affect the efficiency of the road. Therefore, weighing in motion (WIM) method has become the research direction of many scholars. But in the process of passing the dynamic weighing station, the speed and acceleration of the vehicle have a great disturbance to the weight measurement. The average error of the measured axle weight ranges from 5% to 30%, and the corresponding confidence is 90% or 95% [5]. Besides, it is mainly suitable for low-speed occasions. Artur Andrzejczak has designed a loadometer system using piezoelectric modules, the main innovation of whom is to transmit the results of weight measurement through wireless transmission [6]. Similarly, Mohamed Rehan Karim has studied the WIM dynamic weighing system with an error of ±10% of the static weighing [7]. Moreover, Boles"aw Mazurek has designed an inclined plane weighing system based on the principle of deformation-to-piezoelectric signal, which can convert the momentum of the vehicle into the electrical signal by PVDF sensor [8]. Furthermore, Shaokang Xiong has established the relationship between the maximum overshoot in vibration and the static vehicle mass based on BP neural network algorithm [9]. In this paper, a Single-mass vibration model is used, and they assume that the stiffness and damping coefficient of suspension of different vehicles are the same, so the relationship between the maximum overshoot and vehicle mass is linear. In fact, suspension of different automotive are different, ignoring the impact of suspension in practical application is a big deviation.

For vehicle vibration theory is applied to identification of vehicle mass, Benjamin Pence presents a base-excitation model of vertical ride dynamics which treats the unsprung vertical accelerations as the ride dynamics model input [10]. And it employs methods based on polynomial chaos theory and on the maximum likelihood approach to estimate the most likely value of the vehicle sprung mass. It is through the vertical acceleration to extract the characteristic parameters of the vehicle, however, there are few researches on the identification of vehicle mass by the free vibration frequency of suspension and wheel.

Based on the above analysis, a two-axle and two-mass vehicle vibration model is established in this paper. Through this vibration model, the paper establishes a functional relationship between the vehicle mass and the suspension frequency, taking the suspension free vibration frequency as the input, and outputting the vehicle mass. Eventually, the identified mass and the actual mass are compared by kinetics simulation, so as to verify the feasibility of identifying the vehicle weight based the two-axle and two-mass vibration model.

2. Model derivation

Model derivation consists of three parts: postulated conditions, single axle and two-mass vibration model, two-axle and two-mass vibration model. The differential equation considering the coupling vibration of the suspension and the wheel is deduced in the single axle and two-mass vibration model, and the main frequency is solved. Two-axle and two-mass vibration model considers the mass center position of the vehicle, and the mass is obtained eventually.

2.1. Postulated conditions

First, it is necessary to set the postulate conditions to simplify the complex vehicle structure. The postulate conditions are as follows:

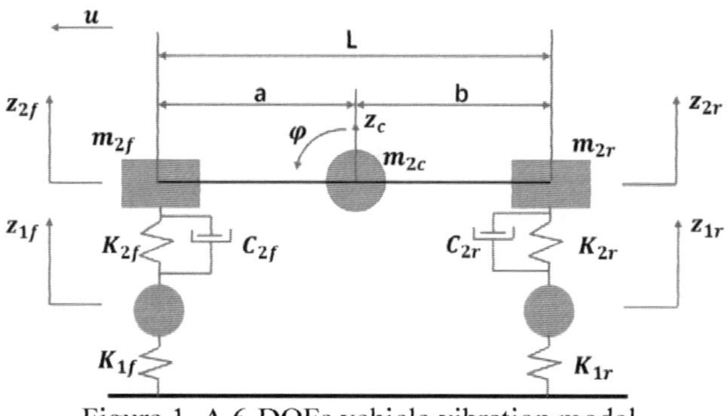

Figure 1. A 6-DOFs vehicle vibration model.

1. The vehicle is simplified into a 6-DOFs vehicle vibration model, including vertical vibration of $m_{2f}, m_{2c}, m_{2r}, m_{1f}, m_{1r}$ and longitudinal vibration of φ. In addition, it is assumed that the irregularity

of the right and left ruts is the same, in other words, the vehicle is symmetric with its longitudinal axis, so the body does not have roll vibration, lateral displacement and yaw vibration. Therefore, only the one side of the vehicle vibration needs to be considered, as shown in figure 1.

2. Ignore the excitation effect of the engine in the vehicle, and only consider the vibration caused by the road surface.

3. The rear wheel runs on a rut, which is the same as the front wheel, so the excitation of the front and rear wheels is the same except for the difference caused by the axle base

2.2. Single-axle and two-mass vibration model

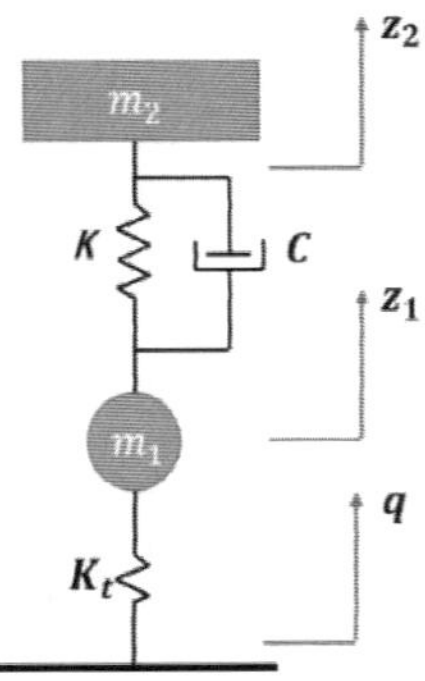

Figure 2. diagram of the single-axle and two-mass vibration model

In figure 2, m_2 is the sprung mass, m_1 is the unsprung mass, K is the suspension stiffness, C is the shock absorber damping coefficient, K_t is the wheel stiffness. The wheel damping is ignored, because that is too small. q is the road roughness function.

In this model, the free vibration generated by passing across a certain excitation is studied, and it can be considered that the roughness of the pavement $q = 0$. The vertical displacement coordinates of the wheel and the body are z_1 and z_2 respectively. The origin of coordinates is selected at their respective equilibrium positions, so the free vibration motion equation is [11]:

$$\begin{cases} m_2\ddot{z}_2 + C(\dot{z}_2 - \dot{z}_1) + K(z_2 - z_1) = 0 \\ m_1\ddot{z}_1 + C(\dot{z}_1 - \dot{z}_2) + K(z_1 - z_2) + K_t z_1 = 0 \end{cases} \quad (1)$$

2.2.1. Suspension vibration model

As can be seen from the equations of motion, the vibrations of m_1 and m_2 are coupled with each other. If m_1 is not moved (z_1=0), we can write:

$$m_2\ddot{z}_2 + C\dot{z}_2 + Kz_2 = 0 \quad (2)$$

This corresponds to the single freedom vibration of the vehicle body m_2. Assume $2n = C/m_2$, $\omega_0^2 = K/m_2$, then equation (2) can be turned into:

$$\ddot{z}_2 + 2n\dot{z}_2 + \omega_0^2 z_2 = 0 \quad (3)$$

The solution of the differential equation is:

$$z = Ae^{-nt} sin(\sqrt{\omega_0^2 - n^2}t + \alpha) \quad (4)$$

This solution shows that, in the case of damped free vibration, m_2 vibrates at the damped natural frequency $\omega_r = \sqrt{\omega_0^2 - n^2}$. This equation can be reduced to $\omega_r = \omega_0\sqrt{1 - \zeta^2}$. The damping ratio of automobile suspension system ζ is usually around 0.25, being small damping. ω_r only decreases by about 3% compared with ω_0, it can be approximately considered that $\omega_r \approx \omega_0$, so the natural frequency $\omega_r = \sqrt{K/m_2}$ [11].

2.2.2. Wheel vibration model

If m_2 does not move ($z_2 = 0$), we can write:

$$m_1\ddot{z}_1 + C\dot{z}_1 + (K + K_t)z_1 = 0 \quad (5)$$

Since both the equation (5) and equation (2) are second order homogeneous differential equations, the solution method is the same, and the natural frequency of the wheel part can be obtained by the same method [11]:

$$\omega_t = \sqrt{(K + K_t)/m_1} \tag{6}$$

2.2.3. Two-mass vibration model

The ω_0 and ω_t obtained above are partial frequencies, and the two vibration are mutually interfering, but if the two vibration are considered comprehensively, the main mode of the two vibration needs to be obtained.

As mentioned above, damping ratio has little effect on frequency, so the damping coefficient C can be ignored. m_1 and m_2 vibrate in simple harmonic mode with frequency of ω_t and ω_0 respectively, amplitude for z_{10} and z_{20}. Because the phase Angle will not have an impact on the calculation results, so set the phase Angle is φ. We can write:

$$\begin{cases} z_1 = z_{10} e^{j(\omega t + \varphi)} \\ z_2 = z_{20} e^{j(\omega t + \varphi)} \end{cases} \tag{7}$$

Because of $\omega_0 = \sqrt{K/m_2}$、 $\omega_t = \sqrt{(K + K_t)/m_1}$, substitute the above two solutions into the differential equations (1) and get:

$$\begin{cases} (\omega_0^2 - \omega^2) z_{20} - \omega_0^2 z_{10} = 0 \\ -\dfrac{K}{m_1} z_{20} + (\omega_t^2 - \omega^2) z_{10} = 0 \end{cases} \Leftrightarrow \begin{pmatrix} \omega_0^2 - \omega^2 & -\omega_0^2 \\ -\dfrac{K}{m_1} & \omega_t^2 - \omega^2 \end{pmatrix} \begin{pmatrix} z_{20} \\ z_{10} \end{pmatrix} = 0 \tag{8}$$

The condition of the matrix equation having a non-zero solution is the determinant of the coefficients of z_{10} *and* z_{20} is zero. We can write:

$$(\omega_0^2 - \omega^2)(\omega_t^2 - \omega^2) - \omega_0^2 \frac{K}{m_1} = 0 \tag{9}$$

Equation 9 is the frequency equation of a two-mass system, whose two roots are the square of the main frequencies ω_1 or ω_2:

$$\omega_{1、2}^2 = \frac{1}{2}(\omega_t^2 + \omega_0^2) \mp \sqrt{\frac{1}{4}(\omega_t^2 + \omega_0^2)^2 - \frac{K K_t}{m_2 m_1}} \tag{10}$$

Low main frequency ω_1 is close to ω_0, high main frequency ω_2 is close to ω_t, and there is a relationship between them:$\omega_1 < \omega_0 < \omega_t < \omega_2$.

Whether the main frequency is ω_1 or ω_2 is determined by the amplitude of the vehicle body and the wheel. In the case of forced vibration, when the excitation frequency ω is close to ω_1, low-frequency resonance is produced and the amplitude of the body mass m_2 is nearly 10 times larger than that of the wheel mass m_1, known as the body type vibration. When the excitation frequency ω is close to ω_2, high-frequency resonance is produced and the amplitude of the wheel mass m_1 is nearly 100 times larger than that of the vehicle mass m_2, known as wheel type vibration.

Because the stiffness of suspension is much smaller than that of wheels, and the excitation frequency is close to 0 in the case of free vibration. Therefore, the vibration amplitude of wheels is smaller than that of the body, generating low-frequency resonance and the main frequency is ω_1. We can write:

$$\omega_1^2 = \frac{1}{2}(\omega_t^2 + \omega_0^2) - \sqrt{\frac{1}{4}(\omega_t^2 + \omega_0^2)^2 - \frac{K K_t}{m_2 m_1}} \tag{11}$$

It can be seen that through the stiffness and free vibration frequency of the suspension and the wheel, as well as the sprung mass and unsprung mass, the main frequency of the free vibration of the single-axle and two-mass vibration model can be obtained. The solution condition is that the sprung mass is known, but the sprung mass is the variable to be solved, so it is necessary to introduce the two-axle and two-mass Vibration Model.

2.3. Two-axle and two-mass vibration model

Since the amplitude of the wheel is much smaller than that of the suspension, and the main frequency of the suspension and the wheel has been calculated, the influence of the wheel is ignored when solving the sprung mass. Therefore, in figure 1, K_{1f} and K_{1r} are ignored. The sprung mass m_2 is decomposed according to equation (12-14) into: concentrated mass m_{2f} allocated to the front axle, connection mass m_{2c} allocated to the center of mass, and concentrated mass m_{2r} allocated to the rear axle, which are connected by a massless rigid rod. In equation (13), a and b are the horizontal distance from the front axle and the rear axle to the center of mass of the vehicle, and L is the axle base of the vehicle. In equation (14), I_y is the moment of inertia of the vehicle body around the Y-axle, and ρ_y is the turning radius of the horizontal Y-axis. The Y-axis passes the center of mass of the vehicle and is perpendicular to the side of the body.

$$m_{2f} + m_{2r} + m_{2c} = m_2 \tag{12}$$

$$m_{2f}a - m_{2r}b = 0 \tag{13}$$

$$I_y = m_2\rho_y^2 = m_{2f}a^2 + m_{2r}b^2 \tag{14}$$

We can write the moment balance equation for the front and rear axles:

$$\begin{cases} m_{2f}\ddot{z}_{2f} + m_{2c}b(\ddot{z}_{2r}a + \ddot{z}_{2f}b)/L^2 + K_{2f}z_{2f} = 0 \\ m_{2r}\ddot{z}_{2r} + m_{2c}a(\ddot{z}_{2r}a + \ddot{z}_{2f}b)/L^2 + K_{2r}z_{2r} = 0 \end{cases} \tag{15}$$

When only the front wheel passes across the excitation, namely, only z_{2f} moves, and $z_{2r} = 0$, the natural circular frequency of the front-end system is:

$$\omega_{of}^2 = \frac{K_{2f}L^2}{m_2(\rho_f^2 + b^2)} \tag{16}$$

K_{2f} is the suspension stiffness of the front axle, L is the axle distance of the vehicle, ρ_f is the rotation radius around the transverse axis y. ω_{of} is equal to the main frequency ω_1 of the two-mass vibration model mentioned above.

Simultaneous equations (11) and (16), and the sprung mass m_2 is obtained. Because the unsprung mass of a kind of vehicle is the same, it can be obtained by referring to relevant automobile manuals. The total mass of the vehicle can be obtained by adding the obtained sprung mass and the checked unsprung mass.

3. Simulation analysis

In order to verify whether the established two-axle and two-mass model can correctly identify the mass of the vehicle, we carry out simulation test by TruckSim. TruckSim is a simulation software for researching the dynamics of commercial vehicles. It has a rich database of vehicles and working conditions, and users can flexibly adjust the structure of vehicles, dynamic performance and working conditions according to their own needs. These characteristics make TruckSim a popular choice for automotive engineers to design their vehicles, as well as helping automobile companies such as GM, Mercedes-Benzes. In this paper, the two-axle commercial vehicle is simulated by TruckSim, and the mass of the vehicle is identified according to the simulation data, and then compared with the actual mass, so as to verify the feasibility of two-axle and two-mass model to identify the mass of the vehicle. Firstly, the vehicle model is established and the commercial vehicle bus is selected as the simulation vehicle. The main parameters of bus are shown in the table 1.

Table 1. The main parameters of bus.

parameter	value
Sprung mass (kg)	6360
Unsprung mass of front axle (kg)	570
Unsprung mass of rear axle (kg)	760
Suspension spring rate (N/mm)	250
Tire spring rate (N/mm)	980

IOP Conf. Series: Materials Science and Engineering **787** (2020) 012024 doi:10.1088/1757-899X/787/1/012024

Radius of gyration (m)	2.20
Axle base (mm)	4490
Distance from the centroid to the rear axle (mm)	1385

Then, road and excitation model are established, and smooth concrete pavement is selected for the road. The properties of the road surface are shown in table 2.

Table 2. Main parameters of road.

Parameters	Value
Friction	0.85
Tire rolling resistance coefficient	1
Path Elevation	Flat

The excitation causes the vehicle to vibrate freely. In this paper, the standard of speed bump is adopted to design the excitation model [12]. The cross section of the speed bumps is shown in figure 3.

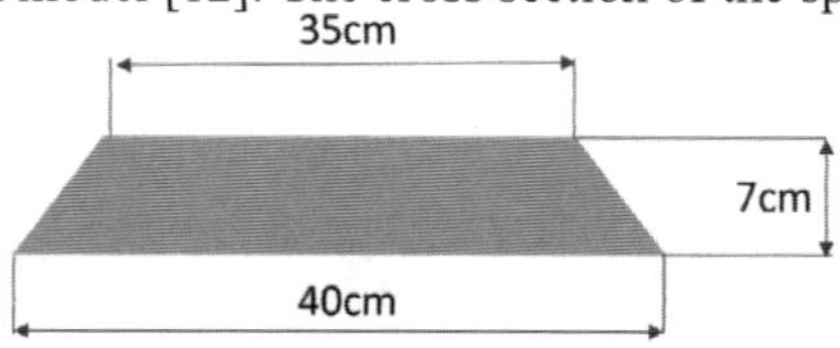

Figure 3. Cross section of excitation during simulation.

In the simulation, the driver is set to drive at a constant speed across the road covered with speed bump. The slow speed will lead to excessive lateral vibration of the vehicle, making the vertical vibration of the left and right suspensions inconsistent. If the vehicle is too fast, the vibration of the front suspension will end prematurely, which will be insufficient, and the fitting effect will be poor. Through several simulation, it is found that a speed of about 10km/h can keep the vertical vibration of the left and right suspension consistent, and the vibration is relatively sufficient. So, the driver is controlled to travel at 10km/h.

The figure of the compression of the front suspension across the excitation is shown in figure 4. Before 3.6s, the front suspension compression remains unchanged, and the vehicle is driving on a flat road. At 3.6s, the front wheel hits the excitation, and the suspension begins to compress. The compression of the suspension becomes larger. At 3.9s, the compression of the suspension reaches the minimum value, which means that the front wheel returns to the ground, and the front suspension begins to vibrate freely. The figure presents a sinusoidal curve of attenuation. At 5.4 seconds, the rear wheel hits the excitation. Obviously, the vibration of the rear wheel will interfere with the vibration of the front suspension and ruin its original sinusoidal attenuation curve.

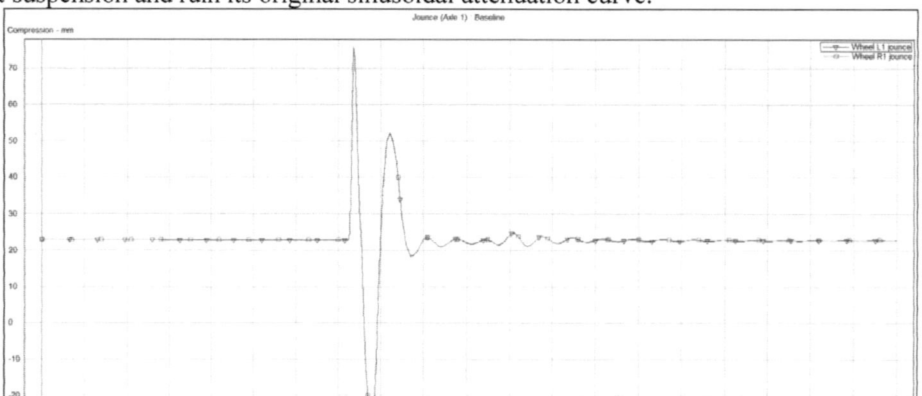

Figure 4. front suspension compression during driving.

Therefore, during 3.9s-5.4s, the front suspension is free to vibrate, and the compression of the suspension presents a relatively regular sinusoidal attenuation curve. The compression data of the front suspension during the free vibration is imported into MATLAB and fitted with equation (4), and the natural frequency of the front suspension ω_0 is 12.63, whose range is [12.15, 13.11], as shown in figure 5.

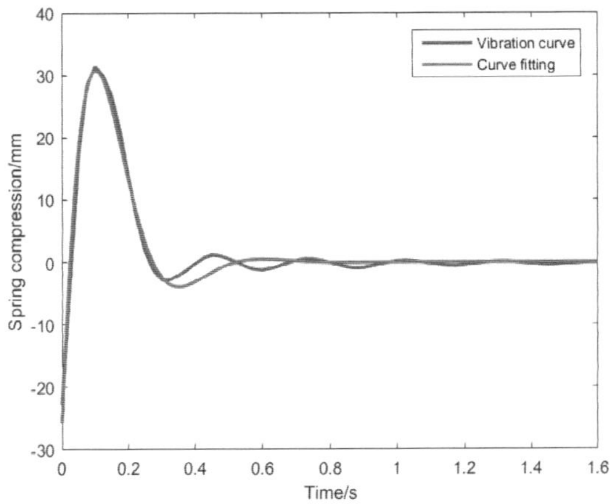

Figure 5. Vibration curve and fitted curve.

Since the unsprung mass of a kind of automobile is determined when it is designed and manufactured, it can be searched from the vehicle database. Thus, the natural frequency of the wheel ω_t can be obtained according to equation 6, which is 46.45.

According to table 1, the position of the center of mass and radius of rotation of the vehicle can be known, and the simultaneous equation (11) and (16) can be used to obtain the main frequency ω_{of} of the front suspension to be 11.18 and the sprung mass m_2. The fitting front suspension frequency ω_0, the corresponding mass m_2 and the relative error are shown in table 3.

Table 3. Mass identification results

	ω_0	Theoretical value of m_2(kg)	Relative error
Optimal fit	12.63	5961.3	6.27%
Left boundary	12.15	6433.7	1.16%
Right boundary	13.11	5540	12.89%
Minimum error	12.22	6361.3	0.02%

It can be obtained from table 3 that the best fit value of ω_0 is 12.63, and the relative error of the corresponding vehicle mass identification value is 6.27%, which can meet the engineering requirements.

The error is mainly caused by the interference effect of tire on suspension vibration. As a result, the actual vibration curve is not the standard sinusoidal attenuation curve. Therefore, there is certain error during the fitting. The following work should be carried out to improve the mathematical model, so that the fitted results better reflect the vibration frequency of suspension.

4. Conclusion

In this paper, a two-axle and two-mass model is proposed to obtain the mass of vehicle. When the front wheel of an automobile passes across a specific excitation, the two-mass model composed of the front suspension and the front wheel generates free vibration. First, the main frequency of the two-mass model is obtained. Then, the mathematic relation between the mass of the vehicle and the main frequency is established through the two-axle vibration model, so as to obtain the mass of the vehicle. The simulation result shows that it is feasible to identify vehicle mass by suspension frequency, and the error is 6.27%.

IOP Conf. Series: Materials Science and Engineering **787** (2020) 012024 doi:10.1088/1757-899X/787/1/012024

It should be pointed out that the application of this theory needs to know the sprung mass, the stiffness of the suspension and the tire, and the location of the center of mass. In practical application, the location of the center of mass is difficult to obtain, and other parameters can be obtained by establishing a database. If the influence of the center of mass is ignored, the mass of front and rear axle can be solved by using the single-axle and two-mass model, and then two mass can be added to be gross mass. In this way, the mass can also be solved, but the error will be improved.

In practical application, the method of image recognition or displacement sensor is used to identify the free vibration frequency of the suspension, and then the mass of the vehicle is obtained through the two-axle and two-mass model. It has a good application prospect and can provide help for intelligent traffic and overload identification.

References

[1] Wilson, A. Vehicle weight is the key driver for automotive composites. Reinforced Plastics, 2017. 61(2): p. 100-102.

[2] Nguyen, P.X., et al. Vehicle's Weight Estimation Using Smartphone's Acceleration Data to Control Overloading. International Journal of Intelligent Transportation Systems Research, 2018. 16(3): p. 151-162.

[3] Zhang R et al. Beidou-based integrated monitoring and management system of vehicle-mounted weight meter. Satellite application, 2015(02): 74-77.

[4] Li H. and li X. Research on vehicle-mounted dynamic weighing algorithm. Computer simulation, 2016.33 (11): 140-143.

[5] Du C. Theoretical analysis of low-frequency dynamic load and algorithm research on improving weighing accuracy. Highway traffic science and technology, 2018.35 (04): 153-158.

[6] Andrzejczak, A, et al. Module hardware structure of wireless vehicle weight measurement system. 2014. Lublin, Poland: IEEE Computer Society.

[7] Karim, M.R., et al. Effectiveness of vehicle weight enforcement in a developing country using weigh-in-motion sorting system considering vehicle by-pass and enforcement capability. IATSS Research, 2014. 37(2): p. 124-129.

[8] Mazurek, B, T. Janiczek and J. Chmielowiec. Assessment of vehicle weight measurement method using PVDF transducers. Journal of Electrostatics, 2001. 51-52: p. 76-81.

[9] Xiong S., et al. Vehicle dynamic weighing technology based on BP neural network. Journal of anhui university of technology (natural science edition), 2014. 31(01): 76-79.

[10] Pence, B., et al. Vehicle sprung mass estimation for rough terrain. International Journal of Vehicle Design, 2013. 61(1-4): p. 3-26.

[11] Yu Z. Automotive theory. machinery industry press, Beijing.

[12] Road rubber speed bumps: JT/T 713-2008[S].

IOP Conf. Series: Materials Science and Engineering **787** (2020) 012025 doi:10.1088/1757-899X/787/1/012025

Research on the development of cold chain logistics of aquatic products in Zhoushan of China

Zhi Han[1], Yunrui Fang[1] , Linhui Hua[1], Qingwen Ma[1], JingXing Wang[2], Junkai Gao[1] and Pengxiang Yuan[3*]

[1] School of Port and Transportation Engineering, Zhejiang Ocean University, Zhoushan, Zhejiang, 316000, China

[2] School of Foreign Languages, Zhejiang Ocean University, Zhoushan, Zhejiang, 316000, China

[3] College of Food and Pharmacy, Zhejiang Ocean University, Zhoushan, Zhejiang, 316000, China

[*]Corresponding author's e-mail: xiangzi661822@163.com

Abstract. Aquatic products are very ideal healthy food. Cold chain logistics is an effective way to ensure the quality of aquatic products in the circulation process. Zhoushan City (China) is situated in Zhoushan archipelago. Zhoushan fishing ground is the largest fishing ground in China, with abundant fishery resources. This paper expounds the theory of cold chain logistics, analyzes the present situation of cold chain logistics of Zhoushan aquatic products, points out the problems existing in cold chain logistics of Zhoushan aquatic products, and puts forward corresponding solutions, which has a positive role in promoting the development of cold chain logistics of Zhoushan aquatic products.

1. Introduction

Aquatic products are the general term of aquatic animal and plant products and their processed products produced by marine and freshwater fisheries. The per capita share of aquatic products in China is twice more than the world average level[1], while the per capita share of aquatic products in Zhejiang Province is 2.22 times more than the national per capita share[2]. Among all the cities in Zhejiang Province, Zhoushan City, located in Zhoushan archipelago, is a significant producing area of aquatic products in Zhejiang Province, and also the national core area of aquatic product catching and distribution. Consequently, it has the reputation of "Fish storehouse in the East China Sea" and "Fishery capital of China".

Aquatic products are nutritious and delicious. Aquatic products are prone to oxidation and bacterial contamination in logistics transportation[3], which will not only damage the taste of aquatic products, but also pose a serious threat to food safety. Cold chain logistics is the key to reducing deterioration of aquatic products in the circulation process. Solving the problems existing in cold chain logistics of Zhoushan aquatic products is not only conducive to the development of local aquatic products industry in Zhoushan, but also will set an example for the development of national aquatic products industry.

2. Summary of cold chain logistics theory

Cold chain logistics refers to the special supply chain system consisting of a series of procedures such as multi-level processing, transportation, storage, distribution and retail of products to maintain a low temperature environment for various kinds of fresh agricultural products such as vegetables, fruits, meat, aquatic products and eggs, so as to ensure quality, reduce deterioration and avoid contamination [4]. Refrigeration technology and other related technologies are the bases of the development of cold chain logistics. With the advance of technology in the industry, cold chain logistics has made substantial progress.

Cold chain logistics of aquatic products is an essential classification of cold chain logistics. Cold chain logistics of aquatic products refers to the system engineering that the cold or frozen aquatic products are in the specified low temperature environment before production, storage, transportation, sales and consumption, to ensure quality and reduce deterioration[5]. The flow chart is shown in Figure 1. In aquatic products industry, it is critical to introduce the advanced idea and technology of cold chain logistics of aquatic products in time to gain the market advantage position.

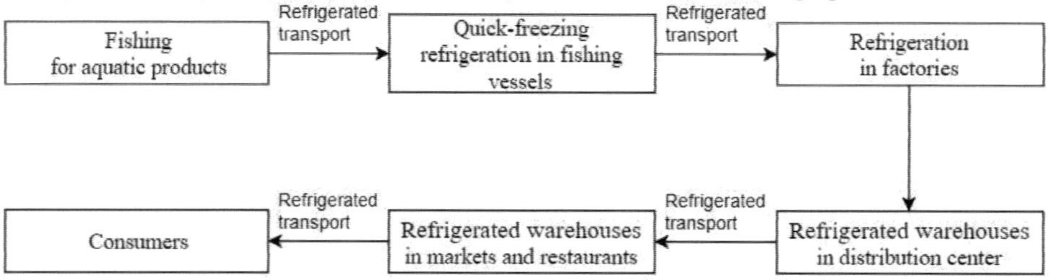

Figure 1. Process flow of cold chain logistics of aquatic products

3. Present situation of cold chain logistics of Zhoushan aquatic products

According to the 2008-2017 National Economy and Social Development Statistical Bulletin of Zhoushan City, the statistics of the total output of Zhoushan aquatic products, pelagic fishery and mariculture in these years are obtained, as shown in Figure 2. It can be seen from the figure that in recent years, the total output of Zhoushan aquatic products has maintained a growth trend, with a rapid growth rate. Except for a slight decline in 2008-2009, there is a great increasing range in the other

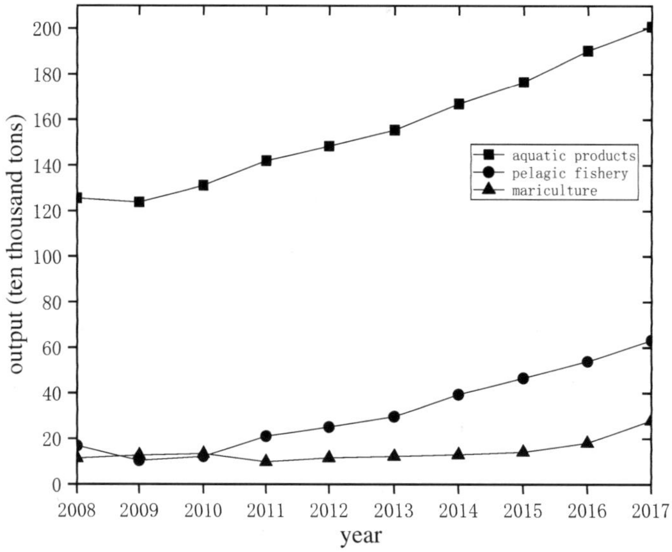

Figure 2 . Yield components and change trend of aquatic products in 2008-2017

3.1. Present situation of cold chain logistics enterprises of Zhoushan aquatic products

There are two characteristics of cold chain logistics enterprises in Zhoushan: one is the rapid development in ladder form. Based on the scale of cold chain logistics enterprises in Zhoushan, they can be divided into three types: small, medium and large. Among them, the small-scale cold chain logistics enterprises are mostly family enterprises, generally with several refrigerated transport vehicles; the medium-sized cold chain logistics enterprises consist of family enterprises and joint ventures, with relatively large scale and with a relatively large number of refrigerated transport vehicles; the large-scale cold chain logistics enterprises are mainly joint ventures, with a large scale and with refrigerated transport fleets. The second is that their distribution presents a concentration effect. In the early stage of cold chain logistics industry in Zhoushan, cold chain logistics enterprises are scattered throughout the archipelago. However, with the development of cold chain logistics in Zhoushan, at present, cold chain logistics enterprises are gradually transferring to important areas of aquatic products production and trade such as Shenjiamen fishing port and Zhoushan International Aquatic Products City[6].

3.2. Present situation of cold chain logistics mode of Zhoushan aquatic products

The cold chain logistics transportation mode of Zhoushan aquatic products can be summarized as follows: after being fished in the fishing ground, aquatic products are transported to ports by refrigerated transport ships, then in refrigerated warehouses. Finally, they are transported to aquatic product markets by refrigerated transport vehicles or flowing water transport vehicles. Ships, refrigerated transport vehicles and flowing water transport vehicles are the main cold chain logistics transport vehicles in Zhoushan.

Zhoushan City is located in the Zhoushan archipelago, and its geographical positon determines that ships are the primary mode of cold chain transportation of local aquatic products. According to the statistics of the 2018 National Economy and Social Development Statistical Bulletin of Zhoushan City, in 2018, there are 1600 marine transport ships in Zhoushan, with a transport capacity of 7.495 million deadweight tons, an increase of 23.2% over the end of last year. Among them, 174 ships have a tonnage of 10000 tons and above, an increase of 24 ships over the end of last year, with a capacity of 4.652 million deadweight, accounting for 62.1% of the total transport capacity of the city.

Refrigerated warehouse, refrigerated container, refrigerated transport ship and refrigerated transport vehicle are the main storage means of cold chain logistics of Zhoushan aquatic products. There are more than 110 refrigerated warehouses in operation in Zhoushan, with a total capacity of 350000 tons. There are 30 aquatic products processing enterprises in Putuo District, Zhoushan City, with an average capacity of 2360 tons for each enterprise, and 38 aquatic products processing enterprises in West Wharf, Zhoushan City. All these processing enterprises have a total capacity of 70000 tons, and 15000 tons for tax declaration [7].

4. Problems in cold chain logistics industry of Zhoushan aquatic products

4.1. Insufficient cold chain logistics equipment

According to the 2016 Distribution Map of National Cold Chain Logistics Enterprises, the refrigerated warehouse capacity of Zhejiang Province is about 1 million tons, ranking 13th in all provinces and autonomous regions in China, and the total number of refrigerated transport vehicles is about 1500, ranking 13th in the country. The per capita refrigerated warehouse capacity is about 200 tons / 10000 people, ranking 15th in the country, and the per capita number of refrigerated transport vehicles is less than 0.5/10000 people, ranking 17th in the country. By contrast, Shanghai's per capita refrigerated warehouse capacity is up to 1243 tons / 10000 people, and the per capita number of refrigerated transport vehicles is 7 / 10000 people, ranking the first in the country. In East China, the per capita refrigerated warehouse capacity is 310 tons / 10000 people, and the per capita number of refrigerated transport vehicles is 0.77/10000 people. It can be seen that each index of cold chain logistics equipment in Zhejiang Province belongs to the intermediate or low level among 32 provinces and autonomous regions in China, which is far behind the national top level and the average level of East China. However, the GDP and per capita GDP of Zhejiang Province have been in the forefront of China for a long time. Insufficient cold chain logistics equipment in Zhejiang Province is not consistent with the situation that Zhejiang Province has a strong economy. Therefore, there is a lot of room for improvement. Zhoushan also faces the problem of insufficient cold chain logistics equipment: the number of refrigerated transport ship is not large, and the technology of refrigerated equipment applied to ships cannot reach the high level of the industry; the number of professional refrigerated transport vehicles is small, and the refrigerated transport vehicles of many small and medium-sized cold chain enterprises are transformed from general vehicles which present poor refrigeration effect; the capacity of refrigerated warehouses is low with old facilities, which cannot give full play to the role of refrigerated warehouse.

4.2. The backward development of third-party cold chain logistics

Third-party cold chain logistics (3PCCL) refers to an operation and management method that frozen and refrigerated food enterprises contract with professional cold chain logistics service providers to entrust them cold chain logistics business in order to concentrate on main businesses and bring down costs. Meanwhile, they keep close contact with cold chain logistics enterprises through information system, so as to achieve whole processing management and control[8]. For Zhoushan aquatic products enterprises, entrusting cold chain logistics business to professional cold chain logistics enterprises can not only help reduce the investment cost of cold chain logistics equipment, but also significantly facilitate the quality and efficiency of cold chain logistics of aquatic products, improve the quality of aquatic products and obtain more profits.

However, there are some problems existing in the development of third-party cold chain logistics of Zhoushan aquatic products: on the one hand, Zhoushan aquatic products enterprises do not pay enough attention to third-party cold chain logistics, and they are not aware of the benefits that third-party cold chain logistics brings to their aquatic product business. On the other hand, Zhoushan lacks sufficient third-party cold chain logistics services and professional third-party cold chain logistics enterprises. There are only 10 registered and existing cold chain logistics companies in Zhoushan, 28 registered and existing cold chain logistics companies in Ningbo, and 42 registered and existing cold chain logistics companies in Hangzhou. What's more, professional third-party cold chain logistics enterprises outside Zhoushan tend to pay more attention to carry out their business in Ningbo and less attention to the demands in Zhoushan. From the perspective of both supply and demand, there is a large gap of third-party cold chain logistics services in Zhoushan.

4.3. Lack of professional cold chain logistics talents

At present, China's cold chain logistics has entered a high-speed development period, the number of cold chain facilities and equipment has risen significantly, and the quality of cold chain logistics has been considerably improved. However, due to the special requirements for cold chain logistics talents, the number of cold chain logistics talents cannot meet the society's needs[9]. According to the Medium- and Long-term Plan for Development of the Logistics Industry (2014-2020), logistics practitioners in China grow at an average annual rate of 6.2%, creating about 1.8 million logistics job vacancies every year. However, there are only 150000 logistics talents from logistics major in colleges and universities every year, which is far from satisfying the society's demand for logistics talents.

Compared with Shanghai, Hangzhou, Ningbo and other developed cities in China, Zhoushan's economic development is obviously not enough, and its attraction to logistics talents is not strong, thus it is difficult to retain cold chain logistics professionals who are originally scarce in China. At present, most of the practitioners of cold chain logistics of aquatic products in Zhoushan have not received professional training, or rarely learned professional knowledge of cold chain logistics. They lack of theoretical consciousness and professional operation skills, which is a restrictive factor that cannot be ignored for the development of aquatic products industry relying on cold chain logistics.

4.4. Poor informatization of cold chain logistics

Logistics informatization refers to a series of processes such as collecting, classifying, transferring, aggregating, identifying, tracking and querying all or part of the information generated in the logistics process by using modern information technology, so as to achieve control of the goods circulation process, thus reducing costs and improving efficiency[10]. Modern logistics cannot be separated from processing information flow. Without information flow, logistics will no longer be smooth[11]. At present, the informatization level of cold chain logistics of aquatic products in Zhoushan is improving, but on the whole, there are still some problems such as backward technology, poor information circulation and inaccurate information transmission, and a lack of widely used logistics information platform. Most of the world's advanced port cities have developed logistics information platforms. For example, in the last century, Singapore has established a comprehensive logistics information conversion platform, TradePalette, which helped Singapore save about $1 billion in logistics. DTTN, the digital trade transport network and data exchange system established in Hong Kong, can be applied to more than 80 trade and logistics services, including aquatic products logistics.

5. Countermeasures for the problems in cold chain logistics of Zhoushan aquatic products

5.1. Increasing investment in cold chain logistics equipment

Logistics equipment is an important tool for the development of modern logistics. The development of cold chain logistics of aquatic products in Zhoushan requires sufficient refrigerated warehouses, and they should be built reasonably and utilized properly. Since 2011, Zhejiang Province has concentrated on building medium and low temperature refrigerated warehouses for aquatic products in important port cities to relieve urgent needs. However, after 2013, due to the soaring costs of pelagic fishery, the prices of aquatic products in international market fell sharply. Many fishery companies' profits fell seriously, and even could not bear the storage costs. In some cities of Zhejiang Province, refrigerated warehouse providers even subsidized the storage costs for fishery companies[12]. The government should draw lessons from above cases. It should review and approve the construction of refrigerated warehouses according to the fishing situation of Zhoushan aquatic products, in order to rationally meet the quantity gap of Zhoushan refrigerated warehouses and ensure the improvement of quality of refrigerated warehouses in Zhoushan at the same time. For the already existing refrigerated warehouses, the proportion of low-level refrigerated warehouses should be minimized. The best refrigerated warehouses with standardized construction and proper operation will be rewarded.

For professional refrigerated transport ships and vehicles, first of all, we should try to reduce the use of ordinary ships and vehicles as refrigerated transport ships and vehicles after a little modification. For this, relevant departments can issue standards to prevent a large number of unqualified modified refrigerated transport ships and vehicles from running. Secondly, we should encourage enterprises to upgrade their equipment and facilities in refrigerated warehouses and try to ensure that the loading and unloading process is completed within the refrigerated warehouses, so as to maintain a whole course cold chain. Finally, Kaili and Dalengwang, located in East China, are the two major manufacturers of refrigeration equipment for refrigerated transport vehicles in China, and the largest suppliers of condensing units for middle and high-end refrigerated transport vehicles. Zhoushan should take advantage of this location advantage to establish its extensive business with such refrigerated transport vehicles and refrigeration equipment giants.

5.2. Developing third-party cold chain logistics actively

On the one hand, Zhoushan's aquatic products producers and sellers should actively cooperate with third-party cold chain logistics enterprises in Shanghai, Hangzhou, Ningbo and other places, release bidding information of third-party cold chain logistics of aquatic products, and attract third-party cold chain logistics enterprises to pay attention to business in Zhoushan. On the other hand, local government of Zhoushan should issue preferential policies for third-party cold chain logistics enterprises, attract third-party cold chain logistics companies in other cities or overseas to carry out businesses in Zhoushan, and promote the development of local third-party cold chain logistics enterprises.

At the end of 2017, Putuo District issued a document named as "Some opinions on accelerating the development of e-commerce", encouraging e-commerce enterprises to actively apply cold chain logistics technologies and schemes of third-party logistics, and providing financial subsidies for e-commerce enterprises that use third-party cold chain logistics. Similar policies can be applied to Zhoushan aquatic products enterprises to promote cooperation between Zhoushan aquatic products enterprises and third-party cold chain logistics companies, so as to achieve win-win results for Zhoushan municipal government, Zhoushan aquatic products enterprises and third-party cold chain logistics companies.

5.3. Optimizing the introduction and training system of cold chain logistics talents

The government and enterprises should establish a reward system for cold chain logistics talents to attract high-quality talents to work in Zhoushan. The organs or organizations where cold chain logistics talents work should carry out safeguard measures and conditions, and establish a reward mechanism for achievements. The high-end management and technical posts related to cold chain logistics will be included in the demand posts for the introduction of high-end talents in Zhoushan Archipelago New Area to attract high-quality cold chain logistics talents to work in Zhoushan.

We need to make good use of local educational resources in Zhoushan, pay attention to the cultivation of cold chain logistics talents in colleges and universities, and establish a vocational education system for cold chain logistics of aquatic products. The government should actively guide Zhejiang Ocean University and Zhejiang International Maritime College to carry out cooperation between school and enterprise. Colleges and universities train professional cold chain logistics talents for enterprises, and enterprises provide practice bases for colleges and universities[13]. At the same time, Zhoushan can attract logistics related students of Zhejiang University to practice in Zhoushan through the medium of the Ocean College of Zhejiang University. In addition, Zhoushan can establish third-party training institutions and public training bases for cold chain logistics, utilizing teaching resources from Zhejiang Ocean University and Ningbo University to train logistics industry employees, unemployed people and job seekers who want to engage in logistics, and to increase the number of middle- and low- end practitioners engaging in cold chain logistics in Zhoushan.

5.4. Enhancing informatization of cold chain logistics

The government should issue relevant logistics information platform construction plans, learn from the logistics information platforms that has been built and operated in advanced port cities such as Hong Kong, Singapore and Shanghai, integrate Zhoushan International Aquatic Products city, major fishing grounds and aquatic products enterprises into a common platform, establish a complete cold chain logistics information platform, and enhance informatization of cold chain logistics.

We should encourage the main information owners of cold chain logistics, such as logistics parks, large- and medium-sized aquatic products enterprises, to extensively apply professional softwares and equipment to cold chain logistics informatization, strengthen the sharing of information with each other, and ensure smooth information flow in the process of cold chain logistics as well as the accuracy and betimes of information transmission. For the small-scale, family township aquatic products producers located in the islands, the relevant departments of the local township government should instruct them to use information softwares and equipment, in order to establish a comprehensive cold chain logistics information platform covering the whole industry of Zhoushan aquatic products.

6. Conclusion

According to the Outline of the 12th Five-Year Plan for Economic and Social Development, which was deliberated and adopted at the Fourth Session of the Eleventh National People's Congress in 2011, it clearly put forward that we should focus on promoting the development of Zhoushan Archipelago New Area in Zhejiang Province. In 2016, Zhejiang Province issued the 13th Five-Year Plan for Open Economic Development in Zhejiang Province, which also pointed out that it is necessary to "further promote the substantial integration of Ningbo and Zhoushan port, as well as the construction of port economy circle and Zhoushan Archipelago New Area ". It can be seen that the State and Zhejiang provincial government are vigorously promoting the construction of Zhoushan Archipelago New Area. Under this favorable environment, we should strive for more beneficial national policy support [14] to promote the development of cold chain logistics of Zhoushan aquatic products.

Aquatic products industry is the characteristic industry of Zhoushan. To solve various problems in aquatic products industry is the key measure for developing Zhoushan's economy. Cold chain logistics is the lifeblood of aquatic products. To improve cold chain logistics of Zhoushan's products, we should take measures based on people, equipment and system, and chase the best in the industry continuously. This is to consolidate the status of Zhoushan as " Fishery capital of China ". It is also of great significance to build Zhoushan into a world-class fishing ground, which is also a critical step for Zhoushan to develop its archipelago new area and implement the national maritime economic strategy.

Acknowledgements
The authors are grateful for the financial supports of the National Sparking Plan Project (No. 2013GA700254), Zhejiang Provincial Public Projects of China (No. 2017C31008), Science and Technology Planning Project of Zhoushan of China (No. 2016C41004), Science and technology innovation program for college students in Zhejiang province of China(2016R411025).

References
[1] Liu, Z. (2019) Logic, effect and direction of China's fishery economy reform -- Commemorating the 40th anniversary of reform and opening up .J. World Agriculture., 01: 41-48.
[2] Le, J. (2011) Current situation, problems and direction of fishery development in Zhejiang Province .J. Heilongjiang Agricultural Sciences., 12: 73-78.
[3] Li, Z., Sun, X. (2015) Research on the application of cold chain logistics of aquatic products in Suzhou .J. Co-Operative Economy & Science., 14: 20-21.
[4] Fu, F., Zhou, W. (2019) Research on real-time monitoring and optimization of cold chain logistics big data .J. Think Tank Era., 34: 18-19.
[5] Li, W., Yu, Y. (2019) Prediction of cold chain logistics demand of aquatic products in China based on grey prediction model .J. Agricultural Technology Service., 36 (06): 103-105 + 107.
[6] Yu, J. (2013) Research on the current situation and application technology of cold chain logistics in northern Zhejiang province -- Taking Zhoushan as an example .J. Private technology., 04: 79+112.
[7] Jin, H. (2013) Research on integration of cold chain logistics of Zhoushan aquatic products.D. Zhejiang Ocean University.
[8] Wu, W. (2014) Research on selection of third-party cold chain logistics service providers.D. Chongqing Technology and Business University.
[9] Hou, X., Ding, H., Huang, Y. (2018) Analysis of cold chain logistics talent training.J. Logistics Engineering and Management., 40 (09): 177-179.
[10] Li, L. (2017) New mode of low-carbon logistics transportation in the era of low-carbon economy .J. Economic Research Guide., 05: 83-84.
[11] Liang, Y., Liu, Z., Li, S. (2016) Analysis of the impact of cold chain logistics cost control on the development of cold chain logistics.J. Journal of Refrigeration Technology., 02: 28-31.
[12] Jing, H., Gao, E., Yu, Z., Li, C. (2015) Chapter 6 Overview of Regional Market of Refrigeration Industry, 2014 Analysis Report of China's Refrigeration Industry Development.J. Chinese Journal of Refrigeration Technology., 35 (S1): 68-82.
[13] Lin, Y. (2018) Research on cold chain logistics development mode of Zhoushan aquatic products .D. Zhejiang Ocean University.
[14] Chen, F. (2013) Research on cold chain logistics development of Zhoushan aquatic products.J. Logistics Engineering and Management., 35 (04): 26-27.

IOP Conf. Series: Materials Science and Engineering **787** (2020) 012026 doi:10.1088/1757-899X/787/1/012026

Research on Influencing Factors of Automotive Body-in-White Laser Welding Quality

Zhang Yue

SiChuan Aerospace Vocational College

Abstract: Laser welding technology is fast, advanced, and flexible. It is the key technology to ensure the development of new automotive products. It is also a technology to improve product quality and reduce costs. However, the application of laser technology in automobile manufacturing still has a certain distance compared with foreign countries. In order to improve the competitiveness of the automotive industry, it is necessary to strengthen the research on the application of laser welding technology in automobile body-in-white. This article will start with the factors that affect the quality of laser welding of automobile body-in-white, and based on the analysis of laser welding mechanism and related characteristics, talk about the influence of laser welding power and give relevant solutions welding speed on the quality of laser welding of automobile body-in-white.

1. Automotive Body-in-White Laser Welding Overview

1.1 Car Body-in-White Overview

The body is an important part of the vehicle, which will have an important impact on the overall safety, power and economic comfort of the car. At the same time, whether a car has a personality and can attract consumers' favor. It is also closely related to the car body design and quality. The car body includes two parts: body-in-white and its accessories. The body-in-white refers to the white leather body that has been welded but not painted. It is a comprehensive welding of the body structure and cover parts which excludes body attachments and trim. The body-in-white can be divided into three types: non-loaded, semi-loaded and load-bearing. The characteristics of the non-loaded body-in-white are that the body is flexible connected by rubber cushions or springs and frames. The body only bears the weight and inertia of the carrying personnel and the cargo itself, and the frame supports the entire vehicle foundation, but the frame's design does not need to consider the auxiliary role of the body in carrying the frame. The characteristic of the load-bearing body is that the car does not have a frame, and the body can bear all the gravity of the car. The body has the bearing capacity of the frame and the body. Because the frame is removed, the mass of the overall vehicle is reduced, and the height is reduced, thereby achieving lighter weight and higher speed of the vehicle body. The body is the bearing foundation, so the performance of the body material can be fully utilized to make the body structure more reasonable. However, the load-bearing body will also increase the bearing capacity of the body due to the cancellation of the frame, which will affect the ride comfort and even reduce the service life of the body. It will be more difficult to modify it later. The structural feature of the semi-loaded body is that the body is rigidly connected to the frame by welding and other methods. The frame bears the load of the main components, and the role of the body is to strengthen the frame and share the advantages of the part of the load of the frame and non-load-carrying body [1].

IOP Conf. Series: Materials Science and Engineering **787** (2020) 012026 doi:10.1088/1757-899X/787/1/012026

1.2 Overview of Automotive Body-in-White Laser Welding Technology

Automotive body-in-white welding is an important part of automobile production. The quality of body welding will directly affect the appearance of the body, the carrying capacity and service life of the body. Adopting scientific and reasonable welding process can improve production efficiency and ensure the quality of body structure. There are many methods of automobile welding, including electric welding, carbon dioxide gas shielded welding, laser welding and so on. The laser welding technology uses a laser mirror to reflect the light beam generated by the laser, so that it is concentrated in the focusing device to generate a beam of huge energy. Once the focus is near the workpiece, it will immediately melt and evaporate. The application of laser welding technology has played a significant role in all aspects of welding schedule, efficiency, reliability and automation. In recent years, because of the research and application of high-performance and high-power laser welding equipment, laser welding technology has become one of the most promising manufacturing welding technologies and has been widely used in welding smelting in various fields. The main application features of laser welding are: firstly, it can reduce heat to a minimum, reduce the range of thermal influence, and minimize deformation problems due to heat conduction. Second, laser welding can reduce the time and materials required for thick plate welding, which has obvious economic advantages. The third laser welding eliminates the need for electrodes and reduces pollution caused by the electrodes. In the fourth laser welding, the laser beam is easy to focus and align, and is also guided by optical instruments. These are characteristics that other welding processes do not have. Generally speaking, the welding technology of the machine is also affected by the welding position, which requires that the laser beam must be in the focus range. In addition, highly reflective and highly thermally conductive materials can be affected by laser welding to cause performance changes, and laser welding equipment is generally expensive.

2. Factors Affecting Automotive Body-in-White Laser Welding Quality

2.1 Effect of Laser Power on Welding Quality

After laser welding of steel and aluminum materials, their quality is mainly affected by factors such as pre-welding processing clamps, welding process parameters, and the physical characteristics of the material being welded. Among them, the welding process parameters have the greatest impact on the quality of laser welding, including laser power, welding speed, high focus, protective gas type and flow, etc. [2]. First of all, for continuous laser welding, the laser power refers to the output power of the laser. Low-power lasers that work continuously generate ordinary and limited heat transfer welds on the sheet at a lower speed. High-power lasers High speeds produce narrower welds, or lower depths produce thicker welds on thick plates. The laser power controls the penetration depth and the welding speed. The laser welding penetration is related to the beam power. When other conditions are unchanged, the penetration depth will increase with the increase of laser power.

2.2 Effect of Laser Welding Speed on Welding Quality

In the case of constant laser power, increasing the welding speed will reduce the heat input and the weld penetration depth. Therefore, appropriately reducing the welding speed can increase the penetration depth. But when the speed is too low, the penetration depth will no longer be increased, the material will be over-melted due to the increase of the melting width, and the weldment will be welded through. This is because during laser welding, small holes are generated and metal gas is in the small holes. When the welding speed is too small, the heat input increases, more and more metals are melted, and the recoil pressure generated by metal vaporization cannot maintain the existence of small holes. The depth of the pinhole no longer deepens, but will deform under the effect of heat conduction. In addition, with the increase of metal vaporization, the temperature in the small holes increases, and the increase of the plasma concentration increases the absorption capacity of the laser, which cannot play the role of laser welding.

2.3 Effect of Spot Diameter on Welding Quality

The effect of the size of the spot irradiated on the welding surface on the quality of welding is that when deep welding is performed, the power density at the laser focus must be high during welding. However, the first method to increase the power density is to increase the laser power, and the second is to reduce the spot diameter. When the spot diameter is reduced, the welding power density is increased. Therefore, we can use a short focal length lens and reduce the transverse mode order of the laser beam to reduce the spot diameter. After the low-order mode is focused, a smaller spot can be obtained [3].

2.4 Influence of Defocus Amount on Welding Quality

Defocus amount refers to the distance from the surface of the weldment to the focal point of the laser. The amount of defocus not only affects the laser spot size on the welding surface, but also affects the incident direction of the laser beam. Therefore, the depth of laser welding, the width of the welding, and the welding The shape of the cross section of the seam will have an effect. When the amount of defocus is large, the penetration depth will be small, but when the amount of defocus is very small, the penetration depth will increase, and a large number of small holes will be generated, making the welding very unstable, which will cause large and small welds, while the seam forming quality is poor. The amount of defocus also affects the radius of the melting spots on the surface of the material and the diameter-depth ratio of the molten pool, which affects the quality of the welding process. When the defocus amount is changed, the size of the laser heating spot and the incident light beam can be changed. When the welding plate is thicker, a reasonable defocus amount can be used to obtain a larger molten pool. However, if the defocus amount is too large, the diameter of the light spot will be increased, which will reduce the power density on the light spot and reduce the penetration depth.

2.5 Influence of Shielding Gas on Welding Quality

The protective gas for laser welding can protect the welding metal from the harmful gases from the outside, avoid the oxidation pollution, and improve the quality of the welded joint. Second, it affects the plasma in welding. During deep-fusion welding, the high-power laser beam heats and vaporizes the metal, forms a metal gas above the molten pool, and dissociates under the action of the electromagnetic field to generate a plasma. It has a certain hindrance to the laser beam. The laser beam is affected and absorbed by the welding. Plasma clouds can be suppressed by blowing a protective gas in laser welding. The specific mechanism is to increase the collision of electrons with ions and neutral atoms by increasing the protective gas, thereby increasing the recombination rate of electrons and reducing the electron density in the plasma. Second, the use of a flowing protective gas to blow out the generated metal gas and ions will also have a certain effect on the gas flow and penetration. When the gas flow rate is too small, the plasma cloud above the molten pool cannot be removed, and the removal effect will increase with the increase of the gas flow rate. However, when the gas flow rate is too large, the problem of collapse under the molten pool will also occur.

2.6 Effect of Pulse Energy and Pulse Width on Welding Quality

In pulse welding, the energy of the pulse will directly affect the heating energy and the amount of metal melting. The width of the pulse determines the heating time during welding and affects the penetration and the size of the heat affected zone. When the pulse energy is constant, different materials have the best pulse width. At this time, the welding penetration depth is the largest. For metals with good thermal conductivity and lower melting point, greater penetration depth will be obtained. There is a certain relationship between the pulse energy and the pulse width during welding. There is a certain change with the thickness and properties of the material. During welding, in order to keep the power constant, as the pulse energy increases, the pulse width must also increase accordingly, so as to ensure the quality of welding.

2.7 Influence of Power Density on Welding Quality
In pulse laser pulse welding, when the power density is small, the welding is performed by heat transfer welding, and the heat conduction directly affects the diameter of the solder joint and the depth of penetration. When the laser spot power density is constant, a pinhole effect will occur during welding, and deep-melt solder joints will be generated. A small amount of metal will evaporate, which will not affect the formation of solder joints. However, when the power density is too large, the metal reacts strongly and the metal gas generated by metal vaporization increases, which will form small holes filled with metal gas in the solder joints, which cannot form strong solder joints, which will affect the quality of the solder.

3. Suggestions on Quality Control Measures for Laser Welding of Automobile Body-in-White

3.1 Laser Welding Equipment Maintenance
Automotive body-in-white laser welding quality defects are most common in voids, fusion-type welds, poor solder connections, single-sided solder joints, rough weld surfaces, and insufficient or excessive weld fill at the beginning and end of the weld. Analyzing the factors that cause these quality defects shows that many quality problems are caused by equipment failures, so it is very important to do daily equipment maintenance and repair. For the maintenance of laser equipment, in addition to daily inspection of glass lenses, the residue generated during welding should be cleaned in time, the bolts on the fixture should be tightened to prevent it from falling off during welding. Besides, the arc lamp in the laser should be checked regularly. We also need to replace aging arc lamps, and robots daily maintenance of auxiliary equipment also needs to be strengthened.

3.2 Dimensional Accuracy Requirements for Laser Welding Workpieces
The main factors affecting the quality of laser welding are process parameters. In order to ensure the accuracy and quality of laser welding, accurate beam positioning and precise clamping technology are required. In addition, controlling the laser power and welding speed, and controlling the gap between the parts to be welded. Generally, the gap between the parts to be welded in the body-in-white laser welding is 0.2mm.

4. Laser Welding Process and its Application Development in Automotive Body-in-White Welding

4.1 Composite Welding
While applying laser welding technology, we are also strengthening research on this technology. In order to solve the disadvantages of laser welding in application, for example, for the base material is easy to be melted and vaporized by heat, creating deep-melted pores, we need to fill the pores with metal gas. These gases will form a plasma cloud under absorption and reflection under the action of laser. After that, the absorption rate of the laser light by the metal material is reduced, thereby reducing the problem of the laser energy utilization rate. In this way, we can use the heating properties of other heat sources to improve the heating of the workpiece by the laser, and combine the laser with other heat sources to achieve composite heat source welding, such as combining laser and electric arc, combining laser and plasma arc, laser and induction heat sources are combined for welding [4]. This composite welding process combines the advantages of multiple welding technologies, which can increase the penetration depth of welding, improve the welding speed and production efficiency, while reducing the cost of equipment and improving the coupling characteristics of laser energy.

4.2 Laser Welding Laser Generator and its Technological Development Trend
The lasers currently used in laser welding technology are generally high-power carbon dioxide lasers and pulsed lasers. The problem we need to solve is how to develop a large-capacity and high-panel light source to improve conversion efficiency and reduce losses, while ensuring the quality of the laser

beam and high-efficiency and high-reliability propagation, while reducing costs. In the future, the output power of single-fiber and single-mode fiber will exceed kW [5], people can use direct diode array laser output wavelength in the near-infrared region to increase the average laser power significantly, and the photoelectric conversion efficiency continues to improve. The application of these devices and technologies in the field of automotive body-in-white welding can improve the quality and efficiency of welding through the improvement of welding processes and materials.

4.3 Development of Automotive Body-in-White Welder Technology

With the mass production of automobiles, the product quality and efficiency have been further improved, so that robot laser welding technology has been widely used in automotive welding. The use of more flexible, highly automated and highly flexible welding equipment can ensure the welding production efficiency while ensuring the high precision and quality of the welding. Automated flexible production system is the main development trend of automotive welding. In welding, the application of robotic laser welding technology and the use of combined lightweight intelligent automatic welding equipment can not only improve product quality, but also improve production efficiency and controlle the material consumption.

5.Conclusion

Laser welding technology has been widely used in modern automobile manufacturing industry because of its high production efficiency and high welding process accuracy. With the continuous development of the automotive industry, the automotive body-in-white welding technology will develop in the direction of automation and informatization. However, the premise of the application of automation technology is that the welding process must be improved. Therefore, continuous research and improvement of laser welding should be conducted. The factors of welding quality are eliminated one by one, making full use of the role of laser welding technology.

References:

[1] Zhang Yan. Application status of laser welding on body-in-white [J]. Electric Welder, 2016 (03): 122-126.

[2] Xu Xinpei, Luo Haiguo, Tang Gaozhong. Research on Advanced Technology of Automobile Body-in-White Manufacturing——Welding Technology [J]. Equipment Manufacturing Technology, 2015 (02): 145-147.

[3] Guo Li. Application of laser welding technology in body-in-white of Shenlong automobile [J]. Metalworking, 2015 (12): 34-36.

[4] Zhang Zhen. Influencing factors of laser welding quality of car body-in-white [J]. Science and Technology Outlook, 2016 (31).

[5] Fu Wenqin, Jia Xiangru. Quality control of laser welding in body-in-white [J]. Quality Exploration, 2014 (03): 55-56.

IOP Conf. Series: Materials Science and Engineering **787** (2020) 012027 doi:10.1088/1757-899X/787/1/012027

A Study on the Network Characteristics of the Road Collection and Distribution System of Tianjin Port

Wang Guanhui*

School of International Business, Tianjin Foreign Studies University, Tianjin, 300204, China

*Corresponding author's e-mail: newhop@126.com

Abstract. The competition among ports is changing from the competition in internal collection and distribution system to that in hinterland, in which road network plays a very important role. Focusing on the hinterland road network of Tianjin Port, this study takes it as a network, and analyzes its network topology feature. The conclusion has guiding significance for the optimization of Tianjin Port road network.

1. Introduction

Tianjin port is the largest port in the north of China. Its port scale and competitiveness are closely related to its own capacity of freight collection, distribution and transportation, which directly affects the development of the port. Generally speaking, good freight collection, distribution and transportation conditions can boost the expansion and development of the port. Otherwise, the development of the port will be negatively affected. Therefore, the importance of freight collection, distribution and transportation system has drawn more and more attention from the academic circle. Meanwhile, with the rapid development of port economy, the competition between ports is more and more fierce, which is extending from one on the level of the collection and distribution system of the port per se to one on the level of port hinterland.

Tianjin Binhai New Area is undoubtedly the hinterland of Tianjin Port. Whether the road freight collection, distribution and transportation network in this area can meet the requirements of port cargo transportation, whether the network is free from network failure, and whether the overall network is robust will directly affect the collection and distribution system capacity of the Port.

2. Model construction of Tianjin Port road collection and distribution system

Hypothesis: N_1 is container throughput, N_2 is highway mileage, both r_1, r_2 are greater than 0. $2<\varepsilon<\eta$,and $0<v<1$. N_1^* is container throughput in balanced state and N_2^* is highway mileage in balanced state. If $when\mu>1, N_2^*\geq lN_1^*; when\mu<1, N_2^*<lN_1^*$, For the convenience of later analysis and comparison, it is considered that when $N_2^*\geq lN_1^*$, then $0<v<1,\ 2<\varepsilon<\eta<\mu+1$.

The model is as follows:

IOP Conf. Series: Materials Science and Engineering **787** (2020) 012027 doi:10.1088/1757-899X/787/1/012027

$$\begin{cases} \dfrac{dN_1}{dt} = -\dfrac{r_1}{K_1}N_1^2 + \dfrac{\mu r_1}{K_2}N_1N_2 + \varepsilon r_1 N_1 \\ \dfrac{dN_2}{dt} = -\dfrac{r_2}{K_2}N_2^2 + \dfrac{v r_2}{K_2}N_1N_2 + \eta r_2 N_2 \end{cases} \tag{1}$$

In this model, considering that the influence of freight collection, distribution and transportation on port development is much greater than that of port development on freight collection, distribution and transportation, because the construction of freight collection, distribution and transportation is mainly restricted by the economic development of hinterland. ε and η are the influence of other factors on the system.

It can be seen from the model that the key to the analysis is that we need to guarantee $\mu > 1$. When container throughput and freeway mileage are in equilibrium, they will not change any more.

Then,. $dN_1/dt = 0$, $dN_2/dt = 0$ Or

$$\begin{cases} \dfrac{dN_1}{dt} = 0 \\ \dfrac{dN_2}{dt} = 0 \end{cases} \Rightarrow \begin{cases} -\dfrac{r_1}{K_1}N_1^2 + \dfrac{\mu r_1}{K_2}N_1N_2 + \varepsilon r_1 N_1 = 0 \\ -\dfrac{r_2}{K_2}N_2^2 + \dfrac{v r_2}{K_2}N_1N_2 + \eta r_2 N_2 = 0 \end{cases} \tag{2}$$

As we made $0 < v < 1,\ 2 < \varepsilon < \eta < \mu + 1$ in the previous assumption, the positive balance condition of this point is $\mu v < 1$. At the same time, because $N_2^* \geq lN_1^*$ needs to be satisfied in the practical sense, the condition that this point needs to meet is $lK_1(\mu\eta + \varepsilon)/(1 - \mu v) \leq K_2(v\varepsilon + \eta)/(1 - \mu v)$, or $(\mu\eta + \varepsilon)lK_1 \leq (v\varepsilon + \eta)K_2$.

Under this equilibrium condition, we can find $N_1^* > K_1$, $N_2^* > K_2$. That is to say, when $N_2^* \geq lN_1^*$, the two populations of container throughput and highway mileage are in a state of mutual benefit and symbiosis. And the number of populations in the equilibrium of both sides is larger than the maximum that could be reached when they develop alone.

3. Analysis of road collection and distribution system network in Tianjin Port

Based on the above formula, 163 nodes and 682 edges of Tianjin Port road freight collection, distribution and transportation network are calculated respectively, which is a relatively complex network. There are no two roads that are disconnected

With a diameter of 9, 9 routes need to be changed from West Ring Road to Huangshan Road. The average path length is 3.64304. That is to say, driving from any road to any other road requires less than four times of changing driving routes on average.

The average degree of each node is 5.96319, so on average each road is connected with nearly 6 other roads, which indicates that the connectivity of the network is still relatively high. Through further statistics, it can be found that the nodes with degrees between 2 and 7 account for 78% of all nodes, but the total degree only accounts for 50% of the total. That is to say, the distribution of the network is relatively uneven, which is related to the traffic capacity of the road and population density.

The degree distribution:

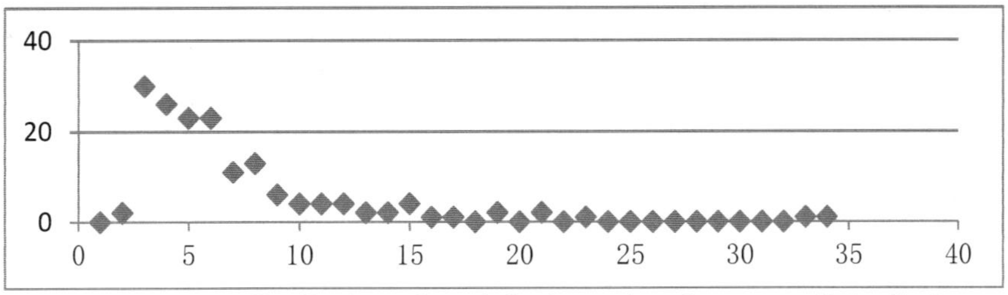

Figure 1 Degree distribution of roads in the main urban area of Tianjin Port

Remove the points with degrees of 0 and 1, and then rearrange the degree distribution function.

Degree distribution function:

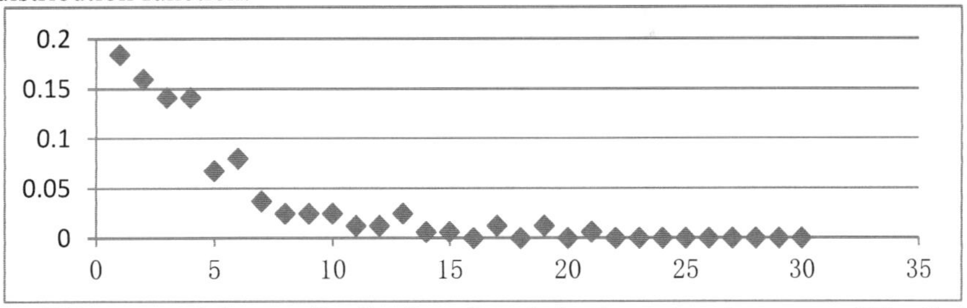

Figure 2 Rearranged degree distribution of roads in the main urban area of Tianjin Port

Function fitting: fitting function $Y = 12.55e^{-0.82x}$, with a goodness of fit of 75%. It passes F test and T test, and the overall fitting effect is good.

The network has a small average path length and a large aggregation coefficient, which shows that the network has the characteristics of a small world. At the same time, because its degree distribution conforms to the exponential distribution, it further shows that the network is a small world network. The characteristics of the small world network proves that the resistance of the highway network in the main urban area of Tianjin Port to random attack is higher, while the resistance to malicious attack is lower.

4. conclusion

The road network in Tianjin Port has moderate diameter and short average path length, and the network is well connected. At the same time, the network has a relatively large average aggregation coefficient. However, compared with other cities in the world, there is still a large gap and development space. The degree distribution of the network basically is in accordance with the exponential distribution. The characteristics of the network shows that it is basically a small world network, which has strong robustness to random attacks, but poor robustness to malicious attacks. The collapse of the road freight collection, distribution and transportation system due to maliciously attacks will lead to the overall collapse of the network.

Acknowledgments

The project is supported by Research project of Humanities and Social Sciences in Tianjin institutions of higher learning (Project No.20132127)

References

[1] YANG S, LIU G, WANG Z, （2018）, Research on coordinated development saturation model of port collection and distribution system. Journal of Ningbo University (NSEE), 31:76-81.

[2] YANG A, (2016), Research on the Container Collection and Distribution System of Xiamen Port. Logistics Sci-Tech, 2:1-3.

[3] KONG L, HE P, (2019), Study on Capacity Coordination of Collecting and Distributing System of Shenhua Heavy Haul Coal Line. Railway Transport and Economy, 39:57-62.

[4] MA L, SONG Y, WANG L, (2018) Logistics Collection and Distribution System of Wuxi Airport under the background of Supply-side Structural Reform. Journal of Wuxi Vocational Institute of Commerce, 18:33-37.
[5] YANG S, LIU G, WANG T, (2017), Study on influential factors of collecting and distributing system for port based on SEM. Journal of Waterway and Harbor, 38:103-108.
[6] KONG L, HE P, (2017), Coordination research on the collection and distribution system of Shenhua heavy haul railway Technology & Economy in Areas of Communications, 19:42-47.
[7] WU Q, (2015), Fuzhou Jiangyin Bonded Port Hinterland Development and Construction of Transportation System. Logistics Engineering and Management, 37:69-71.

IOP Conf. Series: Materials Science and Engineering **787** (2020) 012028 doi:10.1088/1757-899X/787/1/012028

A speed guidance strategy based on cooperative vehicle-infrastructure environment at signalized intersections

Liping Xu, Mingjun Deng

School of Transportation and Logistics, East China Jiaotong University, Nanchang, Jangxi, 330013, China

Corresponding author's e-mail: dmstd98@163.com

Abstract. It is a problem that most traffic scholars have been studying to enable motor vehicles to pass through signalized intersections without parking or with little parking, so as to improve the traffic efficiency of urban road network and reduce congestion. In the vehicle-road collaboration environment, speed guidance strategy has become an effective solution and trend, aiming at the characteristics and shortcomings of the current trunk signal coordination control as a passive response control based on vehicle arrival drive. In this paper, the principle of speed guidance strategy at signalized intersections is systematically proposed, and the corresponding speed guidance strategy model is proposed according to the different traffic conditions when the guided vehicles arrive at the intersections. This strategy can make the guided vehicles pass through the intersections without stopping or stopping time at least, reduce the delay of vehicles and improve the traffic efficiency at the intersections.

1. Introduction

Urban intersections are usually the bottleneck of traffic problems. In particular, vehicles frequently stop, queue or speed up or slow down at signalized intersections, which not only affect the traffic efficiency of the whole road network, but also bring problems such as fuel consumption and exhaust emission. With the development of Internet of vehicles technology, it has become an effective solution and trend for urban traffic to improve traffic efficiency to guide vehicles at the intersection group without stopping or with the least stopping time by virtue of vehicle-road collaboration technology.

Many scholars at home and abroad have studied the speed guidance of intersections under cooperative vehicle-infrastructure environment. QING et al.[1] established a fleet discrimination algorithm based on time headway with real-time traffic data of vehicle-road collaboration technology, and established a signal control optimization model for coordinated control of trunk roads at multiple intersections. RAKHA et al.[2] proposed a speed guidance strategy based on vehicle-road coordination at signalized intersections to improve fuel efficiency. CHEN et al.[3] conducted speed guidance through the roadside variable information board, and proposed a signal control optimization method combining dynamic speed guidance with dynamic signal control, which was used for coordinated control of trunk roads at multiple intersections. Li et al.[4] proposed for individual vehicle speed guide mechanism and model, but the model only for single intersection is analysed, and on the basis of overall benefit in intersection best as the goal, put forward the signal intersection car speed guidance model[5]. However, the influence of signal timing and adjacent intersections was not considered in this study. An et al.[6] proposed a green driving control method for signalized intersections based on multi-stage variable speed limit. However, due to the speed limit and the choice of following model, the optimal results were not obtained. Jing et al.[7] proposed a green wave coordinated control method of double-cycle arterial road based on acceleration and deceleration guidance, but the model did not take into account the coordination between vehicles guided by the upstream intersection and the queued fleet at the downstream intersection and the current signal phase. Long et al.[8] set up the integrated speed guidance and green wave optimization model of trunk line with the maximum green wave bandwidth and the minimum vehicle delay and stopping times as the control variables. But whether there is consistency or mutual influence between multiple objective functions selected in this study requires further study. Liu et al.[9] regarded the guided vehicle and the ordinary vehicle after it as a fleet, and proposed a speed guidance strategy for the signal intersection under the condition of different driving states through the intersection. However, this method does not consider the mutual influence and speed fluctuation between vehicles, and the applicable scene is limited.

Combined with the above existing research and deficiencies, this paper systematically analyses the principle of speed guidance strategy for signalized intersections under the cooperative vehicle-infrastructure environment, and proposes the vehicle guidance speed strategy model under different traffic conditions, in order to provide some references for scholars in this field.

2. Speed guidance principle

In a vehicle speed guidance system at an intersection, we need to collect traffic control information and the running state information of the guided networked vehicle. For the convenience of expression, information about each guided car can be expressed as $\boldsymbol{i=\{E(i),S(i)\}}$, Where, i represents the number of vehicles in the system. E represents traffic environment information, such as current signal phase, remaining time of signal phase, recommended speed of guidance, whether there is a queue ahead. S represents the state of the guided networked vehicle, such as acceleration, deceleration, stop. The main variables and their physical significance are shown in table 1.

Table 1. The physical meaning of the main variables in paper

Variable	Physical meaning
i	Number of guided vehicle
a	Acceleration of the guided vehicle
d	The distance from the guided vehicle to the downstream intersection
b	Deceleration of the guided vehicle
V_{max}	The maximum speed on the road
V_{min}	The minimum speed on the road
V_0	The original speed of the guided vehicle
$V(i)$	Recommended speed of guided vehicle
$P(i)$	The phase of the guided vehicle
$T(i)$	Phase remaining time

The vehicle state $S(i)$ can be expressed as

$$S(i)=[V_0,d,a,b],\quad i=1,2,\ldots,n \tag{1}$$

The current traffic environment can be expressed as

$$E(i)=[R(i),T(i),Q(i),V(i)],\quad i=1,2,\ldots,n \tag{2}$$

With

$$R(i)=\begin{cases}1 & \text{During the red light}\\ 0 & \text{During the green light}\end{cases},\quad Q(i)=\begin{cases}1 & \text{Queue ahead}\\ 0 & \text{No queue ahead}\end{cases} \tag{3}$$

The speed guidance schematic diagram is shown in figure 1. In this paper, the speed guidance strategy model is established according to the green and red traffic conditions when the network vehicles arrive at the intersection.

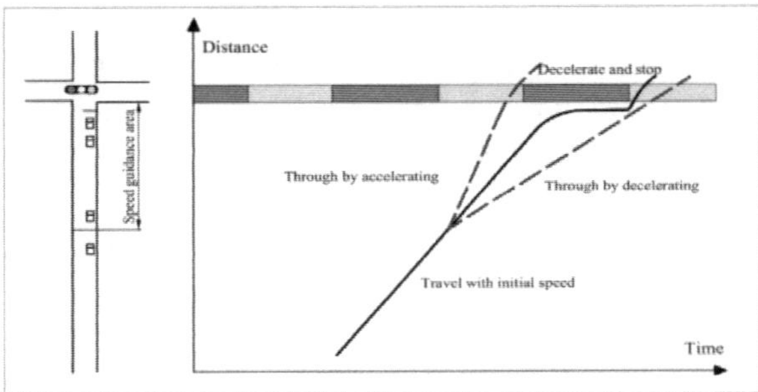

Figure1. Schematic diagram of speed guidance strategy at signal intersection

3. Speed leading strategy discussion

To simplify the research process, the following basic assumptions are made in this research:

- There is no interference from non-motor vehicles, pedestrians and bus stops in the guidance area;
- The vehicles have changed lanes before entering the controlled region, i.e., the vehicles will not change lanes in the controlled region;
- The guided vehicles all strictly follow the speed guidance strategy.

3.1. Speed guidance strategy during red light

Case1: There is no queue at the entrance to the intersection, $S(i)=\left[R(1),T(i),Q(0),V(i)\right]$

(1)Keep going with the original speed

If $\frac{d}{V_0}\succ T(i)$, then the speed guidance strategy of guided car i is "keep the original speed".

(2) Slow down to $V(i)$ and pass

If $\frac{d}{V_0}\prec T(i)$, it indicates that the guided car i will drive at the original speed to the stop line and wait for the green light to start. At this time, the speed of the guided car i adopts a speed guidance strategy of "slow down to $V(i)$ and drive", so that when the guided car i reaches the stop line at the guidance speed, the green light will turn on, and then it can pass the intersection without stopping.

(3) Slow down and stop

If the remaining time of the current red light is too long and the guided car i slows down to the minimum limit speed of the road, it means that the current phase of the guided car to the stop line is still a red light. Then the speed guidance strategy of guided car i is "slow down, stop and wait for the green light".

Case2: There is queue at the entrance of the intersection, $S(i)=\left[R(1),T(i),Q(1),V(i)\right]$

(1)Slow down to follow the line of the last car through

When the guided car i slows down near the end of the queue, the rear vehicle just starts. At this time, the guided car i can follow the rear vehicle. If the time for the guided car i to the stop line is longer than the remaining time for the red light, the speed guidance strategy of guided car i is "slow down and follow the line of the last car through".

(2)Slow down and stop

When the guided car i slows down to reach the end of the queue, the vehicle at the end of the queue still hasn't started, then the guided car i can only stop and wait. At this time, the speed guidance strategy of guided car i is "slow down and stop".

3.2. Speed guidance strategy during green light

Case1: There is no queue at the entrance to the intersection, $S(i)=\left[R(0),T(i),Q(0),V(i)\right]$

(1) Keep going at the current speed

If the remaining time of the green light is relatively long, it means that the signal phase of guided car *i* is still green when it reaches the stop line at the current speed, then the speed guidance strategy of guided car *i* is "keep driving at the original speed".

(2) Accelerate and pass

If the remaining green time is short, it means that the signal phase of guided car *i* has changed to red light when it reaches the stop line at the current speed. If the vehicle accelerates to run, it means that the guided car can pass the intersection within the remaining green time, then the speed guidance strategy of guided car *i* is "accelerate through".

(3) Slow down and stop

When the guided car *i* accelerates to the stop line with the V_{max}, the signal light has changed to red, indicating that the vehicle needs to slow down, stop and wait for the next green light. Then, the speed guidance strategy of guided car *i* is "slow down and stop".

Case2: There is queue at the entrance to the intersection, $S(i)=\left[R(0),T(i),Q(1),V(i)\right]$

(1)Slow down to follow the last car of the line through

When the guided car *i* slows down to near the end of the queue, the vehicle at the end of the queue has started. At this time, the guided car can follow the vehicle at the end of the queue. If the time when the guided car *i* reaches the stop line is less than the rest time of the green light, the speed guidance strategy of guided car *i* is "slow down and follow the vehicle".

(2) Slow down and stop

When the guided car *i* decelerates to the end of the queue with V_{min}, the vehicle at the end of the queue still does not start, indicating that the guided car can only stop and wait. Then the speed guidance strategy of guided car *i* is "slow down and stop".

4. Velocity guidance strategy model

4.1. Model constraints

In actual road traffic, the speed guidance value is not unlimited, and the comfort of driving should also be taken into account. Therefore, all the speed guidance models proposed in this paper are based on the following constraints:

- The range of guidance speed is $V_{\min} \le V(i) \le V_{\max}$;
- Refer to the values of comfort acceleration and comfort deceleration proposed by Yi et al[10],then $a_{\max} = 2.5\,m/s^2$, $b_{\max} = 1.0\,m/s^2$;
- Combined with experience and domestic and foreign literature, the constraint of the speed variation range cannot exceed 60% of the initial running speed, so the guiding speed still needs to be satisfied $0.4V_0 \le V(i) \le 1.6V_0$.

4.2. Model building

4.2.1. During the red light

Case1: There is no queue at the entrance to the intersection

(1)Slow down to *V(i)* and pass

The guided car *i* decelerates firstly and then travels at a uniform speed. The driver's reaction time is t_d, and the deceleration time is t_b, according to the kinematic formula, then

$$L(i) = V_0 t_d + \frac{V^2(i) - V_0^2}{2b} + V(i)\left(T_r(i) - t_d - t_b\right) \tag{4}$$

Guided car *i* to pass through the stop line without parking should satisfy the following optimization function:

$$\begin{cases} F_1 = \max\left(\min\left(0, d - L(i)\right)\right) \\ F_2 = \min\left(\max\left(0, T_r(i) - T_{travel}(i)\right)\right) \end{cases} \tag{5}$$

Where, *Ttravel* is the travel time of the guided car *i* from receiving deceleration information to passing through the stop line.

(2) Slow down and stop

The guided car *i* has slowed down to *Vmin* and is still in the red light when it reaches the stop line. It can only stop and wait for the green light to start, then:

$$T_{travel}(i) = t_d + t_b + \frac{d - V_0 t_d - \left|\frac{V_{min}^2 - V_0^2}{2b}\right|}{V_{min}} \prec T_r(i) \tag{6}$$

At this point, the guided speed is *Vmin*.

Case2: There is queue at the entrance to the intersection

(1)Slow down to follow the line of the last car through

When the guided car *i* is standing at the back of the queue near the entrance of the intersection, the rear car has just started, it should be satisfied:

$$T_{travel}(i) = nt_d + \frac{V(i) - V_0}{b} + \frac{L_{max}}{V_b} + \frac{L_{max}}{V_d} + \frac{d}{V(i)} \succ T_r(i) \tag{7}$$

Where, V_b represents the speed of the wave in front of the queue, V_d represents the starting wave speed of queuing vehicles in front.

(2) Slow down and stop

When the guided car *i* queues up at the rear of the vehicle queue near the entrance of the intersection, the rear car still does not start. The guided car *i* needs to stop and queue up, so it should be ensured that the guided car *i* can stop completely before the collision with the rear car. Then the braking distance should meet equation (8).

$$L_s = V_0\left(t_d + \frac{t_b}{2}\right) + \frac{V_0^2}{2b_{max}} + d_s \tag{8}$$

Where, d_s is the safety workshop distance after parking, take 1.5m.

4.2.2. During the greed light

Case1: There is no queue at the entrance to the intersection

(1) Accelerate and pass

When the remaining green time is short and the guided car *i* can pass the stop line by accelerating, then:

$$L(i) = V_0 t_d + \frac{V^2(i) - V_0^2}{2a} + V(i)\left(T_g(i) - t_d - t_a\right) \ge d \tag{9}$$

Where, t_a is the time of accelerating to $V(i)$

Case2: There is queue at the entrance to the intersection

(1)Slow down to follow the line of the last car through

Network car *i* near the entrance of the intersection to queue at the end of the vehicle line, the rear car has just started, it should meet:

$$T_{travel}(i) = nt_d + \frac{V(i) - V_0}{a} + \frac{L_{max}}{V_b} + \frac{L_{max}}{V_d} + \frac{d}{V(i)} \prec T_g(i) \tag{10}$$

(2) Slow down and stop

the braking distance should meet equation (8).

IOP Conf. Series: Materials Science and Engineering **787** (2020) 012028 doi:10.1088/1757-899X/787/1/012028

5. Solution method of velocity guidance strategy model

At present, the test methods of speed guidance are not mature, and three test methods are mainly adopted: field test, driving simulator and micro-simulation. Among them, micro-simulation software can establish scenes that are difficult to realize in the real world, and it has many advantages such as controllable, lossless, safe, reproducible, which is the most commonly used method in transportation research in recent years.

6. Conclusions and prospects

This paper systematically analysed the principle of speed-guiding strategy, according to the signal status information, the distance between a vehicle and the stop line, vehicle running status information, etc. The kinematics theory is used to construct the vehicle speed guidance model during green light and red light. the guided vehicle can pass the intersection at least without stopping or for a minimum of time under different traffic conditions. This model is the basic model to study and analyse the running state of vehicles that can efficiently pass through the intersections in the vehicle-road collaborative environment, which will provide certain reference value for scholars in this aspect in the future. The next step is to verify and evaluate the proposed model and further study the method of multi-vehicle coordinated speed guidance at signalized intersections under the cooperative vehicle-infrastructure environment.

Acknowledgments

This research was financially supported by the Natural Science Foundation of Jiangxi, China (No.20142BAB201015), and Educational Research Project of Jiangxi, China (No.GJJ160476).

References

[1] Qing, He. (2010) Robust-intelligent traffic signal control with a vehicle-to-infrastructure and vehicle-to-vehicle communication environment D. Tucson: The university of Arizona.

[2] Rakha, H, Kamalanathsharma, R. (2011) Eco-driving at signalized intersections using V2Icommunication. In: Proceedings of the 14th International IEEE Annual Conference on Intelligent Transportation systems. 341-346.

[3] Chen, S, Sun J. (2011) Development and simulation application of a dynamic speed dynamic signal strategy for arterial traffic management. In: 14th International IEEE Annual Conference on Intelligent Transportation systems. 1349-1354.

[4] Li, P. (2012) Modeling vehicle speed guidance at signalized intersections under intellidriversm. J. Traffic information and safety, 30: 136–140.

[5] Li, P. (2013) Modeling and simulation of multi-vehicles cooperative speed guidance and connected vehicles. J. Traffic information and safety, 31: 134–139.

[6] An, S. (2015) Green driving speed control method at signalized intersections. J. Transportation systems engineering and information, 15: 53–59.

[7] Jing, B. (2016) Double-cycle arterial green wave coordination control method based on speed guidance in vehicle infrastructure integration. J. Journal of south China university of technology (natural science edition), 44: 147–154.

[8] Long, K. (2018) Integrated optimization for urban arterial traffic signal coordination and vehicle active speed guidance. J. Journal of chang'an university (natural science edition), 38: 94–102.

[9] Liu, H. (2019) Development of speed guidance strategies at signalized intersections in a cooperative vehicle-infrastructure environment. J. Journal of chongqing jiaotong university (natural science edition), First online paper. http://kns.cnki.net/kcms/detail/50.1190.U.20190822.1522.002.html.

[10] Yi, K, Chung J. (2001) Nonlinear break control for vehicle CW/CA systems. J. IEEE/ASME Transactions on mechatronics, 6: 17–25.

Research and evaluation of one-way traffic setting method

Tengzhou Zhu, Mingjun Deng*, Yongtai Gong, Xinyong Huang

School of Transportation and Logistics, East China Jiaotong University, Nanchang, Jiangxi, 330013, China

*Corresponding author's e-mail: 38232040@qq.com

Abstract. As an effective means of traffic organization, one-way traffic plays an important role in easing traffic congestion. However, the reality is often a one-way traffic organization based on experience, and new traffic problems have emerged during the implementation process. Based on the discussion of one-way traffic setting conditions, this paper models and evaluates the one-way traffic organization scheme that meets the conditions. The fuzzy distribution-analytic hierarchy process is used to evaluate the traffic patency, safety, economy, comfort and environmental impact of the organization scheme. Then, the setting condition-evaluation method is applied to the actual case in combination with VISSIM simulation for inspection. Finally, the experimental results show that the setting conditions-evaluation method can provide a scientific basis for the design and comparison process of one-way traffic organization.

1. Introduction

Compared with the emerging urban areas, the old urban areas have the characteristics of concentrated social public resources, dense road network and obsolete transportation facilities[1], which directly leads to the serious imbalance between traffic demand and supply. Therefore, according to the characteristics of traffic operation in the old urban areas, it is very necessary to find a reasonable way of organization to improve the quality of traffic operation. One-way traffic organization is a measure taken to solve traffic congestion. It is an effective measure taken to improve road transport efficiency or ensure road traffic safety on the premise of no road reconstruction and expansion[2]. At present, the implementation of domestic one-way traffic has largely alleviated the problem of road congestion.

The research on one-way traffic at home and abroad is reflected in three aspects. First, in terms of traffic characteristics, William[3] mainly studied the impact of one-way traffic implementation on traffic safety and regional economy; Liang[4] and Du[5] analyzed in detail the characteristics of one-way traffic, such as road network shape design, road section design and traffic sign design and so on. Second, in terms of organizational scheme optimization, Gao[6] put forward an optimization model combining the optimized site selection of information board with the single-line traffic organization; Jiang[7] focused on the application of one-way traffic in urban traffic microcirculation system, and designed a two-layer planning model for traffic microcirculation network organization optimization. Thirdly, in terms of effect evaluation, Cheng[8] combined the example to analyze the traffic conflict points and complexity of two-way traffic and one-way traffic intersections; Guo[9] used the Processor module in the Q-paramics software to generate an analysis report to compare the traffic conditions of the one-way traffic section; Zhuo[10] developed a specific method for scheme comparison and simulation evaluation of road one-way traffic system by using VISSIM.

In the existing research, the setting method for one-way traffic often relies on empirical analysis and lacks theoretical support. For the evaluation of one-way traffic, or staying in qualitative analysis, or selecting indicators are not perfect, most do not combine the actual operating state of the road network. Based on this, the verified setting method and evaluation model proposed in this paper can effectively solve the problems in the current one-way traffic implementation process.

2. Setting conditions

2.1. Road network conditions

(1) Chessboard road network. The setting method is that two adjacent roads are paired into one-way traffic. (2) Strip-shaped road network. The setting method is to select a road that may be paired locally to form a one-way traffic system. (3) Other road networks. Two adjacent radioactive roads; When there are two adjacent loops in the road network and the length of the road section is short.

2.2. Road conditions

In special cases, when two roads are parallel and not at the same level, consider setting a low-grade road segment as one-way traffic while the other road is still two-way traffic.

2.3. Intersection conditions

The distance between intersections is less than 400 meters, the number of intersecting roads is five or more, and there are many left-turning vehicles, or roads with more difficult arterial coordination control should adopt one-way traffic organization.

2.4. Traffic flow conditions

(1) When the traffic composition is more complicated, it can be set as a one-way traffic type;
(2) Roads with obvious tidal traffic characteristics may be set as reversible one-way traffic;
(3) One-way traffic is implemented by the surveyed traffic that exceeds the road design capacity.

2.5. Other conditions

(1) Environmental conditions: When the two-way traffic road widening is difficult to ensure important buildings, monuments and greening on the road side; When parking space is required on the road but cannot be satisfied. (2) Terrain conditions: One-way traffic can be set up when complicated terrain or on-street traffic will cause great danger on steep slopes. (3) Legal conditions: Must comply with legal requirements such as "*Road Traffic Safety Law of the People's Republic of China*" and "*The administrative licensing law of the People's Republic of China*" .

3. Evaluation method

3.1. Evaluation necessity and principle

The scientific construction of the evaluation system enables real-time tracking and feedback of the one-way traffic implementation plan in the city, and provides an improvement direction for the next step of the one-way traffic planning plan. The evaluation should follow the following principles: scientific principles, dynamic principles and practical principles.

3.2. Select evaluation indicators

On the basis of following the above criteria selection principle, this paper selects relevant evaluation indicators from three aspects: network benefit, bypass cost and environmental impact. In terms of network benefit, the three indicators of average running speed, intersection delay reduction rate and traffic capacity increase rate are selected. In terms of bypass cost, two indicators of bypass distance variation coefficient and node access reduction rate are selected. In terms of environmental impact, two indicators of the reduction rate of parking times and traffic noise pollution are selected. Therefore, this paper selects seven indicators to evaluate the one-way traffic plan.

3.3. Fuzzy distribution method to determine evaluation criteria

The fuzzy distribution method is used to establish the evaluation criteria for the indicators selected in this paper. The evaluation values are listed in Table 1 below.

Table 1. Each index scale and evaluation value

Index	A	B	C	D	E
Average running speed P_1, km/h	>40	30-40	20-30	10-20	<10
Intersection delay decline rate P_2, %	>35	35-25	25-15	15-5	<5
Capacity increase rate P_3, %	>35	35-25	25-15	15-5	<5
Bypass distance variation coefficient P_4	0-0.5	0.5-1	1-1.5	1.5-2	2-2.5
Node access reduction rate P_5, %	0-20	20-40	40-60	60-80	80-100
Stop reduction rate P_6, %	80-100	60-80	40-60	20-40	0-20
Traffic noise pollution P_7, dB	<60	60-65	65-70	70-75	>75
Evaluation value	0.81-1	0.61-0.80	0.41-0.60	0.21-0.40	0-0.2

According to the evaluation criteria of the indicators, the values of the indicators are added up and averaged to obtain the overall evaluation criteria of the one-way transportation organization scheme.

Table 2. Program evaluation standard table

Evaluation score	0.81-1	0.61-0.80	0.41-0.60	0.21-0.40	0-0.20
One-way traffic evaluation rating	A	B	C	D	E

3.4. Analytic hierarchy process to determine the weight of evaluation indicators

This paper uses the analytic hierarchy process to determine the weight. The steps to implement the one-way traffic organization scheme using the analytic hierarchy process are as follows:

(1) Clear questions

To clarify the scope of the problem, this article is to evaluate the single-line program benefits of a single-line area, including the overall goal, the criteria layer and the indicator layer.

(2) Establish a hierarchy

The hierarchical structure of constructing one-way traffic impact assessment is shown in Figure 1. The feature associations between the layers are connected by straight lines.

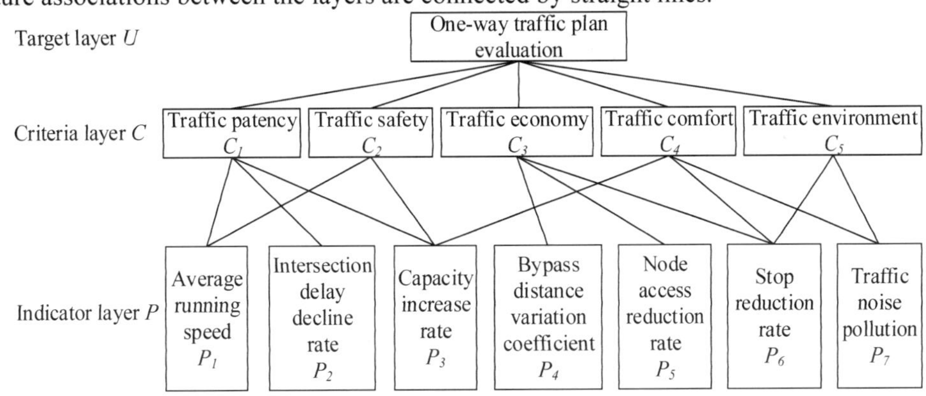

Figure 1. Hierarchy chart

3.4.1. Determining the benchmark. At present, the expert group often uses the nine-point judgment benchmark scale to obtain the ratio of the relative importance between the target criteria, and the judgment benchmark table is shown in Table 3.

Table3. Determine the benchmark table

Two-two comparison	Extremely important	Very important	Important	More important	Equally important	Counter comparison
Scaling	9	7	5	3	1	Reciprocal

3.4.2. Establish a judgment matrix. According to the hierarchical structure of Fig. 1, through the joint investigation and discussion of experts and decision makers, the ninth-rank judgment criterion is used to determine the relative importance of the two indicators and establish a judgment matrix. In the same way, the judgment matrix of the total target and criterion layer, U-C matrix is established, see Table 4 for details.

3.4.3. Calculate relative weight. The orthogonal method is generally used to calculate the orthogonal matrix. The calculation steps are as follows:

(1) Calculate the sum V_i of each row element of the judgment matrix:

$$V_i = \sum_{j=1}^{n} a_{ij} \quad (i = 1, 2, \cdots, n) \tag{1}$$

(2) Normalize V_i, you can get the relative weight of each element under a single criterion W_i:

$$W_i = \frac{V_i}{\sum_{i=1}^{n} V_i} \quad (i = 1, 2, \cdots, n) \tag{2}$$

(3) Calculate the combined weight of the indicator layer for the total target:

$$W_i^{'} = \sum_{j=1}^{7} b_j \bullet W_{ij} \tag{3}$$

Where $W_i^{'}$ is the combined weight of the indicator; b_j is the relative weight of the criteria layer to the target layer; W_{ij} is the relative weight of the indicator layer to the criteria layer.

The calculation results are show in Table 4.

Table 4. Index combination weight calculation table

Criteria layer C / Indicator layer P	Traffic patency C_1	Traffic safety C_2	Traffic economy C_3	Traffic comfort C_4	Traffic environment C_5	Combination weight $W_i^{'} = \sum_{j=1}^{7} b_j \bullet W_{ij}$
	0.2079	0.4157	0.2079	0.0831	0.0854	
Average running speed P_1	0.4000	0.7500				0.3949
Intersection delay decline rate P_2	0.4000			0.1618		0.0966
Capacity increase rate P_3	0.2000	0.2500				0.1455
Bypass distance variation coefficient P_4			0.1618			0.0337
Node access reduction rate P_5			0.3088			0.0642
Stop reduction rate P_6			0.5294	0.3088	0.3333	0.1642
Traffic noise pollution P_7				0.5294	0.6667	0.1009

Therefore, $W^{'} = (0.3949, 0.0966, 0.1455, 0.0337, 0.0642, 0.1642, 0.1009)$

3.5. Building a model

The final selection of the one-way traffic organization plan should be comprehensively calculated.

$$N = \sum_{j=1}^{7} W_i r_i = W_1 r_1 + W_2 r_2 + \cdots + W_7 r_7 \tag{4}$$

Where N is the total score of the one-way traffic organization plan; W_i is the weight of the i-th indicator; r_i is the score of the i-th indicator.

4. Instance verification

4.1. Case overview

This paper takes the Greenland Central area as an example. The Greenland Central is located in the north of the city center of Jinan city. It is adjacent to the historical old street Puli Street in the north, the Gongqingtuan Road in the south, and the Shunhe elevated in the west.

Gongqingtuan Road (Shunhe West Road - Yinhuchi Street) is 520 meters long, with dense commercial areas and heavy traffic. Puli Street is a "golden passage" between the old city and the commercial port. It is 450 meters long and has a very large traffic volume. This small area is suitable for setting one-way traffic conditions as follows: (1) Flow conditions: The area is located in the old city with heavy traffic pressure on road sections and intersections. (2) Road network conditions: the communist youth league road (Shunhe west road - Yinhuchi street section) and Puli street are two adjacent radioactive road. (3) Section conditions: the Gongqingtuan Road (Shunhe west road - Yinhuchi street section) is a slab road. Puli street is a three-board road. One-way traffic can reduce the lateral interference of vehicles and improve vehicle speed and road service level. (4) Intersections conditions: There are five intersecting roads and the distance between intersections is less than 400 meters.

4.2. Design

Gongqing Tuan Road (Shunhe West Street - Yinihuchi Street Section) implements one-way traffic from west to east, with two-way bus and special bus lanes. The Puli Street real motor vehicle runs from east to west in one direction, and the bus has two-way traffic with special bus lanes. After the design of the scheme, place relevant traffic signs and optimize the layout of bus stops.

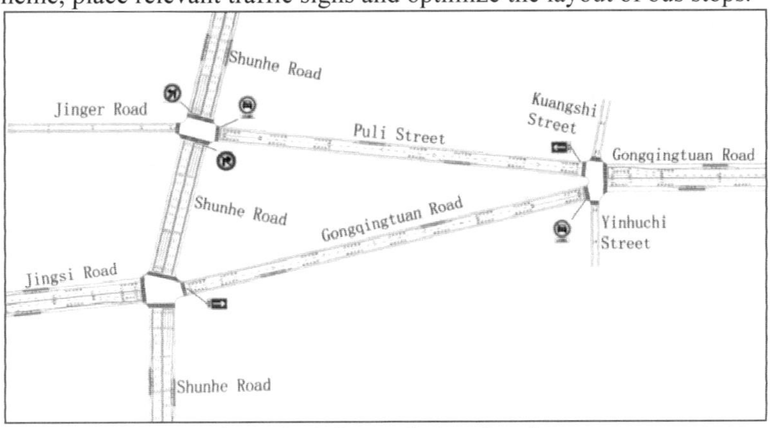

Figure 2. Scheme design drawing

4.3. Scheme simulation

In VISSIM, the flow and signal timing are input into the road network simulation model. Place the detector and get the intersection and road data, such as delay, number of stops, travel speed, etc. Comparing the data with the current survey data, the error is within 8%, indicating that the VISSIM simulation model is well constructed. The resulting data can be used in the next step to calculate values in the evaluate model.

4.4. Scheme evaluation

The score of each evaluation index was obtained from the simulation effect data and membership function: the average running speed score is 0.82, the intersection delay decline rate is 0.80, the capacity increase rate is 0.91, the bypass distance variation coefficient is 0.55, the node access reduction rate is 0.81, the stop reduction rate is 0.75, the traffic noise pollution is 0.49. The one-way traffic plan score is the average value of each evaluation index score, which is 0.78. The evaluation level is B. The traffic effect is good, which proves that the one-way traffic scheme in this place is feasible, and the validity of the model is verified.

5. Conclusion

Aiming at the problems in the setting and evaluation method of one-way traffic, this paper uses VISSIM simulation software and fuzzy distribution-analytic hierarchy process to study the design and evaluation of one-way traffic in detail and establish an evaluation model. The model is applied to the one-way traffic system network in the Greenland Central of Jinan City, Shandong Province, which proves that the evaluation model is reasonable and effective.

Acknowledgement

This research was financially supported by the Natural Science Foundation of Jiangxi Province (No. 20142BAB201015) and Education Commission of Jiangxi Province of China (No. GJJ160476).

References

[1] Li, J.T. (2017) Research on one-way traffic organization optimization of urban old district microcirculation road network [D]. Beijing Jiaotong University.

[2] Gong, W.Y. (2017) Application of one-way traffic system in urban traffic management [J]. Theoretical Research in Urban Construction (electronic version), (10): 228-229.

[3] William, L., William, E., Frawley, W. E. , Park, E. S., et al. (2012) Safety and economic impacts of converting two-way frontage roads to one-way operation [M]// Transportation Research Record: Journal of the Transportation Research Board.

[4] Liang, X.W., Yan, F.F., Hu, Y.L. (2017) Research on one-way traffic organization of urban commercial street[J]. Highways & Automotive Applications, (06): 34-37+41.

[5] Du, X.R. (2018) On the design ideas and advantages and disadvantages of one-way traffic on municipal roads [J].Shanxi Architecture, 44(33):105-106.

[6] Gao, H., Zhang, B. (2016) VMS location model and simulation based on the one-way traffic organization[J]. Journal of Transportation Systems Engineering and Information Technology, 16(06): 54-59.

[7] Jiang, Y.S., Wang, Y., Zhao, B., Han, P., Yao, Z.H. (2019) Optimization design of urban traffic microcirculation network based on one-way traffic [J]. China Transportation Review, 41(01):65-70+126.

[8] Cheng, J.L. (2014) Analysis of influences of one－way traffic on traffic safety at intersections[J]. Technology of Highway and Transport, (01): 126-130.

[9] Guo, D., Zheng, C.Y., Sun, Y.H., Sun, F. (2016) Study on traffic operation benefit evaluation of urban regional one-way traffic transformation [J].Computer Simulation, 33(10):138-141+216.

[10] Zhuo, Y.J., Wang, Y.Q. (2016) Application of VISSIM simulation on evaluation of one-way traffic plans [J].Urban Transport of China, 14(01):72-79.

IOP Conf. Series: Materials Science and Engineering **787** (2020) 012030 doi:10.1088/1757-899X/787/1/012030

Design and Implementation of Vehicle Collision Detection and Alarm System Based on Smartphone

Ding Yang, Mingjun Deng[*], Xueqing Jiang

School of Transportation and Logistics, East China Jiaotong University, Nanchang, Jiangxi, 330013, China

[*]Corresponding author's e-mail: 38232040@qq.com

Abstract. Every year, there are a large number of people died in traffic accidents in our country. If people can get help in time after a traffic accident, the survival rate of the people who are treated can be greatly increased. Therefore, it is very necessary to have a system that can detect traffic accidents and call the police or notify the rescuers in time. According to the survey and analysis, the functions of some similar systems in China are not perfect and most of the cost are high, which cannot be widely promoted in China with a high private car base. Based on the acceleration sequence value obtained by the smartphone accelerometer, this project designs the vehicle collision detection algorithm and realizes the collision detection during the driving process of the vehicle in the android environment. When a collision event is detected, the system can promptly send the collision information to the preset phone number and drive the camera function of the mobile phone, perform instant photo forensics on the scene, and finally send the help information and the accident photo to the preset phone number. The project system can realize the detection and alarm function of automobile collision accidents in time, and provide timely rescue after the collision accident.

1. Introduction

With the rapidly development of the automobile manufacturing industry, the number of motor vehicles in the world has increased substantially. Road traffic problems have become increasingly prominent, traffic accidents have continued to rise, and the number of people who lost their lives due to traffic accidents has increased year by year. Road traffic safety has become a common concern in society. Road traffic accidents have become a serious problem that threatens the safety of national life property. It is particularly necessary to effectively reduce the loss of traffic accidents. Looking up the Statistical Bulletin of the People's Republic of China of the past decade, we can see that the people killed by traffic accidents is increasing every year. The results are shown in the figure 1.

According to the relevant investigations, the following two points can be known. First, most of the victims in traffic accidents dead before they get effective rescue. Second, if the injured person who is equally injured can be sent to the hospital within 30 minutes for treatment, the survival rate can reach 80%. If they have not been effectively treated in more than 90 minutes, the hope of survival is very slim. From this, we know that reducing the waiting time of the injured after the traffic accident is the key to reducing the death rate of traffic accidents [2][3]. Therefore, it is very necessary to develop a system that can report the accident and provide the location of the accident after a traffic accident. It can effectively reduce the waiting time for the injured person after the accident and has good practicability for reducing traffic accident losses.

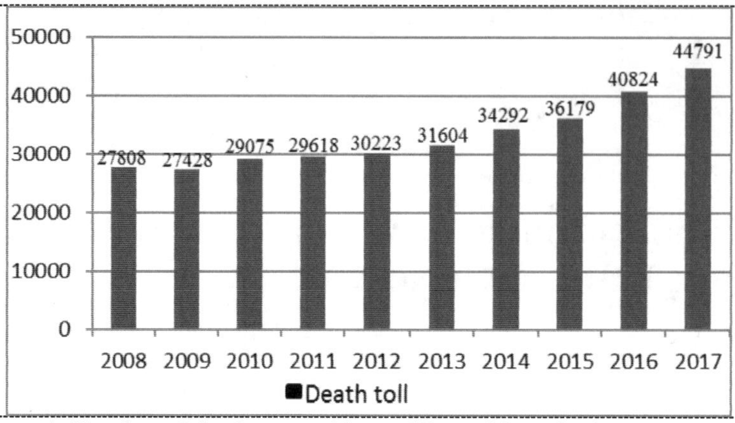

Figure 1. Number of deaths caused by traffic accidents in the past 10 years.

2. Vehicle crash accident characteristic parameters

How to accurately determine the occurrence of collision accidents is the primary task of the whole system and it is also the most important part of ensuring the practicability of the system. After reviewing the relevant literature related to automobile collision accidents, it is found that the following rules existed in the operation of the car, as shown in table 1.

Table 1. The law of car collision speed and time spent.

Collision speed (km/h)	Time spent (ms)
60	80
70	70
80	50
120	40

It can be seen from the table that the time spent by the car in the collision process is between 40 and 80 ms, and gradually decreases with the increase of the collision speed. From The acceleration formula $a=(v-v_0)/t$ can be seen that the greater the collision speed, the greater the acceleration will be at the time of collision, and the greater the threat to the collision of the car. Therefore, according to the speed of the car, there should be an acceleration threshold during the collision of the car, which can be used as the basis for the collision [4]. For example, the acceleration value generated by collision will be large when the vehicle is in a fast speed, and small when the vehicle is in a slow speed. Therefore, it can be comprehensively judged whether the car collision accident occurs according to the driving speed and the acceleration value of the car. According to the relevant research on the theory of automobile collision, it can be known that when the car collides at a speed of 60km/h, the peak value of the acceleration of the car can reach 30g [5], but the average acceleration value is about 11g [6]. Therefore, 11g acceleration value and the speed of 60km/h can be used as two thresholds to judge whether a collision accident occurs. Correspondingly, when the vehicle speed is low, the acceleration threshold as the basis for judgment will be correspondingly reduced. When the vehicle speed is high, the corresponding acceleration threshold should be appropriately increased. The system should increase the frequency of detection when the vehicle speed exceeds a certain maximum value (because the probability of an accident at this time increases). Check the relevant information to know the magnitude of the collision acceleration when a traffic accident occurs at a certain speed.

3. System overall framework and module design

Based on the acceleration sensor of the smartphone, the system uses the obtained acceleration sequence value as a parameter to design the vehicle collision detection algorithm, and realize the collision detection during the driving process of the vehicle in the Android environment. When a collision event is detected, the collision information can be sent to the preset phone number in time. At the same time, the mobile phone camera is driven to take a photo, and the photo is sent to the reserved mobile phone number. The collision detection and alarm function is realized to provide timely support for the alarm when a car collision accident occurs. The smart phone-based vehicle collision accident detection and alarm system should have modules: distress information editing and control module, integrated positioning and position display module, acceleration monitoring module. The system framework is shown in Figure 2 below.

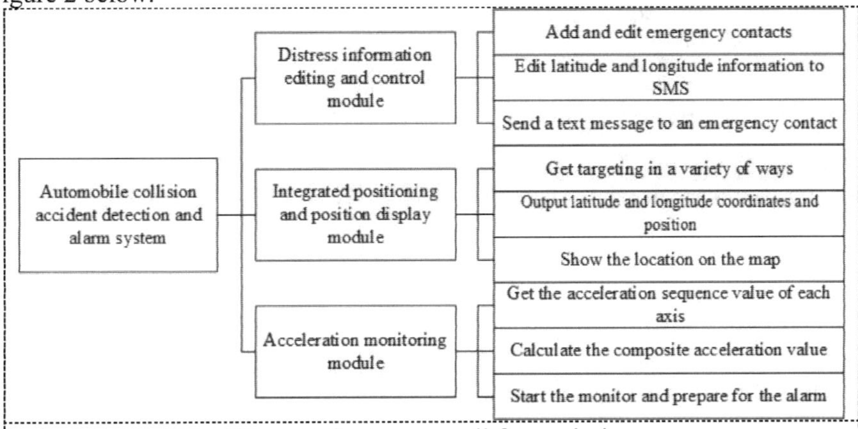

Figure 2. System overall frame design.

3.1. Acceleration monitoring module

The main way to judge the occurrence of automobile collision accident is to detect whether the acceleration value of automobile is equal to the preset threshold (G) in real time. If it exceeds, it is judged that the accident occurs. The value of the acceleration can be obtained by calling the acceleration sensor of the smartphone. The x-axis, y-axis, and z-axis acceleration sequence values are represented by A_x, A_y, and A_z respectively. The z-axis in the acceleration sensor called by this system defaults to 1g (9.8m/s^2) in the horizontal and stationary state of the mobile phone. Therefore, before the synthetic acceleration model is established, the obtained z-axis acceleration value is subtracted from 9.8. For the synthesis of acceleration, the following mathematical model is established: $A = \sqrt{A_x^2 + A_y^2 + A_z^2}$, where A is a parameter directly judged with the threshold G (take 11g). Corresponding to this model, system need to call the mathematical function of JAVA language, where the square operation calls the function: 'Math.pow()'; the square root operation calls the function: 'Math.sqrt()'. To obtain the acceleration value in the smart phone, the API of the acceleration sensor needs to be called. The obtained acceleration values of the x, y, and z axes are used to calculate the combined acceleration and compared with a preset threshold. The implementation process of the acceleration detection module is shown in Figure 3.

3.2. Integrated positioning and position display module

The acquisition of location information is a key step to improve the rescue efficiency. The positioning and location display function of the system is mainly realized by calling the Baidu map API. It can be divided into two parts: latitude and longitude acquisition and map display. Before calling Baidu map, you need to apply for Baidu map AK code and download Baidu map Android positioning SDK package. The SDK parameters used in the positioning process are shown in table 2. The obtained location information is displayed in the form of a map, which helps the user to more intuitively understand where he is. First,add the Baidu map control in the layout file(com.baidu.mapapi.map.MapView). Then obtain the management permission of Baidu map in the main program through 'getMap', and judge the positioning method before displaying the positioning. Finally map will be displayed at a certain multiple of the location.

Table 2. Baidu map SDK parameters.

Call function	Features
.setLocationMode	Set positioning mode
.setCoorType	Set back to latitude and longitude coordinate type
.setScanSpan	Set the interval to initiate a location request
.setIsNeedAddress	Set whether you need address information
.setIsNeedLocationDescribe	Set whether you need a simple location description
.setOpenGps	Set whether to use GPS
.setIsNeedLocationPoiList	Set whether you need peripheral POI information

After the Baidu map SDK is arranged, the latitude and longitude coordinates can be obtained only by calling the corresponding function. The latitude and longitude and position information are displayed on the system interface through 'setText'. Then call the save method declared in advance to save it to the specified txt file.

3.3. Distress information editing and control module

After the acceleration detection module detects that there is a collision accident, the system will enable the distress information editing and control module. The implementation process of the whole module is shown in figure 4.

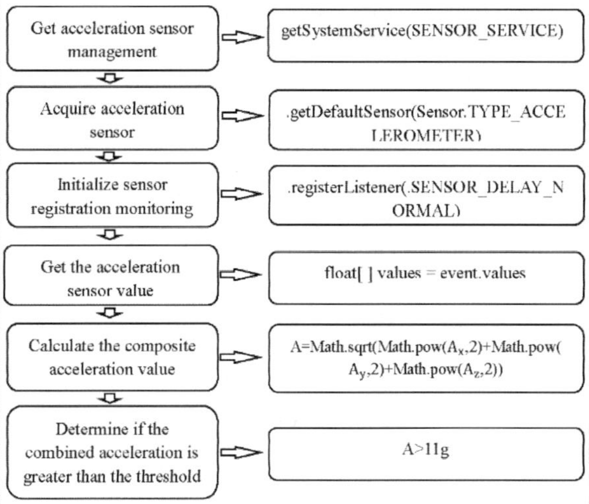

Figure 3. Acceleration detection module flow chart.

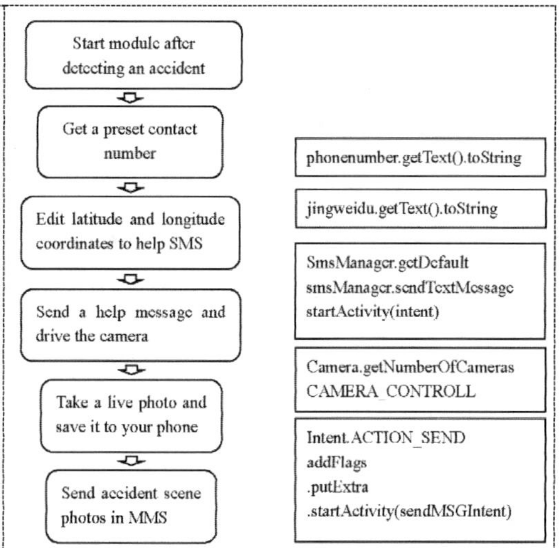

Figure 4 . Distress information editing and control module.

The input of the emergency contact should have very clear prompting and editability, so the form of pop-up input should be added to the main thread. After entering the program, call 'AlertDialog' to get the mobile phone emergency contact mobile number, then display the 'EditText' in the main interface, which is called by 'getText.toString' when a message needs to be sent using an emergency contact.

The content of the help message should mainly include the call for help notification, latitude and longitude coordinates and location information. The implementation of the SMS function in Android needs to call 'smsManager', then use 'sendTextMessage' to edit the content of the SMS and the phone number to complete the sending of the SMS.

After detecting a vehicle collision accident and sending a help message, the system will jump from the main program to the shooting thread that drives the camera. In the jump process of the program, the 'startActivity' is mainly used to realize the jump between activities. After entering the camera activity, the program will drive the mobile phone's own rear camera and automatically save after 6s. The time-lapse shooting is to avoid the blurring of the photos caused by the impact of the collision, and the function used for time-lapse shooting is 'sleep'. Pictures taken by the phone will be saved in the default directory.

In order to let the rescuers know more about the situation of the accident scene, the system will send the photo to the emergency contact through the MMS function after the collision accident is detected. The emergency contact phone is obtained by 'getText.toString'. Different from sending a text message, sending a multimedia message requires defining more parameters through 'putExtra', such as defining a MMS subject through the subject and defining the storage address of the attachment through the 'Intent.EXTRA_STREAM'. The attachment of the MMS is the photo of the accident scene just taken. It's address is obtained by 'getContentResolver( ).insert' and 'Media.EXTERNAL_CONTENT_URI'.

4. System function test

Due to the constraints of the actual situation, it is impossible to test the system in the real situation of car collision. Therefore, the collision detection of the car is tested in a simulated environment, and the electric toy car is used to simulate the collision of the vehicle. Because the speed of the car is slow, the acceleration threshold is lowered, part of the test results are shown in table 3. The test results show that the system has a correct alarm rate of more than 97% when the simulated vehicle collides, and can accurately send help information and accident photos. The interface when the system is running is shown in figure 5, the help information is shown in figure 6.

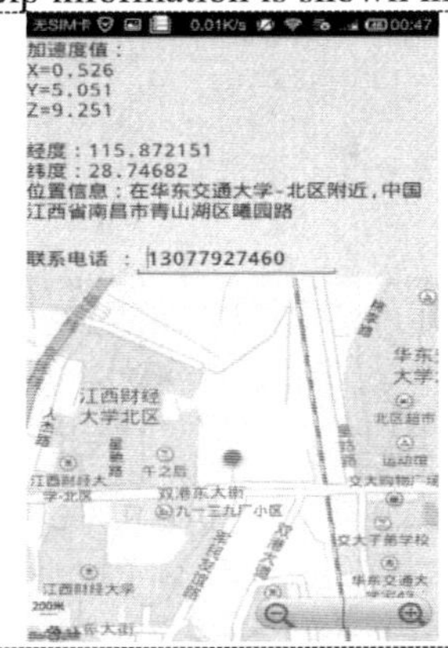

Figure 5. System operation interface.

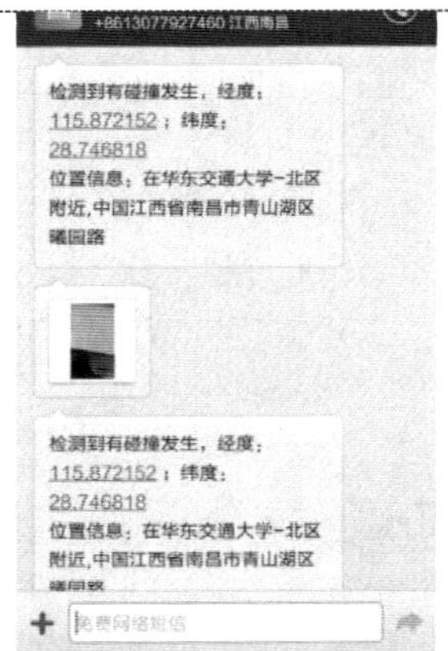

Figure 6. Sending help information.

Table 3. Partial test data.

Time (year, month, day, hour, minute, second)	Simulated vehicle operating status	Acceleration value (m/s^2)			Synthetic acceleration value (m/s^2)	Whether to trigger an alarm
		X axis	Y axis	Z axis		
2019.5.30-11:15:27	normal	2.652	5.099	9.127	5.786695	no
2019.5.30-11:15:29	collision	-19.211	-3.567	3.811	20.43659	yes
2019.5.30-11:16:03	normal	4.276	-4.74	19.614	11.70753	no
2019.5.30-11:17:05	collision	17.909	-10.472	12.502	20.92118	yes
2019.5.30-11:17:55	normal	-5.3	-6.574	10.515	8.474591	no
2019.5.30-11:18:17	collision	18.68	-9.826	12.536	21.28329	yes
2019.5.30-11:19:42	normal	7.676	1.307	11.597	7.991147	no
2019.5.30-11:19:50	collision	17.641	-5.056	4.711	19.04379	no
2019.5.30-11:20:01	normal	13.901	-1.915	6.569	14.39946	no
2019.5.30-11:20:12	collision	15.725	-15.208	19.614	23.97652	yes
2019.5.30-11:21:08	collision	16.75	1.647	-0.924	19.95694	yes
2019.5.30-11:22:00	normal	-6.56	6.814	6.574	9.993562	no
2019.5.30-11:22:22	collision	-19.048	-14.111	1.915	24.98239	yes
2019.5.30-11:23:08	normal	7.714	4.616	8.585	9.071355	no
2019.5.30-11:23:15	normal	6.522	4.352	6.569	8.480315	no
2019.5.30-11:23:28	collision	1.508	4.089	-11.712	21.94903	yes

2019.5.30-11:23:56	collision	2.312	3.912	-11.569	21.84681	yes
2019.5.30-11:24:07	normal	0.478	5.511	11.157	5.695705	no
2019.5.30-11:24:44	collision	18.895	-3.845	14.04	19.74291	yes
2019.5.30-11:25:14	collision	19.613	-1.613	19.614	21.99059	yes
2019.5.30-11:26:23	collision	0.809	19.613	7.838	19.72749	yes
2019.5.30-11:27:14	collision	-6.737	-10.194	-10.903	24.03995	yes

5. Conclusion

The designed system monitors the acceleration value of the mobile phone in real time, and then designs the vehicle collision detection algorithm to realize the collision detection during the running of the vehicle. When a collision event is detected, the alarm message and the live photo can be sent in time. Through the experimental verification in the simulation environment, the system basically realized all the functions designed. The algorithm of automobile collision detection is a very meaningful and complicated topic. In this paper,there are many imperfections in the car collision algorithm. In the future research process, the related research will continue to be improved.

Acknowledgments

Thanks to the support of the Natural Science Foundation of Jiangxi Province (No.20142BAB201015) and the Research Project of Jiangxi Provincial Department of Education (No.GJJ160476).

References

[1] National Bureau of Statistics. Statistical Communique of the 2018 National Economic and Social Development of the People's Republic of China. http://www.stats.gov.cn/tjsj/zxfb/201902/t20190228_1651265.html
[2] Special report. (2016) Current status of emergency rescue of highway traffic accidents abroad. Car and safety.
[3] Li,W.(2016) Design and Development of Automatic Alarm Location System for Car Accident Based on Android Mobile Phone. Hunan University of Science and Technology.
[4] Zhou, H.F. (2016) Research on Abnormal Driving Behavior and Collision Accident Recognition Technology Based on Mobile Terminal. Chongqing Jiaotong University.
[5] Zhu,G.(2011) Research on Emergency Disposal and Command Assistant System of Highway Traffic Emergency in Shaanxi Province. Changan University.
[6] Li,W.J.(2011) Smartphone collision detection and its application in car accident self-rescue. Computer Engineering.
[7] Li,N. (2014)Java Web Programming Practical Warfare. Tsinghua University Press.
[8] Li,G.(2015) Crazy Android Handout (3rd Edition). Electronic Industry Press.
[9] Reto,M.(2013)Professional Android 4 Application Developmen. Tsinghua University Press.
[10] Greg,M, (2013)Adam Stroud. Professional Android Sensor Programming. Tsinghua University Press.
[11] Ian,F.(2013)Android Cookbook.O'Reilly Media.

IOP Conf. Series: Materials Science and Engineering **787** (2020) 012031 doi:10.1088/1757-899X/787/1/012031

Research on Cold Chain Logistics Path Optimization Considering Cascading Failure

Zifu Fan[1], Huilin Ding[1*], Wei Zhang[1]

Chongqing University of Posts & Telecommunications School of Economics & Management, Chongqing 400065, China

*Corresponding author's e-mail: dingzidexiaoyou@163.com

Abstract: Considering that the storage and transportation products in cold chain logistics are perishable and deteriorate easily, in the actual distribution process, the cascading failure of urban roads will affect the transit times and distribution costs of delivery vehicles. According to the cascading failure process under the edge failure condition, a new diversion rule is established, and then the distribution cost is associated with the transportation speed. The cold chain logistics distribution path optimization model is established, with minimal distribution costs as its objective function, and a Dijkstra-GA algorithm is developed. The Dijkstra-GA hybrid algorithm solves the model. The model is verified by simulation examples. The simulation results verify the applicability of the model and the algorithm.

1. Introduction

At present, mobile Internet and mobile app methods make e-commerce purchase methods increasingly convenient, and people's demand for cold-chain foods is also increasing. Cold chain logistics is undoubtedly the key link to ensuring the freshness of refrigerated products. However, cold chain logistics distribution mainly includes agricultural products, processed foods and special commodities, which are perishable, which makes the products more demanding in terms of time and environment during the distribution process and leads to stricter requirements on the timeliness of cold chain logistics distribution. The frequent occurrence of road and traffic congestion incidents has caused the traffic distribution in certain regional road networks to change, which causes successive failures of other road sections and cascading failures, which seriously affect the distribution process and prolongs the transit time of distribution vehicles. The cost of losing the goods in transit and the added energy costs of the distribution vehicle reduce the quality of the delivery service.

Cold chain logistics research focuses on strategic management of cold chain logistics [1-3], security monitoring[4-6], network optimization and so on. The research on the optimization of cold chain logistics networks includes four aspects: the cold chain logistics cargo loss problem, the cold chain logistics distribution centre inventory control problem, the site selection problem and the distribution route planning problem. This paper will start to study the cold chain logistics distribution route planning problem.

There has been some research on the problem of cold chain logistics distribution route planning. The first strain of research seeks to monitor the change of cargo quality and temperature in cold chain distribution. Due to the perishable nature of cold chain products, food quality loss costs should be integrated into production and distribution costs, and models should be built to provide decision support for designing and optimizing product supply chains and distribution routes [7-9]; In addition, in order to reduce the cost of cargo damage in the cold chain distribution process, the cargo temperature is introduced as a factor, using wireless sensor modules and network monitoring technology for real-time monitoring [10-12]. Rong et al. [7] proposed a food quality degradation modelling method, integrating food quality levels into production and distribution costs and using mixed integer linear programming models to provide decision support for designing and optimizing food supply chains. The second strain of research seeks to optimize the Green Vehicle Routing Problem (GVRP) for low carbon emissions. GVRP mainly focuses on reducing fuel consumption and reducing CO2 emissions, and calculates fuel consumption by establishing a fuel consumption calculation model for vehicles, analyses the relationship with carbon emission factors, and rationally arranges the transportation route of vehicles to achieve the effect of green transportation [13-15]. Li et al.[15] developed the Cold Chain Logistics Green Vehicle Path Optimization Model (GVRPCCL) to study the relationship between changing vehicle maximum load and cost and greenhouse gas emissions. The third strain seeks to optimize the cold chain distribution route under certain road conditions. Lan Hui et al. [16] and Ge Xianlong et al.[17] constructed a logistics distribution optimization model based on the traffic conditions of road sections at various times, and developed a hybrid genetic algorithm to solve the problem.

In sum, the previous research on the optimization of cold chain logistics distribution paths mainly focused on the changes of distribution goods quality and temperature monitoring, low carbon emissions, while few studies considered road emergencies in their optimization. Moreover, the dynamic process of roads from normal to cascading failure is not considered. Most of the literature does not consider sudden road situations in the optimization process, and set the speed variable to a certain value, that is, the driving speed of all the paths is set to be the same, which is inconsistent with actual traffic situations. To this end, this paper proposes a cold chain logistics distribution path optimization model in a case of cascading traffic network failure: First, considering the impact of road congestion on traffic flow distribution, the logistic velocity-density model is introduced to develop a new cascading failure diversion rule. Second, the influence of driving speed on the distribution cost of cold chain logistics is considered in the objective function; the impact of the road after cascading failure is analysed and predicted; the sections are predicted according to the degree of failure of each section of the cascade and the traffic speed, a cold chain logistics distribution path optimization model is established, and guidance on the optimization of cold chain logistics distribution path is developed.

2. Problem description and assumptions

2.1 Problem description

Accidents, regulations, congestion and other conditions will affect other road sections when the driving speed of one road section is affected, and cascading failure will occur, which will increase the distribution cost of cold chain logistics and reduce distribution efficiency. After the cascading failure occurs, it is necessary to re-plan the distribution path in time to find an optimal path to provide quality services to customers. The research questions in this paper are described as follows:

The research object is a single distribution centre, single vehicle, and vehicle path problem with a soft time window under cascading failure; that is, the distribution route of a certain vehicle is optimized, and the whole process is delivered. The centre no longer assigns other vehicles to serve the remaining customers; Vehicle fixed cost, unit product cost, unit transportation cost, cargo loss ratio, loading and unloading time, unit energy consumption cost, distance between customers, customer demand and time window restrictions, etc. are known; during the delivery, due to unexpected events, cascading failure of the traffic network occurs; in order to reduce the in-transit time and service delay caused by the cascading failure, the unserved customers need to be re-scheduled, and the remaining paths need to be arranged and re-planned;In the optimization process, take the node where the cascading failure occurs as the starting point, and take the location of the remaining unserved customers as the necessary nodes.

2.2 Research hypothesis

The relevant research considerations in the paper are described as follows:

Hypothesis 1: The vehicle encounters cascading failure during delivery, which makes the original delivery route unfeasible. The distribution centre no longer sends vehicles to complete the service, but continues to plan and distribute the completed service by the original vehicle.

Hypothesis 2: Only the path optimization of the vehicle's distribution service process is studied. The vehicle completes all services, and the path optimization of the process of returning to the distribution centre is beyond the scope of this paper.

Hypothesis 3: The basic information of the customer is known, such as the number of customers, the location of customers, the demand for the goods by the customer, and the time window for the customer to accept the service.

Hypothesis 4: Each customer has one and only one car to serve them. and other factors affecting product quality are not considered.

3. Model formulation

3.1 Analysis of cascade failure process based on edge failure

3.1.1 Measure the importance of the edge

According to the importance of the edge in the transportation network, combined with the inherent attributes of the road given during planning, the importance of different sections (edges) is comprehensively measured. In this paper, the network is represented by an undirected graph $G=\{V,E\}$, where $V=\{v_i\}$ represents the set of nodes in the network and $E=\{e_{ij}\}$ represents the set of edges, $i,j=1,2,\cdots N$.

The initial importance of any node v_i is $S_i^0 : S_i^0 = \frac{k_i}{\langle K \rangle}$ (1)

The traffic impact coefficient δ_{ij} of any node v_i to the adjacent node v_j can be expressed as: $\delta_{ij} = \frac{1}{k_i} \times R_{ij} = \frac{k_i}{(k_i+k_j)\times<K>}$ (2)

The importance of edge e_{ij} is $A_{ij} : A_{ij} = S_i^{0\delta_{ij}} \times S_j^{0\delta_{ji}}$ (3)

The comprehensive importance of edge e_{ij} is Q_{ij}: $Q_{ij} = \alpha A_{ij} + (1-\alpha) B_{ij}$ (4)

In equation (1) (2) (4), k_i represents the degree of the node; $\langle K \rangle$ represents the average of all nodes of the network; $1/k_i$ represents the average partition coefficient; R_{ij} represents the influence weight of any node 1 on the adjacent node 2., B_{ij} represents the degree of importance of edge e_{ij}; α and $1-\alpha$ respectively indicate the importance of the edge in the traffic network and the influence weight of its own importance.

3.1.2 Measure the degree of edge failure

The load capacity is described by the capacity-load model, and the initial load is L_{ij} :

$$L_{ij} = (Q_{ij})^{\varepsilon} * (1 + rand * (P-1) * rt) * A_{ij} \quad (5)$$

The load capacity of edge e_{ij} is C_{ij}: $C_{ij} = P * L_{ij}$ (6)

The introduction of road saturation describes the traffic pressure on the road, and road saturation is represented by the Logistic velocity-density relationship.

At moment *t*, the density $\rho_{ij}(t)$ of the edge e_{ij} : $\rho_{ij}(t)=\dfrac{L_{ij}(t)}{C_{ij}}$ (7)

Degree of failure of edge e_{ij} is M_{ij} : $M_{ij}=\dfrac{1}{u}=\dfrac{1}{u_m e^{-\frac{\rho_{ij}(t)}{\rho_m}}}=\dfrac{1}{u_m e^{-\frac{L_{ij}(t)}{C_{ij}}}}$ (8)

In equation (5)(6)(7)(8), the load capacity of the edge e_{ij} is C_{ij} ,the initial load is L_{ij} , *P* is the network tolerance parameter($P\geq 1$), ε represents an adjustable weight parameter, *rand* is a random number between 0 and 1, *rt* is a parameter that controls the entire network system $(0\leq \mathrm{rt}\leq 1)$, the value is positively correlated with the size of the entire network; the larger the value, the larger the network size. Considering that the load capacity represents the number of vehicles on the side, they are all integers that need to be rounded in the calculation. u_m is the maximum speed limit of the road segment; ρ_m is the maximum congestion density of edge e_{ij} , specifically the density of the load C_{ij} of the road segment; $\rho_{ij}(t)$ is the congestion density of the road segment at time t ; $L_{ij}(t)$ is the load of edge e_{ij} at time t .

3.1.3 Develop a flow distribution rule

If a complete failure occurs, the adjacent edge needs to be shunted. The shunting is based on the importance of the adjacent edge in the network; that is, the more important the adjacent edge, the more it receives of the failed edge e_{ij} 's load. The adjacent edge e_{pq} receives the load distribution $\Delta L_{jp}{}^{*}$ from the failed edge e_{ij} as: $\Delta L_{jp}{}^{*}=\Theta_{jp}L_{ij}=\dfrac{Q_{jp}}{\sum_{e_{p\eta}\in\Gamma_{p\varphi}}Q_{p\eta}}L_{ij}$ (9)

This paper also considers the impact of the degree of edge failure on the driver's path selection. The new load distribution rules are: $\Delta L_{jp}=\dfrac{Q_{jp}}{\sum_{e_{p\eta}\in\Gamma_{p\varphi}}Q_{p\eta}}\left(1-M_{pj}\right)L_{ij}$ (10)

The load $L_{jp}(t)$ of any adjacent edge e_{jp} at moment *t* is: $L_{jp}(t)=L_{jp}(t-1)+\Delta L_{jp}$ (11)

In equation (9)(10)(11), Θ_{jp} is the distribution ratio of the load on the failed edge obtained by the adjacent edge e_{jp} , $\Gamma_{p\varphi}$ is the set of adjacent edges at the v_p end of the node; $L_{jp}(t-1)$ is the real-time load of the adjacent edge e_{jp} at time $t-1$.

To re-determine the relationship between the load distribution $L_{jp}(t)$ of the adjacent edge and the maximum load capacity C_{jp} . If $L_{jp}(t)\geq C_{jp}$, e_{jp} will trigger a new round of failed load distribution for the failed edge; If $L_{jp}(t)<C_{jp}$, the edge e_{jp} does not fail during this period.

3.1.4 Calculate the degree of failure

The loss ratio of the edge is used as an evaluation index of the network damage degree after the cascading network failure. The larger the index value, the greater the destructiveness of the cascading failure towards the network and the lower the network stability. This is calculated by:

$$CF=\frac{\sum_{e_{ij}\in E_f}CF_{ij}}{N_f\left(N_E-1\right)} \tag{12}$$

In equation (12), E_f is the set of completely dead edges, N_E is the number of network edges N_f

is the number of initial dead edges, CF_{ij} is the number of complete failures of other edges caused by the initial failed edge e_{ij} , $0 \le CF_{ij} \le N_E - 1$。

3.2 Analysis of cold chain logistics distribution costs

The model takes the lowest comprehensive cost as the objective function, and the comprehensive cost consists of four costs: transportation cost, cargo damage cost, energy cost and time window penalty cost.

The calculation formula of transportation cost C_1 is: $C_1 = C_a \cdot \sum_{j=0}^{m}\sum_{i=0}^{m} x_{ij} \cdot d_{ij}$ (13)

This paper considers that the cascading failure will indirectly affect the network traffic speed, which in turn affects the cargo damage cost. The cargo damage cost C_2 is expressed as:

$$C_2 = C_b \cdot \left(\begin{array}{l} \sum_{i=0}^{m}\sum_{j=0}^{m}\sum_{p=1}^{n}\left[\xi_1 \cdot \left(t_p + x_{ij} \cdot \frac{d_{ij}}{u_{ij}}\right) \cdot W_p^r\right] \\ +\sum_{p=1}^{n}\left(\xi_2 \cdot t_p \cdot W_p\right) \end{array} \right) \tag{14}$$

The cost of energy consumption refers to the temperature regulation of the vehicle in order to keep the cargo at a suitably low temperature during the distribution process. The energy consumption is the same as the cost of the cargo damage. When the cascading failure occurs, it will indirectly affect the network traffic speed and thus affect the impact. Energy cost C_3 can be expressed as:

$$C_3 = \sum_{i=0}^{m}\sum_{j=0}^{m}\sum_{p=1}^{n}\omega_1 \cdot \left(t_p + x_{ij} \cdot \frac{d_{ij}}{u_{ij}}\right) + \sum_{p=1}^{n}\left(\omega_2 \cdot t_p\right) \tag{15}$$

Penalties include advance delivery, early delivery, timely delivery, delayed delivery, and late delivery of five cases, so the penalty cost $PC_{(p)}$ can be expressed as:

$$PC_{(p)} = \begin{cases} \frac{1}{10}C_b W_p, T_p < AET_p \\ \min\left(\frac{1}{10}C_b W_p, a\left(BET_p - T_p\right)W_p\right), AET_p < T_p < BET_p \\ 0, BET_p < T_p < BLT_p \\ \min\left(C_b W_p, b\left(T_p - BLT_p\right)W_p\right), BLT_p < T_p < ALT_p \\ C_b W_p, T_p > ALT_p \end{cases} \tag{16}$$

Therefore, the time window penalty cost C_4 is: $C_4 = \sum_{p=1}^{n} PC_{(p)}$ (17)

In equation(13)(14)(15)(16), x_{ij} represents the choice of the vehicle path. If the vehicle passes the edge e_{ij}, the x_{ij} value is 1, otherwise it is 0; d_{ij} is the distance of the edge e_{ij} ; u_{ij} is the speed at which the goods are transported from point v_i to point v_j, it is related to the degree of failure M_{ij} of the edge e_{ij} ; C_b is the average price of the product; x_{ij} indicates whether the goods are delivered from point v_i to point v_j ; ξ_1 is the rate of loss per unit time during transportation; ξ_2 is the rate of loss per unit time during loading and unloading. t_p is the time taken for the goods to be loaded and unloaded at the customer node v_p ; W_p^r is the quantity of goods remaining on the vehicle after the completion of point

v_p distribution; W_p is the quantity of goods required at point v_p; ω_1 and ω_2 are the energy costs per unit time of the vehicle during transportation and handling, respectively; T_p is the time when the vehicle arrives at customer p, AET_p is the earliest arrival time acceptable to the customer, BET_p is the earliest arrival time to satisfy the customer, BLT_p is the latest arrival time for customer satisfaction, ALT_p is the latest arrival time that the customer can accept.

3.3 Building a distribution path optimization model

The established cold chain logistics distribution path optimization model with minimum distribution costs as its objective function is as follows:

$$Min\left(C_1 + C_2 + C_3 + C_4\right) \tag{18}$$

$$\text{s.t.} \quad \sum_{i=1}^{n} y_i r_i \le W, i = 1,2,\cdots,n \tag{19}$$

$$x_i = 1, i = 1,2,\cdots,n \tag{20}$$

$$\sum_{p=0}^{m} x_{pa} = \sum_{q=0}^{m} x_{aq}, a = 1,2,\cdots,m \tag{21}$$

Equation (19) is a weight constraint, indicating that the total demand of the transportation line is not greater than the capacity of the transportation vehicle; Equation (20) is a service number constraint, indicating that each customer has one and only one vehicle providing for their delivery service; Equation (21) is the balance constraint, which indicates that the number of times the delivery vehicle arrives and leaves each node is the same.

3.4 Algorithm Design

According to the structural characteristics of the model and the feasible domain of the decision variables, the Dijkstra-GA algorithm is used to solve the model. The overall solution process is based on the genetic algorithm steps. The Dijkstra algorithm participates in the solution of each chromosome in the model. The specific steps of the algorithm's solution are as follows:

Step 1: Encoding to generate the initial population. The starting node of the vehicle is the node in the direction of travel of the vehicle's current path. The coded form is real coded, that is, the real number 1, 2, 3..... n, represents the delivery point number of the unfinished service.

Step 2: Calculate chromosome fitness. First, obtain the distribution cost for each edge in the current situation. Then, determine the initial node to be searched, and generate three arrays of length n for storing the minimum distribution cost lengths, the last node path of the node's minimum distribution cost path, the known node set, the maximum at the beginning, while only the initial node is in the known collection. Finally, calculate the minimum distribution cost, look at the adjacency matrix, and if there is a path from the starting node to the target node, change the length (i) value to the value in the matrix and change path (i) to the destination. Find the minimum unknown length value, add the corresponding node v_i to the known values, and use this node to judge the node v_j value. If the cost to node v_i plus node v_i to node v_j is less than the current record for the cost of node v_j, the node v_j value is updated and repeated until the minimum cost of the initial node to the target node is found.

Step 3: Select a regenerative individual. Individuals with high chromosomal suitability were selected using a roulette and passed on to the next generation as a parent.

Step 4: Crossover. Individuals in the population are randomly sorted into pairs, and some genes in the paired individuals are exchanged with a certain crossover probability to obtain a new generation of individuals.

Step 5: Mutation and two-point exchange operation on the children. The fitness is calculated before and after the exchange. If the fitness is better than before, the value after the mutation is used. If it is worse, the random mutation is performed again.

Step 6: The operation is terminated. After completing the gene generation inheritance, stop the operation and output the optimal solution; otherwise, return to step 3.

4. Model Experiment

4.1 Experimental Design

A city vehicle, A, departs from distribution centre node 22 and serves customers of nodes 4, 6, 15, 26, 31, 36, and 38. It is known that the rated load of A is 7805 kg, the transportation cost per unit of distance is 3.4RMB/km, the time taken by the cargo handling unit per unit of weight is 0.2 h/t, the unit price of goods is 2000RMB/t, the proportion of cargo loss per unit of time in transportation is 0.012, and the unit of time is measured from loading to unloading. The cargo loss ratio is 0.014, the temperature-controlled energy loss cost per unit of time in transportation is 1RMB/h, the temperature-controlled energy loss cost from loading and unloading is 1.2RMB/h, and the penalty cost coefficient exceeding the optimal time window constraint is 100RMB/(h•t). At 4 o'clock, the vehicle found that the road in front of it was seriously congested due to traffic accidents, which began to affect other roads by 5:30. The distribution traffic network structure is shown in Figure 1. The customer demand and time window requirements are shown in Table 1. The distance between nodes in the road network is shown in Table 2. The genetic population size is 40, the genetic frequency is 100, the crossover probability is 0.6, and the mutation probability is 0.05.

Run the Matlab program and simulate under normal conditions. At this time, each road in the network runs well and is in an equilibrium state. Enter the basic information of each node in the traffic network, as shown in Figure 2. The points in the figure indicate that the node is connected to the node. The network node degree distribution is shown in Figure 3. Obtain the initial load and rated load on each side of the network as shown in Table 2.

Table 1 Customer demand and service time window

Customer node	Demand (tons)	[BET,BLT]		[AET,ALT]		Customer node	Demand (tons)	[BET,BLT]		[AET,ALT]	
4	1	4:20	4:50	3:50	5:20	26	0.4	8:10	8:40	7:40	9:20
6	1.2	5:00	5:30	4:30	6:00	36	0.8	9:00	9:30	8:30	10:00
15	1.6	5:50	6:20	5:20	6:50	38	0.6	6:30	7:00	6:00	7:30
31	1.2	7:00	7:30	6:30	8:00						

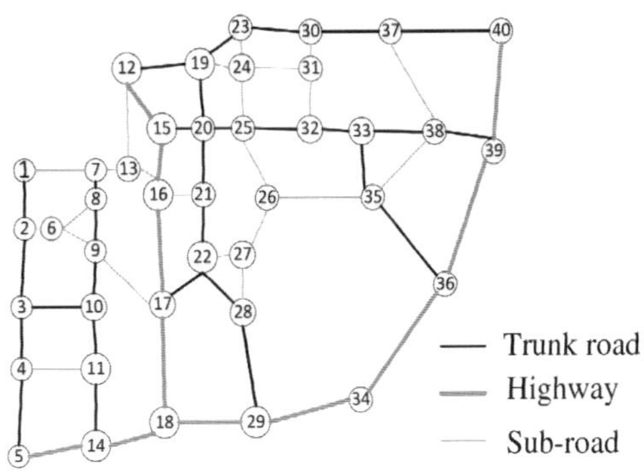

Fig. 1 Traffic network structure within the scope of vehicle A distribution

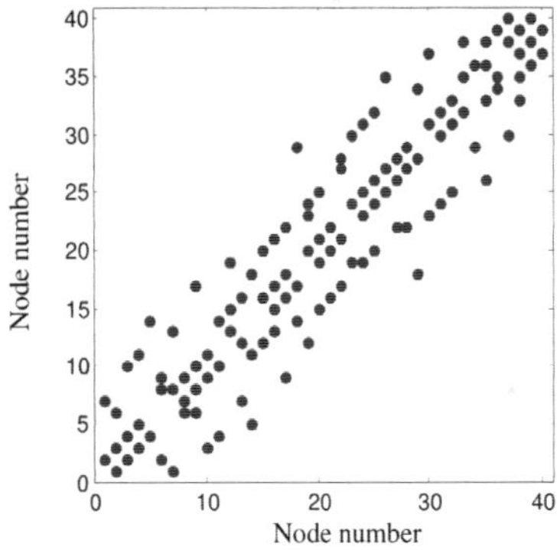

Fig.2 Network connection in Matlab

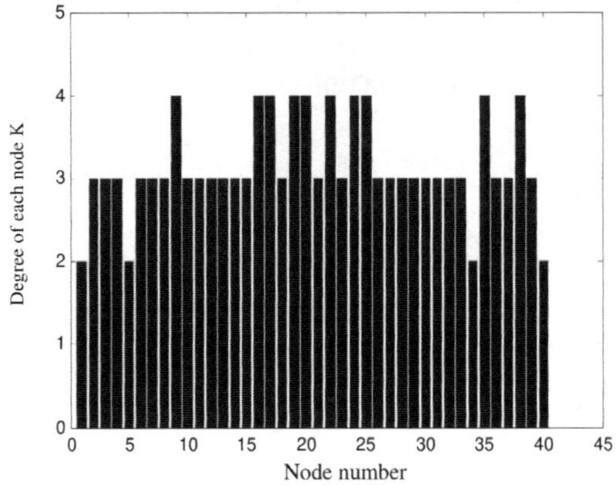

Fig.3 Network node degree distribution

Table2 Road network information table

i	j	d_{ij}	Initial load	Rated load	i	j	d_{ij}	Initial load	Rated load	i	j	d_{ij}	Initial load	Rated load
1	2	1600	1433	1863	13	16	1520	1327	1726	25	26	2240	2027	2636
1	7	2360	1963	2552	14	18	1160	1134	1475	25	32	3040	2961	3850
2	3	2400	2279	2963	15	16	1800	1789	2328	26	27	2000	1828	2377
2	6	920	780	1014	15	20	1000	949	1234	26	35	3200	2904	3776
3	4	2000	1883	2448	16	17	3200	3347	4352	27	28	1680	1494	1943
3	10	2240	2167	2818	16	21	1000	893	1161	28	29	3120	2916	3791
4	5	2640	2275	2958	17	18	3680	3685	4791	29	34	3600	3527	4586
4	11	2240	1956	2543	17	22	2500	2460	3198	30	31	1220	1079	1403
5	14	4000	2275	2958	18	29	2200	2189	2846	30	37	2680	2387	3103
6	8	1680	1454	1891	19	20	1780	1725	2243	31	32	1840	1642	2135
6	9	1680	1478	1922	19	23	1640	1522	1979	32	33	1500	1424	1852
7	8	800	736	957	19	24	1120	1079	1403	33	35	1800	1694	2203
7	13	800	700	910	20	21	1800	1784	2320	33	38	3240	3110	4043
8	9	1600	1486	1932	20	25	1120	1138	1480	34	36	4400	4060	5278
9	10	1600	1496	1945	21	22	1580	1517	1973	35	36	3500	3462	4501
9	17	2400	2258	2936	22	27	1040	940	1222	35	38	3200	3046	3960
10	11	2000	1839	2391	22	28	1840	1765	2295	36	39	4400	4255	5532
11	14	2360	2217	2882	23	24	1280	1181	1536	37	38	3200	2944	3828
12	13	2800	2445	3179	23	30	3240	2950	3835	37	40	3200	2875	3738
12	15	2240	2263	2942	24	25	1820	1648	2143	38	39	2520	2396	3115

Note: The speed limit of the highway is 60 km/h, the speed limit of the trunk road is 50 km/h, and the speed limit of the sub-road is 30 km/h.

4.2 Simulation Analysis

4.2.1 Distribution simulation under normal conditions

Perform the solution under the initial, normal conditions of the network, as shown in Figure 5. The optimal delivery order is 22-4-6-15-8-31-26-36; that is, the delivery vehicle starts from the distribution centre at the 22nd node, and makes deliveries in the order of 4, 6, 15, 38, 31, 26, and 36. At this time, the total distribution cost was 683.44 RMB, and the delivery distance was 49,920 metres. The total time to complete the delivery was 5 hours and 13 minutes. The optimal delivery path is: 22-17-9-10-3-4-3-2-6-8-7-13-16-15-20-25 -32-33-38-33-32-31-32-25-26-35-36.

4.2.2 Distribution Simulation in the Case of Sudden Cascading Failure

The path optimization when cascading failure occurs is divided into two phases. The first phase is before the cascading failure occurs, when the distribution centre plans the distribution route of the vehicle according to normal distribution traffic network conditions, and obtains the normal situation above. The cost, time, route, etc. of the distribution path. The second stage re-plans the distribution route according to the traffic network after the failure. According to the different sources of failure, the distribution path failure and the non-distribution path failure are carried out separately. The algorithm process is the same as it was in the first phase. The final total distribution cost is the sum of the distribution costs before and after the failure. Assume that the failure condition occurs at 5:30: the normal distribution route is calculated, and when the time is known, the vehicle is sent to node 3 and node 2, and the delivery service of the customer point 4 has been completed. 7:00 is set as the failure stop time, and the failure degree of each path at this time is used as the final failure degree to predict the traffic conditions of each path.

It is assumed that the delivery path failure occurs between node8 and node7, the non-distribution path failure occurs between node18 and node 29. According to the cascading failure process and the load diversion rules formulated in this paper, the failure degree of the traffic network cascade is calculated, where CF1 represents the final failure condition of the distribution path, and CF2 represents the final failure condition of the non-distribution path. The results are shown in Table 3.

Table 3 The final failure degree of the network

i	*j*	CF1	CF2	*i*	*j*	CF1	CF2	*i*	*j*	CF1	CF2	*j*	CF1	CF2	*i*	*j*	CF1	CF2
1	2	0.97	0.84	10	11	1.05	1.01	19	24	0.85	0.88	34	0.91	1.21	29	34	0.91	1.21
1	7	1.03	0.87	11	14	1.03	1.08	20	21	0.89	0.91	31	0.89	0.86	30	31	0.89	0.86
2	3	0.9	0.84	12	13	1.18	0.96	20	25	0.8	0.86	37	0.78	0.82	30	37	0.78	0.82
2	6	0.87	0.78	12	15	1	0.84	21	22	0.81	0.83	32	0.77	0.83	31	32	0.77	0.83
3	4	0.88	0.76	12	19	0.98	0.91	22	27	0.84	0.86	33	0.77	0.78	32	33	0.77	0.78
3	10	0.88	0.83	13	16	1.09	0.93	22	28	0.81	0.97	35	0.84	0.83	33	35	0.84	0.83
4	5	0.81	1.03	14	18	1.03	1.06	23	24	0.86	0.8	38	0.93	0.84	33	38	0.93	0.84
4	11	0.8	0.78	15	16	0.88	0.83	23	30	0.9	0.77	36	0.95	1.02	34	36	0.95	1.02
5	14	0.88	1.08	15	20	0.8	0.81	24	25	0.79	0.8	36	0.92	0.89	35	36	0.92	0.89
6	8	1.3	0.86	16	17	0.95	1.08	24	31	0.88	0.73	38	0.92	0.87	35	38	0.92	0.87
6	9	1.19	0.86	16	21	0.81	0.98	25	26	0.85	0.87	39	0.97	1	36	39	0.97	1
7	8	--	0.83	17	18	0.93	1.22	25	32	0.85	0.83	38	0.86	0.89	37	38	0.86	0.89
7	13	1.24	0.88	17	22	0.94	1.13	26	27	0.8	0.86	40	0.84	0.85	37	40	0.84	0.85
8	9	1.21	0.84	18	29	0.81	--	26	35	0.88	0.81	39	0.84	0.87	38	39	0.84	0.87
9	10	1.06	0.9	19	20	0.86	0.91	27	28	0.8	1.01	40	0.9	0.91	39	40	0.9	0.91
9	17	1	0.93	19	23	0.87	0.87	28	29	0.86	1.17							

After the burst cascading failure occurs, the optimal distribution path is first calculated according to the normal situation, and the total cost of distribution, time consumption, optimal delivery sequence, driving distance, and optimal path are obtained, as shown in Table 4.

Then, optimize again to obtain the optimal distribution path under cascading failure. According to the failure prediction, the delivery route is changed, and the network is updated with the final failure result. The current location of the vehicle is the starting point, the current load is the initial weight, and the current time is the departure time. The customer nodes for which delivery has been completed are deleted, and the remaining nodes are distributed. The task is performed, and the delivery order of the remaining nodes is finally obtained. The results of the total distribution cost, time consumption, optimal delivery sequence, driving distance, and optimal path of the final delivery route and the non-distribution route failure are shown in Table 4.

Table 4 Comparison of experimental results

	Normal	Delivery path failure	Non-delivery path failure
Distribution cost as normal	683.44(RMB)	2532.56(RMB)	2078.62(RMB)
Optimized distribution cost	-	1787.89(RMB)	1785.01(RMB)
Time-consuming as normal	313(min)	422(min)	399(min)
Optimized time-consuming	-	377(min)	364(min)
Travel distance as normal	49.92(km)	55.63（km）	52.35(km)
Optimized driving distance	-	52.88(km)	50.4(km)
Optimal delivery order	22-4-6-15-38-31-26-36	6-15-31-26-36-38	6-15-31-26-36-38
Optimal path	22-17-9-10-3-4-3-2-6-8-7-1 3-16-15-20-25-2-33-38-33-3 2-31-32-5-26-35-36	22-17-9-10-3-4-3-2-6-9-1 7-16-15-20-25-32-31-32- 25-26-35-36-39-38	22-17-9-10-3-4-3-2-6-8-7-1 3-16-15-20-25-32-31-32-25 -26-35-36-39-38

Experiments are carried out in the above situation. It can be seen from Table 4 that, after the failure of the distribution route and the non-distribution route, the distribution cost, delivery time, and delivery distance greatly increased, indicating that cascading connection failure will occur after the cold chain distribution occurs. This will have a substantial impact. However, through the optimization of the model, the goals of reducing the distribution cost, shortening the driving distance and reducing the delivery time are achieved.

It can be seen from Tables 3 and 4 that the final failure degree of the network when the distribution path fails is greater than the final failure degree of the network when the non-distribution path fails, and the delivery time, delivery cost, and delivery distance of the delivery path failure are greater than those of the non-distribution path. From the distribution time, distribution cost and distribution distance, it can be seen that the impact of the delivery path failure on the traffic along other distribution paths is greater than the impact caused by non-distribution path failure.

5. Conclusion and future work

In this paper, first, road congestion is judged according to the driving speed of the vehicle, and then the road is judged to be invalid. In this paper, firstly, according to the vehicle speed to judge the road congestion, and then judge whether the road is invalid, according to the importance of the edge, make the diversion rules, better depict the cascading failure effect. and according to the importance degree of the edge, the diversion rules are formulated to better describe the impact of the cascading failure. Second, considering the distribution cost, not only the transportation cost, the distance is related to the delivery time. When cascading failure occurs, it will affect the network traffic speed indirectly, which will affect the delivery time and distribution cost. The model's cost function is fully considered to affect the cost of the vehicle. Finally, according to the particularity of the cold chain distribution requirements, when constructing the model with minimum distribution costs, in addition to considering the transportation cost of the vehicle, various cost factors, such as the soft time window limitation and the cargo damage cost during the transportation loading and unloading process, are also considered.

The model established by this paper not only considers the characteristics of urban road network cascading failure but also considers the characteristics of urban road congestion speed change. Compared with the traditional cold chain vehicle path planning problem, this approach is closer to the real urban distribution environment. In the future, the distribution process can be further optimized and improved by combining the dynamic process of cascading failure, and the adaptability of the distribution path to the real-time change of vehicle transportation status can be improved.

Acknowledgement

The authors gratefully thank the policy-making consulting project of Chongqing university of posts and telecommunications (K2017-112), the project of humanities and social sciences Ministry of Education in China(18YJC790224), the planning project of Chongqing social science (2016BS057), The Ministry of Education humanities social sciences study the youth fund project (17YJC630189), Chongqing University e-commerce and modern logistics Key Laboratory of open fund project (ECML201708).

References

[1] Xiao X ,Li Z ,Matetic M ,et al. Energy-efficient sensing method for table grapes cold chain management[J]. Journal of Cleaner Production, 2017, 152:77-87.

[2] Wang X , He Q , Matetic M ,et al. Development and evaluation on a wireless multi-gas-sensors system for improving traceability and transparency of table grape cold chain[J]. Computers and Electronics in Agriculture, 2017, 135:195-207.

[3] Hsiao YH,Chen MC,Chin CL. Distribution planning for perishable foods in cold chains with quality concerns: Formulation and solution procedure[J].Trends in Food Science&Technology,2017,61： 80-93.

[4] Yu Y,XiaoT.Pricing and cold-chain service level decisions in a fresh agri-products supply with&logistics outsourcing[J].Computers＆Industrial Engineering,2017,111： 56-66.

[5] Thakur M , For?S E . EPCIS based online temperature monitoring and traceability in a cold meat chain[J]. Computers and Electronics in Agriculture, 2015, 117:22-30.

[6] Gogou E,Katsaros,G,Derens,et all. Cold chain database development and application as a tool for the cold chain management and food quality evaluation[J]. International Journal of Refrigeration, 2015, 52:109-121.

[7] Rong Ai-ying,Akke Rman R,Grunow M. An optimization approach for managing fresh food quality throughout the supply chain[J]. International Journal of Production Economics,2011,131(1):421－429.

[8] Nakandala D , Lau H , Zhang J . Cost-optimization modelling for fresh food quality and transportation[J]. Industrial Management & Data Systems, 2016, 116(3):564-583.

[9] Azadeh A , Elahi S , Farahani M H , et al. A genetic algorithm-Taguchi based approach to inventory routing problem of a single perishable product with transshipment[J]. Computers & Industrial Engineering, 2017, 104:124-133.

[10] KUO Ju-chia, CHEN Mu-chen. Developing anadvanced multi-temperature joint distribution system for the food cold chain ［J］. Food Control,2010,21(4):559－566.

[11] Xiao X , He Q , Fu Z , et al. Applying CS and WSN methods for improving efficiency of frozen and chilled aquatic products monitoring system in cold chain logistics[J]. Food Control, 2015, 60:656-666.

[12] Thakur M , For?S E . EPCIS based online temperature monitoring and traceability in a cold meat chain[J]. Computers and Electronics in Agriculture, 2015, 117:22-30.

[13] Yoshinori Suzuki. A dual~objective metaheuristic approach to solve practical pollution routing problem [J]. International Journal of Production Economics,2016,6（176）: 143~153.

[14] Çağrı Koç, Karaoglan I . The green vehicle routing problem: A heuristic based exact solution approach[J]. Applied Soft Computing, 2016, 39:154-164.

[15] Li Yan,Lim Ming K,Tseng Ming-Lang .A green vehicle routing model based on Modified particle swarm optimization for cold chain logistics[J].Industrial Management &Data Systems.2019,119(3) : 473-494.

[16] Lan Hui, He Qinfei, Bian Zhan, et al. Optimization of Cold Chain Logistics Distribution Routing Considering Road Traffic Conditions[J]. Journal of Dalian Maritime University, 2015, 41(4).

[17] Ge Xianlong, Xu Weiping, Wang Weixin. Research on Urban Logistics Transfer and Intermodal Transportation Based on Vehicle Cooperation under Traffic Restriction[J]. Chinese Management Science,2017,25(10):130-139.

IOP Conf. Series: Materials Science and Engineering **787** (2020) 012032 doi:10.1088/1757-899X/787/1/012032

Research on Electric Vehicles Considering V2G Technology in Photovoltaic Residential Areas

Shanting Ma*, Haifeng Su and Yibing Yu

School of Electrical and Electronic Engineering, North China Electric Power University, Baoding, Hebei Province, 071003, China

*Corresponding author's e-mail: aurora_ting@163.com

Abstract. With the emergence of global energy crisis, electric vehicles and photovoltaic power generation have been widely used as an effective way. However, the connection of large-scale electric vehicles and the uncertainty of photovoltaic power generation also do harm to the operation of the power distribution system. In order to solve the problem that electrical load of resident varies big between peak and valley caused by disordered charging of electric vehicles, this paper constructed a cooperative scheduling model for V2G electric vehicles and photovoltaic power generation. Taking residential community as research background, this paper try to solve the SOC threshold of electric vehicle discharge by reducing the grid load variance and the objective function. The reasonable SOC threshold is optimized by the improved particle swarm optimization algorithm, and the reasonable SOC threshold can be compared and analyzed to reduce the load variance value more effectively. Finally, the paper takes one residential area as an example to prove the rationality and practicality of the proposed model.

1. Introduction

In recent years, energy issues have become one of the most important issues facing the world. A series of renewable energy sources such as photovoltaic power generation and wind power generation are gradually slowing down the dependence on traditional energy sources. EVs(electric vehicles) are powered by electric energy and do not produce greenhouse gases. With the in-depth study of electric vehicle V2G (vehicle-to-grid) technology, electric vehicle batteries can be used as a load or as a power source[1]. Therefore, making full use of the characteristics of electric vehicle charging and discharging and renewable energy is of great significance to the reduction of grid load fluctuations.

The literature [2] discussed that the charging time of electric vehicles is mostly the same as that of grid load, and the phenomenon of peak load on the grid load occurs. Therefore, a reasonable load management scheme is needed to distribute electric vehicle charging. In [3], the random charge and discharge behaviors of hybrid electric vehicles and the rules of wind power generation and photovoltaic power generation are analyzed. The cooperative scheduling model considering V2G is constructed to optimize the system network loss. In the above literature, the disordered charging of electric vehicles and the original load of the distribution network are discussed, which makes the load peak-to-valley difference increase, especially in the residential area load. At the same time, although the V2G strategy can reduce the load network loss, it does not specifically study the impact of different electric vehicle discharge states on the system.

Based on the research background of residential community, this paper constructs a V2G electric vehicle charging and discharging and photovoltaic output collaborative scheduling model to reduce the total load peak-to-valley difference of residential areas. Firstly, according to the probability density distribution of the electric vehicle V2G and the photovoltaic power generation system, the driving law and photovoltaic power generation characteristics of the electric vehicle are obtained. Secondly, with the load variance and minimum as the objective function, a reasonable SOC threshold is optimized to respond to the discharge state of the electric vehicle. Finally, the results of the optimization by the improved particle swarm optimization algorithm are compared.

2. Probability distribution of electric vehicles and photovoltaic power generation systems

2.1. Electric vehicle slow charging and discharging power model

At present, the battery of electric vehicles is mainly a lithium battery[4]. The charging and discharging powers $P_C(t)$ and $P_F(t)$ of electric vehicles are normally distributed within 3-4kW[5].

2.2. Electric vehicle charging and discharging time

The approximate charging and discharging time calculation formula for electric vehicles is as follows:

$$T_C = \frac{LW_{100}}{100P_C\eta} \tag{1}$$

Where T_C is the charging time, the unit is h; L is the daily driving mileage, the unit is km; W_{100} is the power consumption per 100 km, the unit is kWh; η is the charging efficiency, which is 0.95.

In 5 minute intervals, the day is divided into 288 time periods. Then the charging period of each electric vehicle T_n is (The time period T_n is rounded up):

$$T_n = \frac{T_C}{5} \tag{2}$$

2.3. Driving rules of electric vehicles

According to the US Department of Transportation's survey of household cars in the United States, more than 85% of EVs are not used in one day, and there is plenty of time to charge at home.

(1)Travel/return time probability density distribution function:

$$f_S(x_t) = \begin{cases} \frac{1}{\sigma_S\sqrt{2\pi}}\exp[-\frac{(x_t-\mu_S)^2}{2\sigma_S^2}], & (\mu_S-12) < x_t \le 24 \\ \frac{1}{\sigma_S\sqrt{2\pi}}\exp[-\frac{(x_t+24-\mu_S)^2}{2\sigma_S^2}], & 0 < x_t \le (\mu_S-12) \end{cases} \tag{3}$$

Where x_t is a moment between [0, 24], $\mu_s = 17.6$, $\sigma_s = 3.4$.

(2) Probability density function for daily mileage:

$$f_D(x_L) = \frac{1}{x_L\sigma_D\sqrt{2\pi}}\exp[-\frac{(\ln x_L - \mu_D)^2}{2\sigma_D^2}] \tag{4}$$

Where x_L is the daily mileage (unit: mile); and μ_D=3.20, σ_D=0.88 .

2.4. Probability distribution of photovoltaic power generation system output

Photovoltaic has a strong intermittent nature. Studies have shown that the sun's light intensity $r(t)$ at a certain moment satisfies the Beta distribution.

$$f(r(t)) = \frac{\Gamma(\alpha+\beta)}{\Gamma(\alpha)\Gamma(\beta)}\left(\frac{r(t)}{r_{max}(t)}\right)^{\alpha-1}\left(1-\frac{r(t)}{r_{max}(t)}\right)^{\beta-1} \tag{5}$$

Where: $\Gamma(\cdot)$ is the Gamma function; $r_{max}(t)$ is the maximum illumination intensity at time t; α and β are the Bate distribution shape parameters.

Photovoltaic power generation system output function[6]:

$$P_G(t) = r(t)A\eta \tag{6}$$

Where: A is the total area of the photovoltaic power generation array; η is the photoelectric conversion efficiency.

Photovoltaic power generation probability density function[7]:

$$f(P_g(t)) = \frac{\Gamma(\alpha+\beta)}{\Gamma(\alpha)\Gamma(\beta)}\left(\frac{P_G(t)}{P_{Gmax}(t)}\right)^{\alpha-1}\left(1-\frac{P_G(t)}{P_{Gmax}(t)}\right)^{\beta-1} \tag{7}$$

Among them, $P_{Gmax}(t)$ is the maximum output value of the photovoltaic generator set at time t:

$$P_{Gmax}(t) = A\eta\, r_{max}(t) \tag{8}$$

3. Based on V2G of electric vehicle optimization scheduling mode

3.1. Objective function

According to the conventional V2G, every electric vehicle energy storage battery can discharge a certain amount of remaining SOC. According to the driving law of electric vehicles, the number of EVs returned at time 19:00 and after is larger. If the vehicle with the remaining value of the stored value of the electric vehicle at this time satisfies the conventional set value (0.2), the vehicle selects to discharge, and the load peak sharply drops during the late peak period, resulting in a larger peak-to-valley difference. Therefore, a reasonable SOC threshold is determined, and the SOC value of the electric vehicles discharge is controlled to change the number of electric vehicles. This paper chooses to use the total variance of the load to measure the peak-to-valley difference of the whole system. The variance load is minimized by optimizing a reasonable overall SOC threshold.

Use the grid total load variance and minimum to optimize the objective function, namely:

$$\min F(t) = \sum_{j=1}^{100}\left[\frac{1}{288}\sum_{t=1}^{288}\left(P(t)-\frac{1}{T}\sum_{t=1}^{T}P(t)\right)^2\right] \tag{9}$$

Where: $P(t)$ is the total load of the residential area of the system during the t period of the day. It includes the initial load $P_l(t)$ of the entire residential area, the power $P_G(t)$ of the photovoltaic output, and the charging and discharging powers $P_C(t)$ and $P_F(t)$ of the electric vehicle participating in the V2G strategy, and the set power consumption is positive.

$$P(t) = P_l(t) - P_G(t) - P_F(t) + P_C(t) \tag{10}$$

3.2. Restrictions

(1)Electric vehicles charging and discharging time constraints

According to the daily load law of the residential area, it is assumed that the peak load time of the system is the noon time period $[t_{f1}, t_{f2}]$, the evening time period $[t_{f3}, t_{f4}]$, and the rest is the valley period. The charging time of the electric vehicle at the peak time and the starting time of the discharge are t_{fc1} and t_{ff1}, and the charging time and the discharging time at the end are t_{fc2} and t_{ff2}. The initial charge and discharge start time in the valley is t_{gc1} and t_{gf1}, and the charge and discharge time at the end of the valley is t_{gc2} and t_{gf2}.

Charging time constraint:

$$\begin{cases} t_{fc1} < t_{f1} < t_{f4} < t_{fc2} \\ t_{fc2} - t_{fc1} \le 24 - T_C \\ t_{fc2}、t_{fc1} \in [0,24) \end{cases} \tag{11}$$

Discharge time constraint:

$$t_{f4} \le t_{ff1} \le t_{ff2} \le t_{fc2} \tag{12}$$

(2)Electric vehicle battery storage capacity constraints

IOP Conf. Series: Materials Science and Engineering **787** (2020) 012032 doi:10.1088/1757-899X/787/1/012032

$$S_{\min} \le S(t) \le S_{\max} \tag{13}$$

Among them, $S_{\min}$ is the minimum storage capacity of the electric vehicle battery, and $S_{\min}=\varepsilon S$, S is the rated capacity of the electric vehicle battery storage, ε is the minimum retained electricity ratio. $S_{\max}$ is the maximum storage capacity of an electric vehicle battery. Due to battery loss, it cannot reach 100%, that is $S_{\max}=(1-\delta)S$, and δ is the percentage of battery loss.

3.3. Algorithm features and processes
The main flow of this algorithm is as follows:

Step 1: According to the probability density function, sample the PV output, the return time of each electric vehicle, the charging and discharging power of EVs, the mileage of each electric vehicle, and the electric vehicle number that responds to V2G in each period, and extract 100 samples respectively;

Step 2: Using the improved particle swarm optimization algorithm to find the optimal SOC threshold for the whole day and the optimal SOC threshold for the time period, and make the load variance of 100 samples minimum;

Step 3: Compare the results of the three SOC thresholds to verify the effectiveness of the optimization.

4. Case analysis

4.1. Conditional assumption
(1) The users of electric vehicles are rational people. Users have all the information about charge and discharge electricity prices and can respond to V2G strategies.

(2) With the popularity of electric vehicles, it is assumed that 80% of the users in this community own electric vehicles.

(3) The charging device for electric vehicles fully meets the charging needs of residents.

4.2. Research object and parameter settings
This paper selects a total of 750 housing units in a residential area in a certain city in China. The number of EVs is 600. The model of the EV selected in this paper is Nissan Leaf 2, and it is $S=24kWh$, $\delta=8\%$, $\varepsilon=20\%$. Investigation of the collected light intensity characteristics residential city, the actual situation of photovoltaic function and output density residential area of 2.4 part. And then, $A=3042m^2$, η=0.15. Take the photovoltaic power of Figure 1 as the basic data.

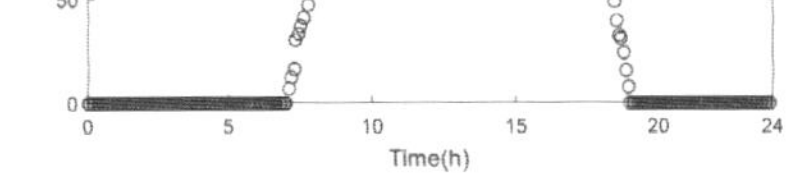

Figure 1. Photovoltaic power generation output graph.

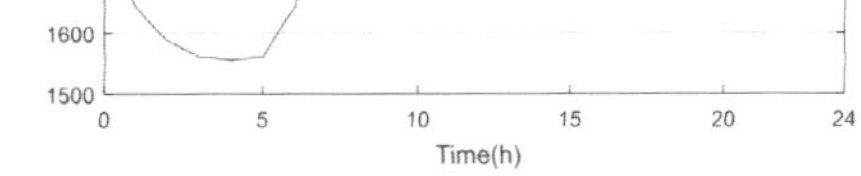

Figure 2. Residential area daily load curve.

The electricity consumption of residential areas has a strong regularity. This paper conducts statistics on the daily load of residential areas. At the same time, the electric vehicle random access cell is recorded for disorderly charging, and a typical daily maximum load curve is obtained as shown in Figure 2. The peak daily load of this cell is 3,080 kW, and there are two distinct peak load periods, [11:00-13:00] and [19:00-22:00].

4.3. Case study analysis

4.3.1. Comparison of daily load of residential area without optimized SOC threshold

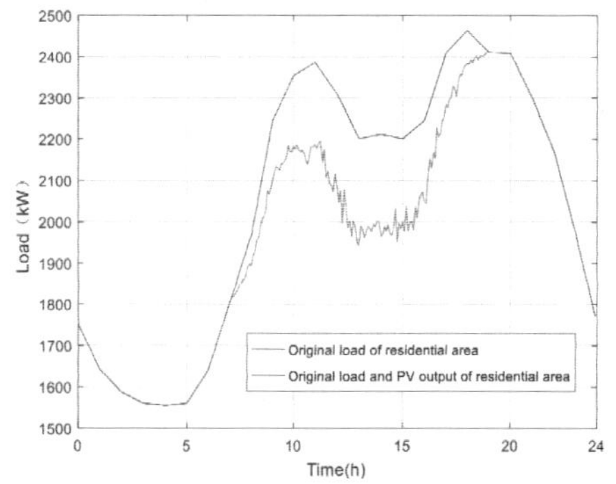

Figure 3. Comparison of residential area load and original load of photovoltaic after superposition .

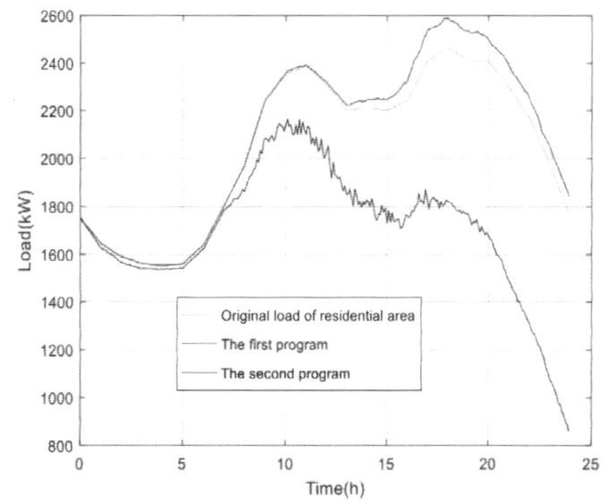

Figure 4. SOC load curve of residential area load in three programs.

It can be seen from Figure 3 and Figure 4 that when 600 EVs are disorderly charged, an obvious "peak peaking" phenomenon occurs in the daily load peak of the original residential area. Adding photovoltaics can effectively alleviate the load demand of residential areas during the day. During the 10:00-16:00 period, the peak hour load is significantly reduced. In order to study the influence of the SOC threshold on the load peak-to-valley difference when the EV is not optimized, this paper designs two programs for comparison. The first program is the disordered charging of electric vehicles. The second program is that the EV accesses the photovoltaic community and participates in the V2G strategy. It can be discharged when it meets the traditional setting value of 0.2.

Since the peak of photovoltaic output is around at 13:00, the peak load at noon is significantly lower than the original. At the same time, according to the daily driving law of electric vehicles, since there are more vehicles coming back after 16:00, there are relatively many vehicles that choose to discharge at this time, and the second load peak is particularly obvious.

4.3.2. Optimized SOC threshold

The third program is to set a reasonable discharge SOC value based on V2G and photovoltaic power generation. In this paper, the improved particle swarm optimization algorithm and optimization model are programmed and solved by MATLAB software. In the program, the number of particle groups is set to N=50, and the maximum number of iterations is 500.

Table 1. Load variance value under different conditions.

Different schemes	Load variance（GW^2）
Original load	10.170
The first program	11.241
The second program	6.929
The third program	4.026

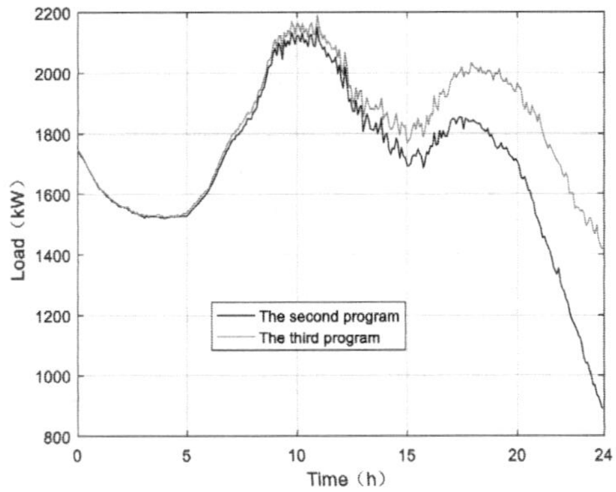

Figure 5. Load curve of SOC in different programs.

Through simulation optimization, the SOC threshold of the whole day is 0.72. At the same time, calculate the variance of the load under different programs, as shown in Table 1. It can be seen from Table 1 that the implementation of the V2G technology can reduce the load peak-to-valley difference. By determining the SOC threshold of the discharge, the load peak-to-valley difference can be more effectively reduced.

5. Conclusion

(1)This paper establishes a collaborative scheduling model for electric vehicles and photovoltaic power generation that takes into account V2G for residential communities. In this paper, the SOC threshold of electric vehicle discharge is solved by reducing the variance of grid load as the objective function, and compared with other methods, the rationality of the model is verified.

(2)In practical applications, when building a new residential area or when the residential power distribution is not expanded, adding a photovoltaic system and optimizing a reasonable SOC threshold will not only effectively increase the access volume of the system electric vehicle, but also effectively reduce peak-to-valley difference and reduce system cost.

(3)The SOC threshold at the time of electric vehicle discharge can be optimized for multiple periods. This will further reduce the peak-to-valley difference of the load, and also select the optimal optimization scheme based on actual economic and technical considerations.

References

[1] Johannes Kester,Gerardo Zarazua de Rubens,Benjamin K. Sovacool,Lance Noel. Public perceptions of electric vehicles and vehicle-to-grid (V2G): Insights from a Nordic focus group study[J]. Transportation Research Part D,2019,74.

[2] GAO Ciwei, ZHANG Liang. A Survey of Influence of Electrics Vehicle Charging on Power Grid[J]. Power System Technology, 2011,35(2):127-131.

[3] Mohsenzadeh A, Pazouki S, Haghifam M R, et al. Optimal planning of parking lots and demand response programs in distribution network considering power loss and voltage profile[C]// Innovative Smart Grid Technologies Conference. IEEE, 2015:1-5.

[4] LIU Xiaofei,ZHNAG Qianfan,CUI Shumei. Review of Electric Vehicle V2G Technology[J].Transactions Of China Electrotechnical Society, 2012,27(2):121-127.

[5] SU Haifeng, LIANG Zhirui. Household electric vehicles ordered charging Control method in residential quarter based on peak-valley TOU power price[J]. Electric Power Automation Equipment, 2011, 35(6): 17-22.

[6] ATWA Y, EL-SAADANY E, SALAM A, et al. Optimal renewable resources mix for distribution system energy loss minimization[J]. IEEE Trans on Power Systems, 2010, 25(1):360-370.

[7] TIAN Liting, SHI Shuanglong, JIA Zhuo.A Statistical Model for Charging Power Demand of Electric Vehicles[J]. Power System Technology 2010, 34(11):126-130.

Evaluation of Induction Heating Effects on Asphalt Concretes Containing Waste Steel Shavings and Iron-based Particles

Qi Pang[1*], Jiangbo Zhang[1], Lei Gao[1], Bo Sun[1], Shenglei Yao[1]

[1] China Changjiang Construction Investment. Corp. Ltd, Chengdu, 610000, P.R. China

[*]Corresponding author's e-mail: 1163637410@qq.com

Abstract. The self-healing ability of asphalt has great benefits for the service of asphalt pavement, while this ability is not significant at room temperature. This paper utilized waste steel shavings (WSS) and iron-based particles (IBP) as conductor materials, and induction heating technology to heat asphalt concretes. High temperature rutting test, low temperature cracking test and freeze-thaw splitting test were carried out to evaluate the effect of WSS and IBP on the pavement performance of asphalt concretes, and induction heating test was conducted to analyse the heating efficiency. Results shows that both WSS and IBP are beneficial to high temperature performance of asphalt concrete, but not good for low temperature performance. Besides, WSS and IBP has no significant impact on the water stability. Considering the pavement performance and induction heating efficiency, 6%-8% WSS with 2% IBP is recommended.

1. Introduction

As a kind of viscoelastic material, asphalt has the ability of self-healing after being damaged, which has been widely recognized by researchers [1]. Researchers are also trying to make use of this characteristic of asphalt to realize self-healing of the asphalt pavement. The representative technologies include microcapsules [2], electromagnetic induction heating [3-5], microwave heating [6] and so on. Compared with the other methods, induction heating has the advantages of simple process, multiple repairs and uniform heating [7].

The induction heating technology is to add some conductor materials to the asphalt mixture and heat the mixture by the alternating magnetic field (AMF) [3]. Commonly used conductor materials are waste steel shavings (WSS) [3], steel wool fibres [8], iron-based particles (IBP) [5] and iron power [9].

Liu et al. studied the heating performance of asphalt mixture with different WSS content, heating power and frequency, and found that the induction heating uniformity (IHU) was affected by these three factors and the heating effect was obvious [3]. Jeoffroy et al. utilized iron-based particles as the conductor material and investigated the effect of sizes and chemical compositions of iron-based particles to heal the crack. The results showed that the particle size under 1 mm had uniform heating effect and the crack healed with 80% performance recovery by heating the asphalt sample to 104°C [5]. Apostolidis et al. studied asphalt mastic and mortar containing steel fibers and iron power. He reported that the electrical conductivity of asphalt material was improved by steel fibers and iron power, but the effect of steel fibers is better [9].

Considering the high induction heating efficiency of WSS and good heating uniformity of IBP, this paper utilized WSS and IBP as the asphalt induction heating materials and explored both heating effect and pavement performance of asphalt concretes with conductive additives. The addition of IBP is expected to enhance the induction heating uniformity and efficiency of asphalt concrete containing WSS. This paper aims to provide an optimum content of the conductive additives.

2. Materials and methods

2.1. Materials

An SBS modified asphalt supplied by Sinopec Group was used. The aggregates were limestone gravel and the mineral powder was made of high quality limestone. The WSS and IBP were purchased from Guangzhou West tier Industrial Products Co., Ltd. The basic technical indexes of all the materials were listed in Table 1 and Table 2.

Table 1. Basic technical indexes of materials

Materials	Properties	Test result	Specifications
SBS modified Asphalt	Penetration (25°C) / 0.1 mm	57.7	40-60
	Softening point / °C	66.9	≥60
	Ductility (5°C) / cm	25.3	≥20
	Viscosity (135°C) / (Pa·s)	2.19	≤3
Waste steel shaving	Length / mm	3-6	-
	Density / (g·cm^{-3})	7.7	-
Iron-based particle	Diameter / mm	1.0	-
	Density / (g·cm^{-3})	7.4	-

Table 2. Basic properties of aggregates and mineral powder

Specification	15~25 mm	5~15 mm	3~5 mm	0~3 mm	Mineral powder
Bulk density	2.721	2.729	2.720	2.733	2.714
Water absorption / %	0.69	0.71	0.55	-	-

2.2. Specimen preparation

AC-13 is selected as the gradation of SBS modified asphalt mixture in this paper, and the mix proportion is designed according to the specification requirements. See Figure 1 for the gradation curve. The asphalt content is 4.6% by mass. WSS and IBP were mixed with aggregates for 90 s before adding the asphalt and mineral powder. The WSS contents (by mass) were 0 %, 4%, 6%, 8% and 10%, and each content had two specimens (with and without 2% IBP for comparison).

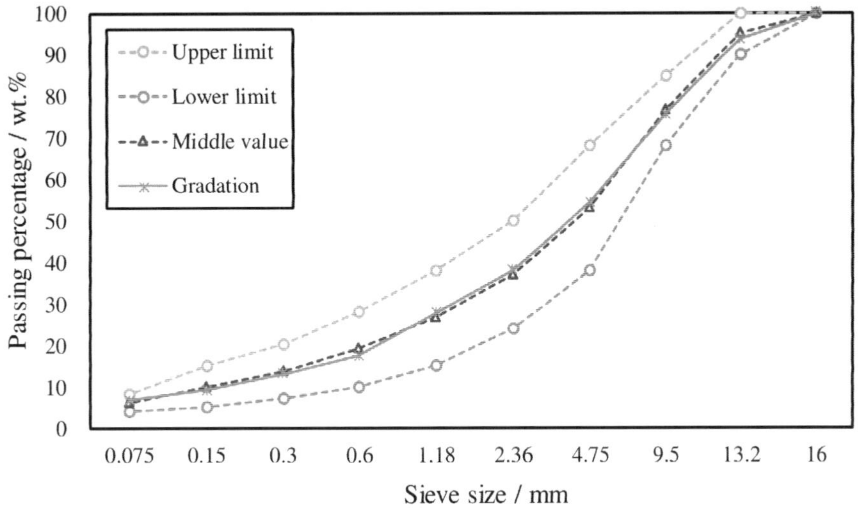

Figure 1. Gradation curve of AC-13 SBS modified asphalt mixture

2.3. Pavement performance test

High temperature rutting test, low temperature cracking test and freeze-thaw splitting test were employed to evaluate the high temperature performance, low temperature performance and water stability of asphalt concretes containing conductor materials. Dynamic stability, flexural failure strain and TSR of asphalt concretes with different content of WSS and IBP were measured and calculated.

2.4. Induction heating test

As shown in Figure 2, the induction heating tests were carried out with electromagnetic induction coils, a high frequency AC power supply and a cooling device. The power and frequency were set to 8 kW and 120 kHz, respectively. Marshall samples with different content of WSS and IBP were prepared based on the mix proportion. The gap between top of Marshall samples and induction coils was 1 cm. The temperatures of asphalt concretes were measured by an infrared imager (FLIR E6-XT). All the samples were placed below the middle of the electromagnetic induction coils to ensure the same alternating magnetic field.

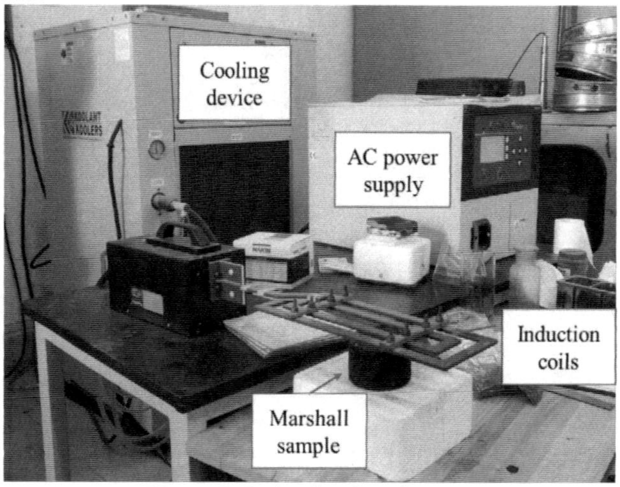

Figure 2. Induction heating test

3. Results and discussion

3.1. High temperature rutting test

Rutting tests were conducted at 60°C under 0.7 MPa of the wheel pressure, and dynamic stability (DS) was gathered to investigate the effect of WSS and IBP on high temperature rutting resistance of asphalt concretes. From Figure 3, it is obvious that the addition of WSS and IBP increases the DS of asphalt concretes, indicating that the effect is positive and high temperature rutting resistance is improved. When WSS content raised up to 10%, DS of asphalt concretes increased 33% and 40%, respectively. Besides, the effect of the same quality of WSS and IBP on the increase of DS is not significant, which means the improvement of high temperature rutting resistance caused by WSS and IBP is almost the same. And the high temperature stability of all the asphalt concretes satisfied the specification.

IOP Conf. Series: Materials Science and Engineering **787** (2020) 012033 doi:10.1088/1757-899X/787/1/012033

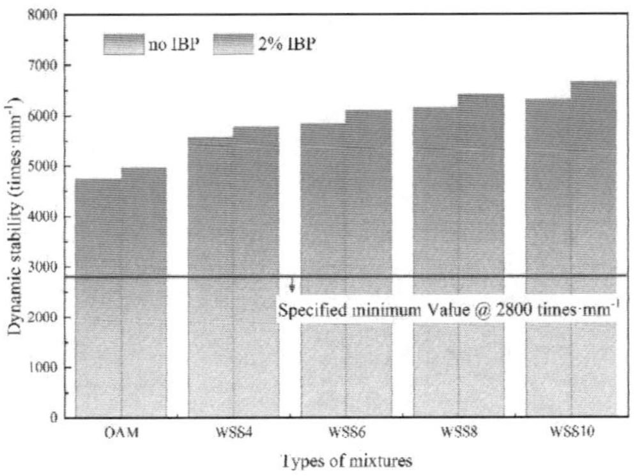

Figure 3. High temperature rutting test results

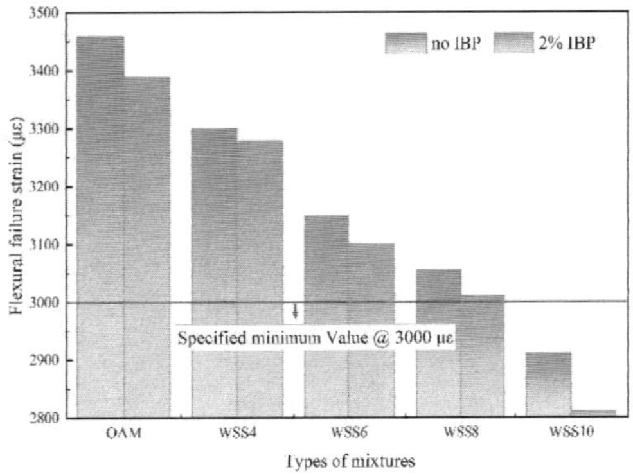

Figure 4. Low temperature cracking test results

3.2. Low temperature cracking test

Cracking tests were carried out at -10°C and flexural failure strain was measured to evaluate low temperature cracking resistance of asphalt concrete containing WSS and IBP. As shown in Figure 4, flexural failure strain decreased with the increasing WSS content and the addition of IBP. The variation trend is obvious and basically linear. This result proves that the addition of WSS and IBP is negative to low temperature performance of asphalt concretes. Moreover, when the WSS content raised up to 8%, flexural failure strain of the asphalt concrete with 2% IBP was very close to the specified minimum value of 3000 με. While when the WSS content was 10%, both samples with and without IBP cannot satisfy the specification. This result proves that when the WSS content is increased to more than 8%, the deformation ability of asphalt concrete is too weak to resist low temperature cracking. Thus, considering the cracking resistance of asphalt concrete at low temperatures, it is suggested that the content of WSS should not exceed 8%.

3.3. Freeze-thaw splitting test

Freeze-thaw splitting tests were employed to evaluate the water stability of asphalt concretes. After Marshall samples totally absorbed water in vacuum, they were placed in a refrigerator at -18°C for 16 h. Then these samples were put into constant temperature water tank of 60°C for 24 h. The ratios of Marshall stability of the samples after and before freeze-thaw process were calculated, which is TSR. From Figure 5, it can be seen that the variation of TSR shows low significant correlation with WSS and IBP. It is hard to say the effect of WSS and IBP on water stability of asphalt concretes. Besides, the fluctuation range of TSR is relatively small. In addition, TSRs of all the specimens were higher than 90%, which indicated good water stability.

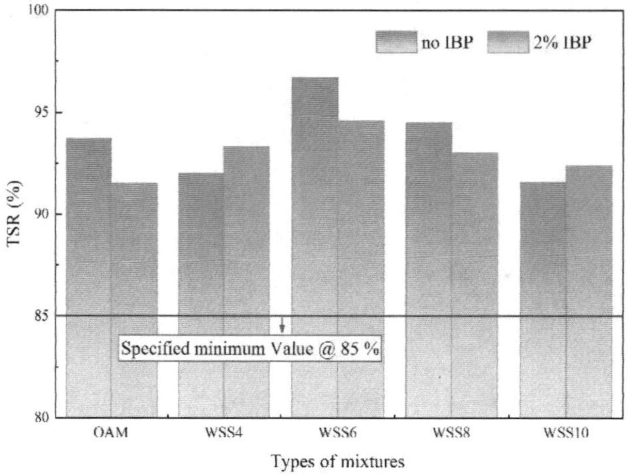

Figure 5. Freeze-thaw splitting test results

3.4. Induction heating test

Heating efficiency was measured by induction heating test. According to the previous literature [1], asphalt mixture has strong self-healing ability above 100°C. While the temperature should not be too high to prevent the mixture from loosening. In this test, the upper temperature limit was set to 160°C to get the heating trend. But when choosing the best content of WSS, the temperature is still 100°C. As shown in Figure 6, the temperatures of asphalt concretes increased with the addition of WSS and IBP. Induction heating efficiency significantly increased with the increasing WSS content. Even the induction heating effect of 2% IBP is significant, and the temperature of asphalt concrete can reach 70°C. Besides, when WSS content was higher than 6%, the asphalt concretes can be heated up to over 100°C in 60 s with the help of IBP. When WSS content is increased to 10%, the temperature can even reach 100°C in 30 s, which shows very high induction heating efficiency. Thus, it is suggested that WSS content is not less than 6% with the addition of IBP, according to the test results.

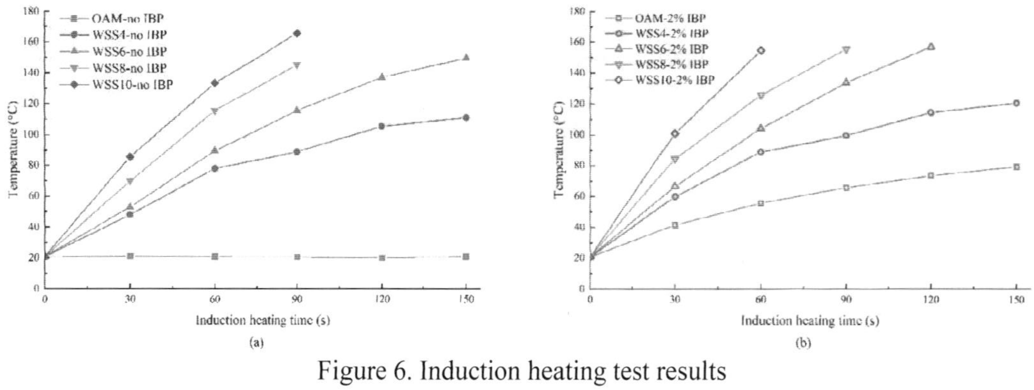

Figure 6. Induction heating test results

4. Conclusions

In this paper, SBS modified asphalt concretes containing different contents of WSS and IBP were prepared and tested. The pavement performance and induction heating efficiency were measured and evaluated. The main conclusions are shown as follows:

(1) WSS and IBP have positive effect on the high temperature performance of asphalt concretes, but not good for low temperature performance. Besides, the water stability of asphalt concretes shows low significant correlation with WSS and IBP.

(2) According to induction heating test results, the heating efficiency is positively correlated with the content of WSS and IBP. Asphalt concrete can be heated up to over 100°C in 60 s with WSS content of 6% and IBP content of 2%.
(3) Considering both low temperature performance and heating efficiency, 6%-8% WSS and 2% IBP are recommended.

Acknowledgments
Thanks for the assistance and support from China Changjiang Construction Investment. Corp. Ltd.

References

[1] Sun, D., Sun, G., Zhu, X., Guarin, A., Li, B., Dai, Z., & Ling, J. (2018). A comprehensive review on self-healing of asphalt materials: Mechanism, model, characterization and enhancement. Advances in colloid and interface science, 256: 65-93.
[2] Sun, D., Lu, T., Zhu, X., Li, B., & Tian, Y. (2018). Optimization of synthesis technology to improve the design of asphalt self-healing microcapsules. Construction and Building Materials, 175: 88-103.
[3] Liu, K., Dai, D., Fu, C., Li, W., & Li, S. (2020). Induction heating of asphalt mixtures with waste steel shavings. Construction and Building Materials, 234, 117368.
[4] Dinh, B. H., Park, D. W., & Le, T. H. M. (2018). Effect of rejuvenators on the crack healing performance of recycled asphalt pavement by induction heating. Construction and Building Materials, 164: 246-254.
[5] Jeoffroy, E., Bouville, F., Bueno, M., Studart, A. R., & Partl, M. N. (2018). Iron-based particles for the magnetically-triggered crack healing of bituminous materials. Construction and Building Materials, 164: 775-782.
[6] Norambuena-Contreras, J., & Garcia, A. (2016). Self-healing of asphalt mixture by microwave and induction heating. Materials & Design, 106: 404-414.
[7] García, A., Norambuena-Contreras, J., & Partl, M. N. (2013). Experimental evaluation of dense asphalt concrete properties for induction heating purposes. Construction and Building Materials, 46: 48-54.
[8] García, A., Norambuena-Contreras, J., Bueno, M., & Partl, M. N. (2014). Influence of steel wool fibers on the mechanical, termal, and healing properties of dense asphalt concrete. Journal of Testing and Evaluation, 42(5): 1107-1118.
[9] Apostolidis, P., Liu, X., Scarpas, A., Kasbergen, C., & van de Ven, M. F. C. (2016). Advanced evaluation of asphalt mortar for induction healing purposes. Construction and Building Materials, 126: 9-25.

A Real-time Traffic Sign Detection Model Based on Improved YOLOv3

XiongFei Liu [1], Fan Xiong [2]

[1]School of Physics and Electronics, Central South University, Changsha, Hunan, 416000, China

[*]Corresponding author's e-mail: 809649938@qq.com

Abstract. Advanced driver assistance system, which can quickly and accurately detect driver assistance personnel, ensure traffic safety, and play an important role in guiding road traffic signs. However, traditional detection models often have false detection, missed detection, and predicted low accuracy of the target box, especially for the Identification of small targets, its accuracy is difficult to meet the requirements. Therefore, based on the target detection algorithm model YOLOv3, this paper replaces the common two-dimensional convolution with dilated convolution and expands the receptive field, At the end of the model, a three-layer spatial pyramid cascade structure is added to make the multi-scale feature map deep fusion. finally, the improved model is tested for performance on the GTSDB dataset. The results show that the improved model implemented in this paper can effectively solve some problems in the actual road traffic sign detection, and meet the requirements of real-time detection .

1. Introduction

Whether it is an automated driving assistance system or the current hot driverless field, traffic sign detection is undoubtedly very important. However, among a large number of traffic sign detection methods, algorithms based on deep learning have drawn great attention from scholars[1-2]. the improvement of computer computing performance has made convolutional neural network(CNN) models widely used in the field of object detection. CNN have become an application trend in the field of computer vision with their high efficiency. Such as R-CNN[3-4], YOLO[5-6]. Many researchers have also applied it to the field of traffic sign detection, and have improved them.

YOLOv3 algorithm obvious deficiencies, such as the target prediction box accuracy is not high, small target detection scenario occurs erroneous detection, missed detection[7-8]. As for these problems, this paper improves YOLOv3, and proposes a fast traffic sign detection model YOLOv3-pro, which is a deep fusion of feature maps. Its structure shown in Figure 3: Dilated convolutional in deep residual networks; 3 layers convolution of cascade structure. Together, they form a feature pyramid containing 6 convolutional layers of different scales; Meantime, perform an upsampling operation on the feature pyramid in 2 times steps and fuse with the previous deep residual network to form a deep fusion detection model. This paper uses the GTSDB[9] dataset to compare and test different algorithms. Experimental results show that our method achieves good results in small target detection.

IOP Conf. Series: Materials Science and Engineering **787** (2020) 012034 doi:10.1088/1757-899X/787/1/012034

2. YOLOv3 model

YOLOv3 the latest version of the YOLO series is an end-to-end one-stage universal object detection algorithm. It uses the K-means clustering to automatically select the best initial regression box for the dataset, and uses a multi-scale anchor structure to improve the accuracy of target detection.

$$confidence = p_r(object) \times IOU_{pred}^{truth} \quad , \qquad p_r(object) \in \{0,1\} \tag{1}$$

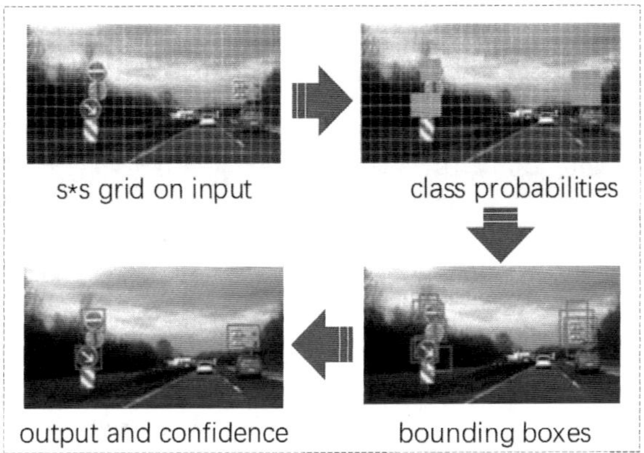

Figure 1. Principle of YOLO algorithm

Figure 1 shows the processing flow. It divides each input image into s*s grids. If the center of the real target box falls on a grid, the grid is responsible for detection the goal. Each grid needs to predict the probability value of the class C target, B boxes, and corresponding confidence score. Formula (1) shows the confidence score calculation, where $p_r(object)$ is 1 for the grid containing the target, IOU_{pred}^{truth} represents the error between the prediction box and ground truth, it reflects the grid about the accuracy of the prediction. If a target has multiple prediction box, NMS is used to select the best.

	Type	Filter	Size	Output
	conv	32	3*3	256*256*16
	conv	64	3*3/2	128*128*32
1x	conv	32	1*1	
	conv	64	3*3	
	res			128*128*32
	conv	128	3*3/2	64*64*64
2x	conv	64	1*1	
	conv	128	3*3	
	res			64*64*64
	conv	256	3*3/2	32*32*128
8x	conv	128	1*1	
	conv	256	3*3	
	res			32*32*128
	conv	512	3*3/2	16*16*256
8x	conv	256	1*1	
	conv	512	3*3	
	res			16*16*256
	conv	1024	3*3/2	8*8*512
4x	conv	512	1*1	
	conv	1024	3*3	
	res			8*8*512
	avgpool		global	
	conected		1000	
	softmax			

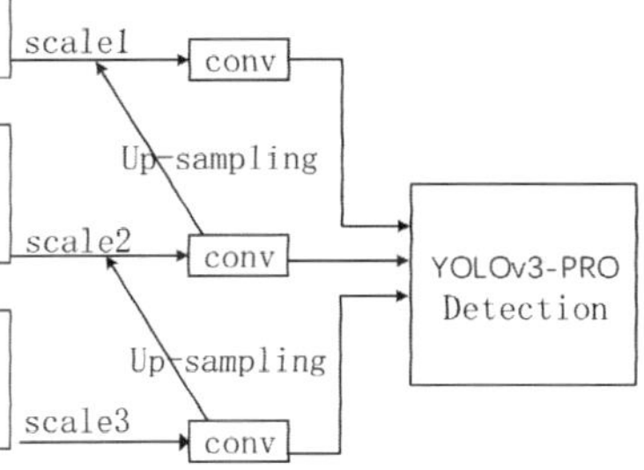

Figure 2. YOLOv3 model structure

The structure YOLOv3 model is shown in Figure 2. Backbone is Darknet-53. It uses upsampling and residual networks. In the feature maps of the last 3 scales, 3 channels are used to perform target detection independently.

3. Improved YOLOv3 model

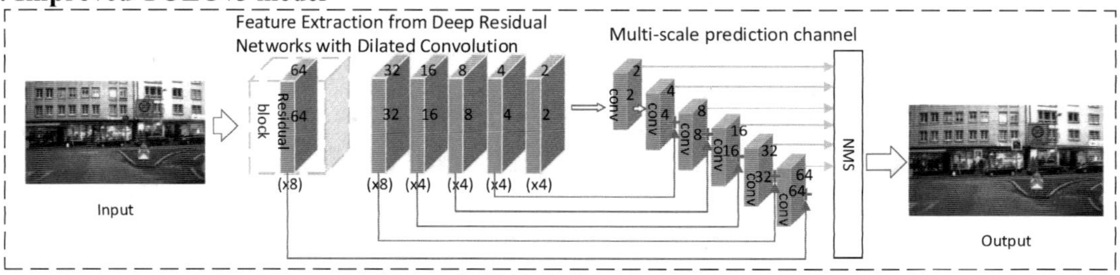

Figure 3. YOLOv3-pro structure

3.1. Candidate box generation strategy based on K-means

Anchor box mechanisms have been proposed in target detection models such as Fast R-CNN[4], but the size of the prediction box is manually set based on experience, which will cause the network to converge slowly during training and be prone to local optimization. The traditional K-means clustering method uses the Euclidean distance function, which makes larger boxes have more error clustering than smaller boxes, and the clustering results may be biased. For this reason, we use IOU[10] scores to evaluate the clustering results, thereby avoiding errors caused by the scale of the box. The formula for the distance is as follows:

$$d(box, centrod) = 1 - IOU(\text{box, centrod}) \qquad (2)$$

The central particle is the center point of the cluster; $IOU(\text{box, centrod})$ is the overlap ratio of the cluster box and the central box. The K-means clustering method is used to compare the IOU scores of different k values to find the best number and size of anchor boxes.

3.2. Backbone Network Based on Dilated Convolution and FPN Structure

For the detection of small targets, high-resolution feature maps are required. In order to obtain more detailed semantic features and accurately determine the location of the target, a wider receptive field and more global information are required. This paper combines dilated convolution and FPN structure, and proposes a high-resolution, large receptive field deep residual backbone network for feature extraction.

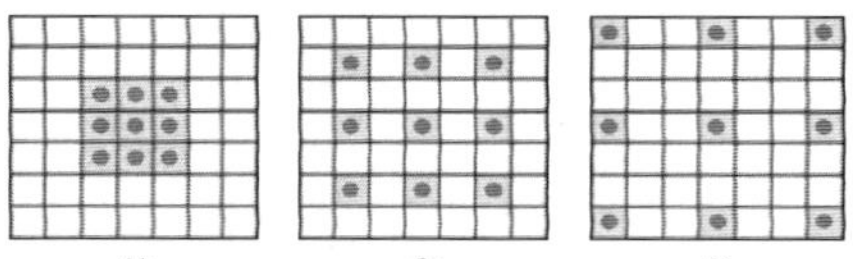

(a)Original; (b) Double interval; (c) Triple interval

Figure 4. dilated convolution

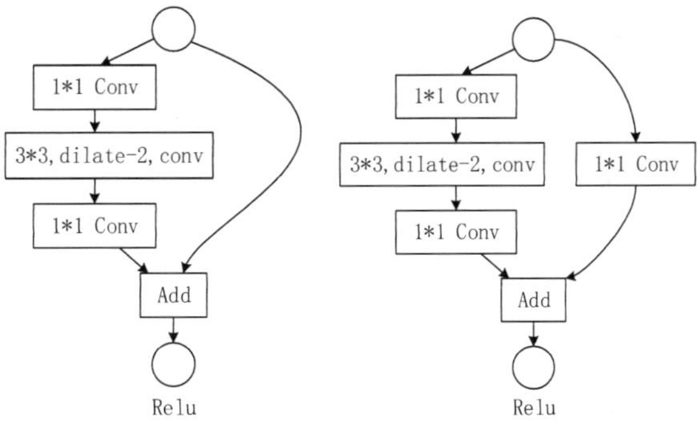

Figure 5. Two types of residual unit with dialated convolution

3.2.1. Dilated convolution structure

Dilated convolution is to expand the convolution kernel convolution kernel by changing the internal spacing receptive field. Figure 4 shows three differently spaced convolution kernels, where rate represents the interval of holes in the convolution kernel. (a) is 3*3, rate = 1 dilated convolution, the convolution kernel receptive field range is 3*3. (b) is 3*3, rate = 2 dilated convolution, convolution receptive field increases to 7*7. (c) is a range of 3*3 and rate = 3. The receptive field of the convolution kernel increases to 15*15, which ensures that the convolutional network can extract feature information in a larger field of view [11-13]. Figure 5 shows two application types.

IOP Conf. Series: Materials Science and Engineering **787** (2020) 012034 doi:10.1088/1757-899X/787/1/012034

	Type	Filter	Size	Output
	conv	16	3*3	512*512*16
	conv	32	3*3	256*256*32
	conv	16	1*1	
1x	conv	32	3*3	
	res			256*256*32
	conv	64	3*3/2	128*128*64
	conv	32	1*1	
2x	conv	64	3*3	
	res			128*128*64
	conv	128	3*3/2	64*64*128
	dilated-conv	64	1*1	
8x	dilated-conv	128	3*3	
	res			64*64*128
	conv	256	3*3/2	32*32*256
	dilated-conv	128	1*1	
8x	dilated-conv	256	3*3	
	res			32*32*256
	conv	512	3*3/2	16*16*512
	dilated-conv	256	1*1	
4x	dilated-conv	512	3*3	
	res			16*16*512
	conv	1024	3*3/2	8*8*1024
	dilated-conv	512	1*1	
4x	dilated-conv	1024	3*3	
	res			8*8*1024
	conv	2048	3*3/2	4*4*2048
	dilated-conv	1024	1*1	
4x	dilated-conv	2048	3*3	
	res			4*4*2048
	conv	4096	3*3/2	2*2*4096
	dilated-conv	2048	1*1	
4x	dilated-conv	4096	3*3	
	res			2*2*4096
	avgpool		global	
	connected		4	
	softmax			

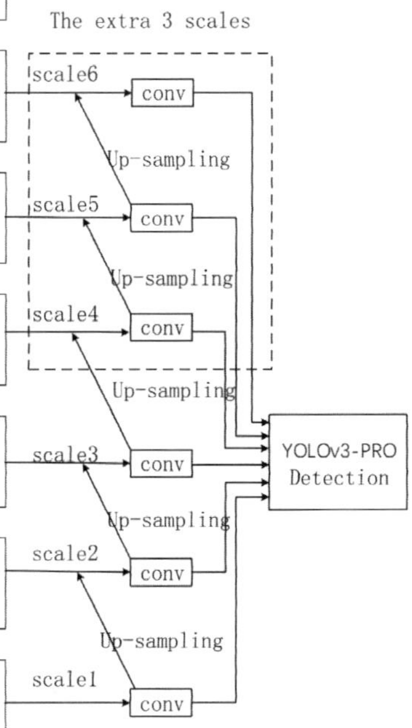

Figure 6. Improved YOLOv3 model structure (YOLOv3-pro)

3.2.2. Multi-scale spatial pyramid structure

For error detection, undetected presence, analyze the reasons for the lack of features for the target model of learning. Therefore, here we try to add 3 scales of target detection channels after the main network of YOLOv3 (as shown by the dotted line in Figure 6), together with the original 3 convolutional layers of YOLOv3 to form a convolution feature pyramid with 6 different scales. That is: 64*64, 32*32, 16*16, 8*8, 4*4, and 2*2 resolutions; at the same time, perform upsampling operations on the feature pyramid in 2 times steps with the previous residual network layer to form a deep feature fusion detection model[14-15],and the detection is performed independently on the final 6 independent fusion feature maps.

Based on the above improvements, the final YOLOv3-pro model feature extraction network structure shown in Figure 6.

4. Experiments and results

Follow the idea YOLOv3, the target detection task as an end to end regression model using stochastic gradient in training drop the loss function to achieve the optimal solution. From actual training dataset, generated a priori box, training model and the final results of the analysis elaborated on.

IOP Conf. Series: Materials Science and Engineering **787** (2020) 012034 doi:10.1088/1757-899X/787/1/012034

4.1. GTSDB dataset

The model in this paper uses German Traffic Sign Detection Benchmark (GTSDB) dataset for training and testing. The data set were 900 road traffic signs panoramas, contains a total of 1206 traffic signs, each picture contains 0, 1 or more traffic signs, of which the first 600 as a training set (including 846 traffic signs), 300 as a test set (including 360 traffic signs). The data set labels are divided into 4 categories: mandatory, prohibition, danger, and other. Figure 7 shows some.

Figure 7. Some examples in the GTSDB dataset

4.2. Generation of prior box

For the selected anchor frame, is generated by K-means clustering, When doing clustering,The width and height of all ground truth boxes in the dataset are used as feature inputs for K-means clustering,we use the aforementioned IOU to measure the distance between two bounding boxes, and when calculating the iou of two bounding boxes, we only need to use their 4 position parameters (xmin, ymin, width, height).

According to the magnitude of the GTSDB and the experimental results of K-means, it is finally determined that the position coordinates are predicted using convolution feature maps of 9 scales.The prior boxes are: 10*13, 16*30, 33*23, 30*61, 62*45, 59*119, 116*90, 156*198, 373*326.

4.3. Model training

YOLOv3-pro training decreased by stochastic gradient algorithms end. In order to make the training converge quickly, the pre-trained model Darknet-53 is used to initialize the shared convolutional layer in this model. In training, we need to match the bounding box anchor predict the real box in order to establish their correspondence. If the predicted bounding box a priori has the highest overlap with the real box, mark the traffic sign as a positive sample while ignoring the non-optimal bounding box; it is worth noting that for each real box, only one bounding box is assigned to Reduce duplication in your target audience. Training process parameters set as shown in Table 1.

Table 1. Initialization parameters of YOLOv3-pro network

Size of input	Batch	Momentum	Initial learning rate	Decay	Training steps
512*512	64	0.9	0.001	0.0005	20000

Figure 8 shows the corresponding Loss curve of the method in the training process. the number of training iterations, and the ordinate represents the Loss value during training. It can be seen that the model will achieve a convergence effect. That is, at the beginning of training, the method in this paper decreases rapidly; after 10,000 iterations of training, the Loss curve declines steadily; when iterating to 20000 times, Loss converges to 0.2 and ends the training.

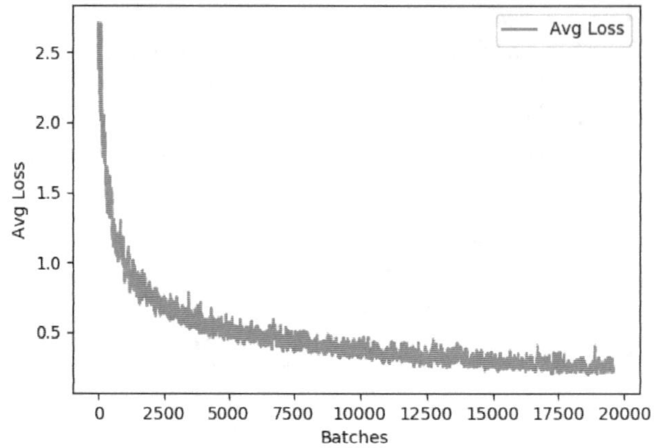

Figure 8. Loss curve during training

4.4. Results and analysis

4.4.1. Test on small targets

From the GTSDB dataset, a sample of small target traffic signs at 6 scales is selected, and its ground truth is around 10*20, 20*40, 30*60, 40*80, 50*100, and 60*120. Select 50 pictures each scale together to form a small target sample test set, and use Faster-RCNN, YOLOv3 and this method for comparison experiments.

Figure 9. YOLOv3-pro for small target detection results

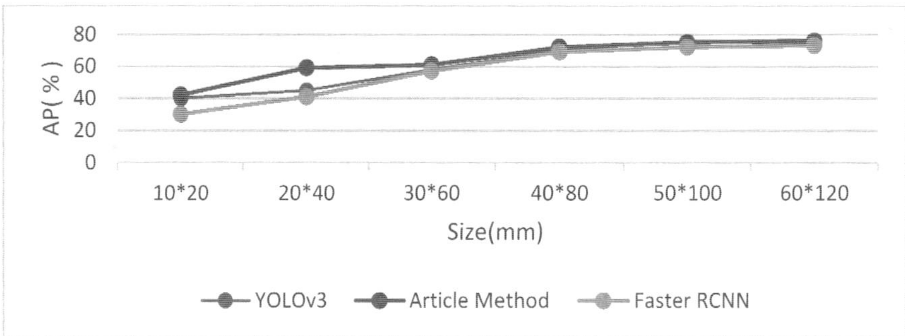

Figure 10. different models for small target detection performance

Figure 9 shows YOLOv3-pro detects an actual effect on small targets, from an image point of view, all of the small target are properly marked and classified, no missed or false detection cases, to achieve the desired. The results of different model algorithms for small target detection are shown in Figure 10. It is obvious that the average accuracy of this model is higher than the other two models in the small target data sample. As the target size increases, the AP value also increases. When the traffic sign box is about 60*120, the average accuracy of YOLOv3 is 77.3%, while the method in this paper can reach 81.6%, an increase of 4.3%, and performance has improved.

4.4.2. Performance comparison of different detection algorithms

Table.2 Performance comparison of different models

method of detection	mAP（%）	FPS（f/s）
YOLOv3	80.44	46.55
Faster R-CNN	78.42	2.51
Article method	85.26	38.52

Table 2 shows the results of the three models on a test dataset GTSDB after training. As can be seen from the table, the model proposed in this paper obtained 85.26% mAP, which is superior to Faster R-CNN in terms of accuracy. Compared with the original YOLOv3, the detection accuracy is improved by 4.82%. This indicates that cascade through multi-scale features can get more detailed target feature, the detection accuracy is significantly improved, to some extent, solve undetected, error detection of the problem. Of course, can also be found from the table, YOLOv3 detection rate improved compared to the decline before the improvement, mainly due to the deeper layers of the iterative model, bulky structure of the cascade model of training time has increased, In addition, Dialted convolution is more time-consuming and computationally intensive than 2D convolution. But from the perspective of detection speed, 38.52 f / s has reached the requirements of real-time detection.

5. Conclusion

This paper proposes an improved real-time traffic sign detection model based on YOLOv3. The extended convolution method is applied to the backbone network to effectively expand the receiving domain and improve the recognition effect of small target traffic signs on the road. Multi-level spatial pyramid The feature fusion backbone network can obtain more detailed semantic features during feature extraction, which effectively improves the problem of missed detection and false detection in the road traffic sign recognition scene. The experimental results show that combined with dilated convolution and multi-level spatial pyramid feature fusion, the performance of the final model target detection has been improved. mAP is 4.82% higher than YOLOv3 and 36.01 fps faster than Faster R-CNN.

References

[1] A. De La Escalera, L.E. Moreno, M.A. Salichs, J.M. Armingol, Road traffic sign detection and classification, IEEE Trans. Ind. Electron.44 (6) (1997) 848 – 859,doi:10.1109/41.649946.

[2] Á. Arcos-García, J.A. Álvarez-García, L.M. Soria-Morillo, Deep neural network for traffic sign recognition systems: an analysis of spatial transfor.

[3] Girshick R.Fast R-CNN[C]//ICCV，2015：1440-1448.

[4] Ren S Q，He K M，Girshick R，et al.Faster R-CNN：towards real-time object detection with region proposal networks[J].IEEE Transactions on Pattern Analysis and Machine Intelligence，2017，39（6）：1137-1149.

[5] Redmon J，Divvala S，Girshick R，et al.You only look once：unified，real-time object detection[C]//CVPR，2016.

[6] Redmon J，Farhadi A.YOLOv3：an incremental improvement[J].arXiv：1804.02767v1，2018.

[7] Gong An, Li Chengqian, Niu Bo. Real-time pedestrian detection method based on convolutional neural network [J]. Journal of Computer System Applications, 2017, 26 (9): 215-218.

[8] Guo Zhi, Song Ping, Zhang Yi, et al. Aircraft target detection method for remote sensing image based on deep convolutional neural network [J]. Journal of Electronics & Information Technology, 2018, 40 (11): 149-155.
[9] S. Houben, J. Stallkamp, J. Salmen, M. Schlipsing, C. Igel, Detection of traffic signs in real-world images: the German traffic sign detection benchmark, in:Proceedings of the 2013 International Joint Conference on Neural Networks,IJCNN, IEEE, 2013, pp. 1 – 8.
[10] F. Ahmed, D. Tarlow, D. Batra, Optimizing expected intersection-over-union with candidate-constrained CRFs, IEEE International Conference on Computer Vision (2016).
[11] Zhang Fukai, Yang Feng, Li Ce. Fast vehicle detection method based on improved YOLOv3 [J] .Computer Engineering and Applications, 2019, 55 (02): 18-26.
[12] Jianming Z , Manting H , Xiaokang J , et al. A Real-Time Chinese Traffic Sign Detection Algorithm Based on Modified YOLOv2[J]. Algorithms, 2017, 10(4):127-.
[13] Guo Jinxiang, Liu Libo, Xu Feng, Zheng Bin.Aeroplane scene detection method based on YOLO v3 [J / OL] .Laser and Optoelectronics Progress.
[14] Yi,Z.,Yongliang, S., & Jun, Z. (2019). An improved tiny-YOLOv3 pedestrian detection algorithm.Optik.doi:10.1016/j.ijleo.2019.02.038
[15] Yunong Tian,Guodong Yang,Zhe Wang,Hao Wang,En Li,Zize Liang. Apple detection during different growth stages in orchards using the improved YOLOv3 model[J]. Computers and Electronics in Agriculture,2019,157.

IOP Conf. Series: Materials Science and Engineering **787** (2020) 012035 doi:10.1088/1757-899X/787/1/012035

Bridge Deterioration Analysis Based on Censored Data: A Case Study in Yunnan Province

Xin He[1*], Wei Wang[1,2], Weidi Hu[1] and Hanyu Zhang[1]

[1] Software School, Yunnan University, Kunming, Yunnan Province, 650500, China

[2] Key Laboratory for Software Engineering of Yunnan Province, Kunming, Yunnan Province, 650500, China

*e-mail: 2205494777@qq.com

Abstract. Bridge deterioration reflects the change of bridge condition and it is caused by many factors. Predicting bridge deterioration as a research problem has attracted increasing attention around the world. Most methods contrived in previous work and the results are satisfactory. Despite their considerable success, these methods are weighed down by some problems. First, nature environment and climate factors are not considered. Second, they neglected the existence of censored data. Censored data means that the event outcome of interest can't be observed exactly because of some reasons. It is usual in bridge deterioration prediction and will cause deviation between the prediction result and the actual situation. In this paper, we propose a method to tackle the censored data issue and add nature environment and climate factors into deterioration analysis. This method uses an indicator variable to identify censored data in parameter estimation. In order to ensure the objectiveness of experiment, data collected from the Yunnan Communications Investment & Construction Group LTD are used in this study. Empirical results show that different structure and material of bridge may have different effect on bridge condition. In addition, we confirmed that nature environment and climate can affect the deterioration rate of bridge comparing with other factors in Yunnan Province. The bridges in the areas with moderate temperature and large precipitation may deteriorate more slowly. These findings are important for bridge management and maintenance.

1. Introduction

Bridge deterioration analysis has attracted considerable attention for past ten years. Success in such analysis leads to the development of bridge maintenance research[1, 2].

In recent years, various machine learning algorithms adopted in deterioration analysis, such as Artificial Neural Network model[3], Markov chain model[4], Linear Regression model[5] and fuzzy model[6], and produced many encouraging results. However, the advance of the deterioration analysis is slowed down by the two issues, without considering the nature environment factors and the nature of censoring of training data. These issues are described below.

(1) Natural environment and climate factors were not included in the study. The difficulty of bridge deterioration analysis is caused by two aspects: First, bridges are composed of several components, superstructure, substructure, decks, bearings, foundation, etc. Different components have different contribution on the bridge deterioration. Second, bridge deterioration is not only due to the age and structure, but also depends on complex environmental and climate factors. In order to get a more comprehensive analysis, all these factors should be considered in the study as far as possible.

(2) The nature of censoring of training data. Many models based on machine learning assume the data used to construct the deterioration analysis model is non truncated. However, the hypothesis is not always true. Censored data means that data was truncated because of some reasons. In bridge deterioration analysis, these reasons could be summarized as follows. First, some bridges were damaged by unnatural factors during the observation interval, such as artificial reconstruction, demolition, etc. Second, some bridges stop regular inspection during the observation interval. Third, the bridge did not deteriorate until the end of the observation interval. Fourth, the standard range of scores limits the assessment of bridges. The condition of bridge does not get a proper numerical value until the upper limit of the scoring criteria. The fourth case of censoring is usually neglected by previous studies. In statistics, censoring arises when an event of interest cannot be directly observed in the observation interval, which causes incomplete data[7]. Most common problems encountered in the models are caused by the unobserved events, such as a large deviation between the prediction result and the actual situation because of data censoring[8]. Censored data can't be removed as exception values. Because it contains some important information, removing it will result in loss of sample size and information. If the information of these objects is lost, one-sided or unstable analysis results will be produced[9]. Censored data also can't be regarded as complete data. Otherwise, the bridge condition will be underestimated. If the bridge with very good condition in the censored data is lost, the prediction result will produce deviation.

In this paper, we add the data of nature environment and climate factors into the training data. Besides, aiming at the problem of censored data, Cox regression model is used to predict major factors that lead to the deterioration of bridge. Using Cox regression model, we can handle this incomplete data as censored data in the analysis instead of removing it. The partial regression coefficients show the magnitude and direction of the influence of the variable on the dependent variable. Cox regression also can filter variables and select variables that have more significant influence on the dependent variable by using stepwise regression.

The rest of paper is organized as follows. First, the current research states and progress of bridge deterioration analysis are reviewed. Second, necessary statistical background is briefly presented for censored data and Cox regression model. Then, the proposed method is presented. Afterward, the case study is described, followed by a presentation of empirical results, including the experimental setup and results analysis. Finally, the paper concludes with summaries and discussions.

2. Literature review

Numerous methodologies have been applied to bridge deterioration analysis. All these methods can be divided into two types: deterministic and stochastic[10].

The Markov chain is one of the stochastic methods which was used extensively in forecasting the performances of bridge infrastructures. As early as 1992, Cesare proposed methods[11] for utilizing Markov chain in highway bridge deterioration research and bridge condition ratings were considered as Markovian states. Markov-chain model assumed that the future state is based on the present condition[12], each state can transfer to another at each time step according to the fixed transition probabilities. In bridge deterioration analysis, transition probability matrix was expressed as a matrix because that bridge condition is evaluated using several rating levels. The transition probability matrix is commonly calculated by statistical data of bridge conditions. Therefore, if the present bridge condition is known, the future condition at a certain time can be obtained. Up to recent years, Markov chain is still the mainstream model for bridge deterioration prediction. Li Li[13] used a group of bridge condition data about ten years in Shanghai as training data, and proposed a Markov-chain model to predict the deterioration tendency of urban bridges in Shanghai. Although Markov chain is widely used in bridge deterioration analysis, most studies predict the transition of bridge condition, focus only on a particular bridge component or directly consider the bridge as a whole[14] not consider the influence of nature environmental and climate factors to bridge, which may affect the accuracy of decision-making in bridge maintenance. In addition, previous work assumed that the data used to construct the deterioration model without censoring. This hypothesis will lead to deviation in the experimental results.

Regression model is a deterministic method can quantify the impact of different factors on the final results. It is a predictive modelling technology, which is primarily used to find the best-fitted mathematical model, so that a dependent variable (objective) can be predicted from independent variables (predictors). It also serves a major purpose that describing the cause and effect relationship between dependent variable and independent variable. Generally, the potential determinants of bridge deterioration are identified first in bridge deterioration analysis, which were inputted as independent variables. And the condition rating was used as the dependent variable. The values of the coefficients estimated by regression analysis for the independent variables represent the influence extent to bridge deterioration. In reference[15], Yanev used regression model to analysis bridge deterioration under the only factors bridge age. Chase considered more factors including age, average daily traffic, precipitation, frequency of deicing, temperature range, freeze thaw cycles and type of bridge construction in reference[16], and they applied three different regression methods to capture the relationship between bridge condition and impact factors. They neglected the nature environment and climate factors. In addition, both of them also assumed that there is no censoring in the data used to construct the regression model.

Cox regression is a multivariate statistical method and has an ability to handle censored data[17]. Yang Ying Nan and Kumaraswamy[18] used Cox regression analysis to predict the distribution of durations between changes in condition states. Censored data was mentioned and handled in their study, but they didn't consider the case of censored data caused by score range limitation. Reference[19] described the data limitation in their study. There are two major problems. First, because of the subjective nature of bridge inspections, the same bridge was inspected by different inspectors could result in the same bridge got different ratings. Second, some newly constructed bridges are not given an initial condition rating of the best rating. They solved these problems by training inspectors, review of inspection reports, and spot checks of individual bridges. Similarly, reference[20] also found that bridge inspection observations are innately imprecise or fuzzy because of the subjective nature of bridge inspections and linguistic descriptions. A matrix-driven multiple fuzzy linear regression model was proposed to deal with a mixture of fuzzy data and crisp data. The two studies thought that these data problems were due to misapplication of rating criteria by bridge inspectors. However, when inspectors could assess bridge condition correctly according to rating criteria, we can find that the rating also can't represent the actual bridge condition because of the score range limitation. This is one of the censoring cases, which is common in bridge deterioration analysis. However, the present studies neglected this case of censored data and the influence on bridge deterioration model.

Therefore, the aim of our study is to handle the censored data caused by score range limitation and an analysis method based on Cox regression was used in this paper.

3. Background
In this section, backgrounds are briefly presented for censored data and Cox regression model. More technical details are presented in [21] for more statistical treatments of this model.

3.1. Censored Data
Censoring represents an incomplete data structure, which means that the event outcome of interest can't be observed exactly because of some reasons during the observation interval[7]. Censoring falls into three main categories, right censoring, left censoring and interval censoring[22]. Interval censoring means that the event outcome of interest is known only to lie within an interval [L, R] instead of being observed exactly. Left censoring means that the event outcome of interest occurred before the initial point of observation interval, but exact time is unknown. Right censoring is the most common type of censoring data in practical research. In right censoring, the initial point of observation interval is known. Generally, according to the difference of end point, right censoring can be divided

into three types: Type-I censoring, Type-II censoring and Type- III censoring. Type-I censoring means that the initial point is same. Except for those whose event outcome have already occurred, the end point of the rest of research objects is unified to a fixed time, and the event outcome of these objects cannot be observed, because they occur after the end point of the observation. For example, in bridge deterioration analysis, the minimum limit of bridge scoring standard is 0. Due to the limitation of the scoring criteria range, researchers cannot observe the outcome events of all bridges. The outcome event is that bridge score gets a proper numerical value. Higher numerical values cannot be given after reaching the maximum limit 100 of the scoring standard. The true values of bridge scores whose outcome event have not occurred are unknown, but not less than 100. The data type studied in this paper is Type-I censoring. When the observation does not end until enough event outcomes occur, the survival time of the individual without event outcome is unknown, which is called Type-II censoring. The type that different individuals have different initial point and censoring point is called Type-III censoring.

3.2. Cox Regression Model

Cox regression model was first proposed by Cox[23]. It is a semi-parametric regression model. Compared to parametric method, Cox regression model has an advantage that we don't need to consider the distribution of data when we use it. Cox regression belongs to one of the analytic techniques used in survival analysis. At present, the application of survival analysis is no longer limited to the analysis of survival, but as an ideological method of dealing with censored data and has gradually formed a relatively complete new theoretical system[24]. The event outcome in survival analysis can be death or something we concern. Cox regression model has been extensively used in the medical field[25, 26], economics[27] and management field[28] since its introduction. It has shown the power of regression analysis for censored data[17] and achieved considerable success. Cox regression model takes event outcome and survival time (bridge score in this study) as dependent variables. An indicator variable was set to identify whether the event outcome occurs. It is easy to determine if the event outcome occurs, and it is customary to set the indicator variable as 1. If the event outcome does not occur, it will contain at least two situations. One is that the bridges have not had an outcome event until the end of the observation interval. The other one is that in the middle of the observation interval, the observation of some bridges were stopped for various reasons, resulting in the failure to observe the outcome. As long as the event outcome does not occur, the indicator variable of these bridge data was set as 0. For the incomplete data structure in bridge deterioration analysis, using cox regression model avoids the deviation caused by censored data and ensures the integrity of information.

4. Proposed Method

In this section, we introduce the process of bridge deterioration model construction based on cox regression model. It comprises three components: variable selection, model construction and parameter estimation. The whole process is shown in Fig. 1.

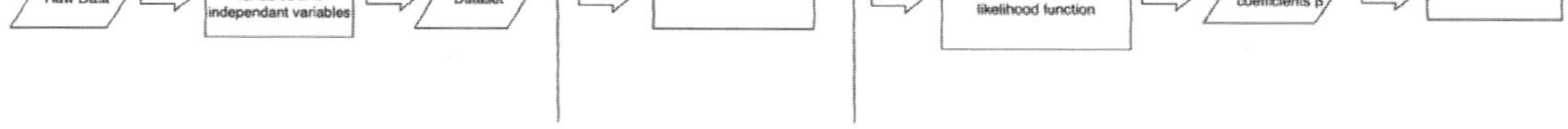

Fig. 1. the process of bridge deterioration model construction.

4.1. Variable Selection

Cox regression takes two variables as dependent variables. One is a classification variable. And another is a continuous variable. In this study, the event outcome and bridge score were taken as dependent variables.

An indicator variable c was set to represent the event outcome. If the event outcome occurs, then c=1, otherwise c=0. This indicator variable c also identifies whether the data is censored data. Then, the bridge condition score was defined as t. For a given individual i, we mark its score as t_i.

In order to analyse the influence of impact factors on bridge condition score, all the impact factors were defined as independent variables. Assuming that the number of impact factors for score is k, all factors can be defined as $x_1, x_2, \ldots, x_k$, respectively, and note as covariate vector form is X = $(x_1\ x_2 \ldots x_k)$.

4.2. Model Construction

Cox regression is different from conventional regression analysis. It does not directly use bridge condition score as the dependent variable of regression equation. The influence of covariate variables on bridge condition score is reflected by the ratio of hazard function and baseline hazard function, in which the hazard function and baseline hazard function are unknown. The hazard function was defined as h(t), it measures the probability that the bridge gets a proper score. We can describe it as Eq.(1).

$$h(t) = \lim_{\Delta t} \frac{P(t \le t_i < t + \Delta t | t_i \ge t)}{\Delta t} \tag{1}$$

At the same time, the baseline hazard function can be defined as $h_0(t)$. It represents the form of a bridge's hazard function without effect of covariate variables. $h_0(t)$ can be any function related to t. There is no any assumption for it in Cox regression model.

Supposing that the hazard function of bridges with various factors X = $(x_1\ x_2 \ldots x_k)$ represented by $h_i(t, X)$. According to the proportional hazards and logarithmic linear hypothesis, the Cox regression model proposed for bridge deterioration analysis is defined as Eq.(2).

$$h_i(t, X) = h_0(t) exp\,(\beta_1 x_{i1} + \beta_2 x_{i2} + \cdots + \beta_k x_{ik}) \tag{2}$$

where $h_0(t)$ is the equal of $h_i(t, X)$ when X = (0 0 ...... 0); exp $(\beta_k x_{ik})$ denotes the effect of the X, the effect is the mortality rate from $h_0(t)$ to $h_0(t)$exp $(\beta_k x_{ik})$, exp $(\beta_k x_{ik})$ can be regarded as a proportional constant. So Eq.(2) was called proportional hazards model. β_k are parameters (estimate value based on sample) describing the impact of various factors on hazards. When β_k=0, factor x_k has no effect on hazards; when β_k>0, factor x_k increases hazards; when β_k<0, factor x_k reduces hazards and becomes protective factor .Eq.(2) also can be converted into Eq.(3).

$$ln\left[\frac{h_i(t,X)}{h_0(t)}\right] = \beta_1 x_{i1} + \beta_2 x_{i2} + \cdots + \beta_k x_{ik} \tag{3}$$

The left part of the equation is the natural logarithm of the relative risk degree, and the right part of the equation is the linear function of the corresponding regression coefficients with covariate variables.

4.3. Parameter Estimation

Cox regression model does not make any assumptions about $h_0(t)$. For two individuals, the hazard ratio of them is irrelevant to $h_0(t)$. We can find that $h_0(t)$ will cancel out in this expression from Eq.(4).

$$\frac{h_i(t,X)}{h_{i'}(t,X)} = \frac{h_0(t)\, exp(\beta_1 x_{i1} + \beta_2 x_{i2} + \cdots + \beta_k x_{ik})}{h_0(t)\, exp(\beta_1 x_{i'1} + \beta_2 x_{i'2} + \cdots + \beta_k x_{i'k})} = \frac{exp(\beta_1 x_{i1} + \beta_2 x_{i2} + \cdots + \beta_k x_{ik})}{exp(\beta_1 x_{i'1} + \beta_2 x_{i'2} + \cdots + \beta_k x_{i'k})} \tag{4}$$

Even in the case that $h_0(t)$ is unknown, we can still estimate parameters. The radio dependents only on regression coefficient β. This is the reason that Cox regression model is a semiparametric model. One of the advantages of Cox model is that it can obtain reliable estimation of regression coefficients and related statistical inferences without any assumptions about the functional form of the baseline hazard function h_0. We defined partial likelihood rather than traditional likelihood function when learning the model. A censoring indicator δ_j was used to identify the censored data, which equals 0 if the j-th bridge data is censored, otherwise is 1. This is the method of dealing with censored data. In addition, we set R(t) to represent the hazard set of score t. Supposing that there are m events. Then we can define the corresponding conditional partial likelihood function as Eq.(5).

$$L_p(\beta) = \prod_{i=1}^{m}\left(\frac{e^{\beta x_j}}{\sum_{j\in R(t_i)} e^{\beta x_j}}\right)^{\delta_j} \quad (5)$$

We also can write the log-likelihood function as follows:

$$log\, L_p(\beta) = \sum_{i=1}^{m} \delta_j[\beta x_j - log(\sum_{j\in R(t_i)} e^{\beta x_j})] \quad (6)$$

The numerical methods such as iteration algorithm is usually used to derive the Eq.(6)[29], and obtain the maximum likelihood estimation value of the parameter β and its standard error, then the likelihood ratio test, score test or Wald test are performed which determine whether each covariate has statistical significance. We can judge whether these factors are dangerous factors or protective factors for bridge deterioration according to the value of β. The statistical significance can help us to find the impact factors that have more significant influence on bridge deterioration from all factors.

5. Case Study

The implement of bridge deterioration analysis in Yunnan Province using Cox regression is described in this section. The data we use in experiment are processed and coded. The sample size is 2793, of which 746 data are right-censored. And the percentage of censored data is 27%. The experiment is implemented in R language.

5.1. Data Preparation

Multi-factors analysis on bridge deterioration need to consider more factors as much as possible. In this paper, we considered that the factors causing bridge deterioration are not only the age and structure of bridge, but also the environmental factors and climate. Yunnan Province is located in the southwestern frontiers of China. The climate of Yunnan Province is diverse, it crosses different climatic zones including temperate monsoon climate, subtropical monsoon climate and tropical rainforest climate. Most districts of Yunnan belong to the subtropical monsoon climate. The climate has remarkable features that small annual temperature difference, large daily temperature difference and distinct dry and wet seasons. The bridge data of Yunnan provide more comprehensive information for bridge deterioration analysis. Furthermore, we find that the bridge data collected from Yunnan have significant censoring, which is accord with our research purpose. Next, data preparation before experiment will be described.

5.1.1 Data Collection. Real world bridge condition data we use in this study were obtained through contacting directly the Yunnan Communications Investment & Construction Group LTD, China. The data include bridges on 14 highways and cover 16 districts. These districts are distributed in three climate zones, temperate monsoon climate, subtropical monsoon climate and tropical rainforest climate. In this study, bridges in subtropical monsoon climate were selected as representatives for Yunnan Province with 2793 data. The basic data is a sample of all relevant information for each bridge, each sample includes 12 attributes, there are underpass, deck pavement, deck width, bridge type, etc. In addition, considering the effect of temperature and precipitation, we add 6 attributes. There are annual average temperature, average minimum temperature, average maximum temperature, annual average precipitation, rainy days and snow days. Furthermore, an indicator variable c was set to define the status of censoring. If the data is censoring, then c=0, otherwise c=1.

5.1.2 Data Processing. First of all, there are some missing data in raw data. If we simply discard missing data, valuable information may be lost, and inferential power compromised[30]. In doing so, random forest (RF)[31] was applied to interpolate missing data. Then, we merge similar attribute values and remove unreasonable attribute values after discussion with experts in bridge engineering. In order to facilitate computer processing, we use real-number encoding. For example, in the attribute values of "Bridge Type", "super bridge" is coded as 1, "large bridge" is coded as 2, "middle bridge" is coded as 3, "small bridge" is coded 4.

5.1.3 Data normalization. Because different evaluation indicators have different dimension and dimension units, which will affect the result of data analysis. In order to eliminate the dimension effect between indicators, it is necessary to normalize the data to solve the problem of comparability between data indicators. So, we normalize the attributes whose attribute values are numerical values. The normalization method we used is linear normalization. Conversion function is shown in Eq.(7).

$$x' = \frac{x - min\,(x)}{max(x) - min\,(x)} \tag{7}$$

After the processing, the overview of data pattern is shown in Table 1.

Table 1. The overview of data pattern.

Underpass	Deck Pavement	Deck Width	Bridge Type	…	Annual Average Temperature	Snow Days	Score	c
1	4	3	3	…	0.2105263157895	0.0612244897959	100	0
1	4	3	3	…	0.2105263157895	0.0612244897959	100	0
1	5	1	4	…	0.2105263157895	0.0612244897959	100	0
1	5	1	4	…	0.2105263157895	0.0612244897959	100	0
1	5	1	4	…	0.2105263157895	0.0612244897959	100	0
2	5	1	3	…	0.2105263157895	0.0612244897959	100	0
…	…	…	…	…	…	…	…	
3	3	1	4	…	0.1973684210526	0.1020408163265	63.9	1
3	3	3	4	…	0.1973684210526	0.1020408163265	60	1

5.2. Experimental Setup

5.2.1 Single Factors Analysis. In order to understand the determinant effect of each covariate variable on the model and which variable has impact on the bridge condition score, we do single factor analysis of 17 covariate variables by using Cox single factor analysis first. And we remove two covariate variables that fail to pass the significance test after the single factor analysis. The analysis results of covariate variable Deck Width are shown in Table 2 and Table 3.

Table 2. Parameter estimation of Deck Width.

Variable name	Coefficient(β)	Standard Error(se)	z	Pr(>|z|)
Deck Width-wide	-0.05168	0.11947	-0.433	0.665
Deck Width-narrow	-0.12599	0.09297	-1.355	0.175

Table 3. Hypothesis Test of Deck Width.

Name	Value
Likelihood ratio test	p= 0.4
Wald test	p= 0.4
Score (logrank) test	p= 0.4

We do hypothesis test on the model and suppose the model is invalid, $H_0: \beta = 0$. According to the hypothesis test result of the model, with the significance level $\alpha = 0.05$, null hypothesis is invalid, i.e., the model only considering the impact factor Deck Width is not significant and doesn't have statistical significance(when $P<0.05$, a model can be regarded as having statistical significance). Besides, according to the result of parameter estimation, variable Deck Width-wide does not pass the test of significance (P=0.665>0.05), so does variable Deck Width-narrow (P=0.175>0.05), which means that this factor has little effect on bridge score. Therefore, the covariable variable Deck Width should be removed from the data set.

The analysis results of covariate variable Rainy Days are shown in Table 4 and Table 5.

Table 4. Parameter estimation of Rainy Days.

Variable name	Coefficient(β)	Standard Error(se)	z	Pr(>|z|)
Rainy Days	-0.10458	0.06544	-1.598	0.11

Table 5. Hypothesis Test of Rainy Days.

Name	Value
Likelihood ratio test	p= 3e-12
Wald test	p= 9e-14
Score (logrank) test	p= 5e-14

Suppose the model only considering the impact factor Rainy Days is invalid, $H_0: \beta = 0$. According to the hypothesis test result of the model, with the significance level α =0.05, null hypothesis is valid, i.e., the model is significant and has statistical significance (P<0.05). But we find that the covariable variable Rainy Days does not pass the test of significance (P=0.11>0.05) in parameter estimation. Therefore, the covariable variable Rainy Days should be removed from the data set.

In single factor analysis, 15 variables have significant influence on bridge score include Underpass, Deck Pavement, Bridge Type, Superstructure Form, Superstructure Material, Substructure Form, Substructure Material, Fundamental Form of Substructure, Support Form, Age, Annual Average Temperature, Annual Average Precipitation, Average Minimum Temperature, Average Maximum Temperature and Snow Days. All of these factors will be considered in multivariate Cox regression analysis.

5.2.2 Multi-factors Analysis. Cox single factor analysis has a problem that it ignores the possible interaction between variables. Synergistic effect could exist between some variables. When other factors are considered in the model, the influence of the factor on the score will be weakened or even become insignificant. So, multivariate Cox regression analysis is used in this paper. 15 variables are considered in Cox regression model.

The comprehensive evaluation index of the model and the result of goodness of fit test are shown in Table 6. The comprehensive evaluation index Concordance is 0.713. It shows that it has satisfactory indexes of discrimination. And the value of R-square that represents the goodness of fit is 0.315, which means that the expression equation has high reference value.

Table 6. The value of comprehensive evaluation index and goodness of fit.

Name	Value
Concordance	0.713
R-square	0.315

Next, the hypothesis test of model is presented. Suppose the model is invalid, $H_0: \beta = 0$. The result of hypothesis test is shown in Table 7.

Table 7. The result of hypothesis test.

Name	Value
Likelihood ratio test	p=<2e-16
Wald test	p=<2e-16
Score (logrank) test	p=<2e-16

According to the hypothesis test result of the model, with the significance level α =0.05, null hypothesis is invalid, i.e., the model is significant and has statistical significance (P<0.05). Then, the parameter estimation is presented in Table 8.

Table 8. Parameter Estimation of Cox Model.

Parameter	Variable name	Coefficient(β)	Exp (coefficient)	Standard Error(se)	z	Pr(>|z|)	Signif. codes
X_1	Underpass-river	4.002e-01	1.492e+00	5.545e-02	7.218	5.29e-13	***
X_2	Underpass-valley or gully	2.681e-01	1.307e+00	8.788e-02	3.050	0.002286	**
X_3	Deck Pavement-asphalt concrete	-1.323e+00	2.664e-01	3.298e-01	-4.011	6.05e-05	***
X_4	Deck Pavement-cement concrete	-1.528e+00	2.170e-01	3.046e-01	-5.016	5.28e-07	***
X_5	Deck Pavement-cement asphalt concrete	-1.169e+00	3.107e-01	3.031e-01	-3.857	0.000115	***
X_6	Bridge Type-middle bridge	-8.775e-01	4.158e-01	4.369e-01	-2.009	0.044590	*
X_7	Bridge Type-small bridge	-1.009e+00	3.644e-01	4.458e-01	-2.264	0.023570	*
X_8	Superstructure Form-I type composite beam	1.319e+00	3.739e+00	3.710e-01	3.554	0.000379	***
X_9	Superstructure Form-hollow slab	4.030e-01	1.496e+00	1.876e-01	2.149	0.031658	*
X_{10}	Superstructure Form-box type	5.674e-01	1.764e+00	2.636e-01	2.153	0.031334	*
X_{11}	Superstructure Form-shape beam	1.299e+00	3.667e+00	4.252e-01	3.056	0.002245	**
X_{12}	Superstructure Form- prestressed hollow slab	6.893e-01	1.992e+00	1.616e-01	4.265	2.00e-05	***
X_{13}	Superstructure Material- masonry coarse stone	3.803e+00	4.486e+01	1.391e+00	2.734	0.006263	**
X_{14}	Superstructure Material-prestressed reinforced concrete	-4.511e-01	6.369e-01	1.667e-01	-2.706	0.006804	**
X_{15}	Substructure Form-thin-walled abutment	-6.794e-01	5.069e-01	3.196e-01	-2.126	0.033530	*
X_{16}	Substructure Form-frame abutment (multi-column)	-1.368e+00	2.547e-01	6.098e-01	-2.243	0.024901	*
X_{17}	Substructure Form-ribbed column pier	-1.539e+00	2.145e-01	2.512e-01	-6.128	8.88e-10	***
X_{18}	Substructure Form-buried abutment column pier	-7.249e-01	4.844e-01	1.938e-01	-3.739	0.000185	***
X_{19}	Substructure Form-light abutment	-8.680e-01	4.198e-01	1.544e-01	-5.621	1.90e-08	***
X_{20}	Substructure Form-gravity	-4.276e-01	6.521e-01	1.419e-01	-3.012	0.002592	**

	abutment						
X_{21}	Substructure Form-gravity abutment column pier	-3.513e-01	7.038e-01	1.464e-01	-2.400	0.016404	*
X_{22}	Substructure Form-pile column (T-shaped pier)	-1.017e+00	3.616e-01	2.587e-01	-3.932	8.43e-05	***
X_{23}	Substructure Form-pile column (column pier)	-8.979e-01	4.074e-01	1.548e-01	-5.801	6.57e-09	***
X_{24}	Substructure Material-reinforced mortar concrete	-5.567e-01	5.731e-01	1.296e-01	-4.296	1.74e-05	***
X_{25}	Fundamental Form of Substructure-expand foundation	-1.784e+00	1.680e-01	7.184e-01	-2.483	0.013034	*
X_{26}	Fundamental Form of Substructure-expend foundation & digging pile foundation	-1.576e+00	2.067e-01	7.364e-01	-2.141	0.032296	*
X_{27}	Fundamental Form of Substructure-embedded foundation	-1.519e+00	2.189e-01	7.701e-01	-1.973	0.048523	*
X_{28}	Fundamental Form of Substructure-gravel stone cushion	-2.161e+00	1.152e-01	8.207e-01	-2.633	0.008461	**
X_{29}	Support Form-Φ250×300×57	1.260e+00	3.526e+00	4.552e-01	2.768	0.005639	**
X_{30}	Support Form-rectangular plate rubber bearing	-5.745e-01	5.630e-01	2.590e-01	-2.218	0.026543	*
X_{31}	Support Form-rubber bearing (plate, basin)	-6.035e-01	5.469e-01	2.945e-01	-2.049	0.040443	*
X_{32}	Support Form-round plate rubber bearing	-6.026e-01	5.474e-01	1.650e-01	-3.653	0.000259	***
X_{33}	Annual Average Temperature	5.538e+00	2.542e+02	1.642e+00	3.374	0.000742	***
X_{34}	Average Minimum Temperature	-1.538e+00	2.148e-01	3.245e-01	-4.740	2.14e-06	***
X_{35}	Annual Average Precipitation	-4.533e+00	1.075e-02	8.522e-01	-5.319	1.04e-07	***
X_{36}	Snow Days	1.979e+00	7.237e+00	6.082e-01	3.254	0.001138	**
X_{37}	Age	2.407e-01	1.272e+00	1.070e-01	2.250	0.024449	*

On the premise that the Cox regression model is valid, we can find that there are 37 variables have statistical significance (P<0.05) from the output of parameter estimation. On observing Table 8, it is clear that all the variables have stars in the last column. The star is significance code, where "***", "**", "*" represents P<0.001, P<0.01, P<0.05, respectively. The significance code indicates that a variable with more stars have more significant influence on the bridge score in the model.

In summary, empirical results show that the Cox regression model can generate in term of Table 8, and therefore, the model can be defined as Eq.(8).

$$h(t,X) = h_0(t)\,exp(4.002e-01X_1 + 2.681e-01X_2 + \cdots + 2.407e-01X_{37}) \quad (8)$$

We can estimate whether the variable is a hazard factor or a protective factor according to Eq.(8). And the magnitude of the influence of the variable on the bridge score also can be presented. Because the influence of covariate variables on bridge score is reflected by the ratio of hazard function and baseline hazard function, the exp(coefficient) in Table 8 represents the change in the hazard ratio caused by a unit change of a variable. The higher the hazard, the higher the probability of low bridge score, and vice versa.

5.3. Result Analysis

On observing the empirical results, we can get some useful information for bridge management department. According to the influence on bridge score, these 17 factors selected by experts in bridge engineering are divided into three types, hazard factors, protective factors and factors without significant influence.

As we all know, the bridge condition will deteriorate rapidly with the passing of time. In other words, the longer the bridge is used, the lower the bridge score will be. This is also reflected in the experimental results. In addition to the age of bridge, other factors also have impact on bridge deterioration. First, we introduce the influence of bridge structure on bridge deterioration. In terms of Bridge Type, the deterioration rate of middle bridge is slower than large bridge and small bridge. Deck Width is a factor that doesn't have significant influence on bridge score. Then, we find that Support Form is particularly special compared with other factors. When the attribute value of Support Form is Φ250×300×57, Support Form is a hazard factor which may cause faster deterioration of bridge. But when the Support Form is rubber bearing, it plays a protective role on bridge condition. In particular, using round plate rubber bearing as Support Form can significantly reduce the hazard of bridge deterioration. And the Superstructure Material is as special as Support Form. When the Superstructure Material is prestressed reinforced concrete, it is a protective factor. But when Superstructure Material is masonry coarse stone, it is a hazard factor. If Superstructure Form is shape beam or hollow slab, it will increase the hazard of deterioration. And using I type composite beam and prestressed hollow slab has a greater impact on deterioration hazard. In addition, we find some protective factors for bridge. When cement concrete and asphalt concrete are used on Deck Pavement, it has significant effect on reducing the hazard of deterioration. In order to reduce the hazard of bridge deterioration, these measures also can be adopted. Select reinforced mortar concrete as the Substructure Material. The Substructure Form can be ribbed column pier, buried abutment column pier, light abutment, pile column (T-shaped pier), pile column (column pier) or gravity abutment. And the Fundamental Form of Substructure can be gravel stone cushion, expand foundation, digging pile foundation and embedded foundation.

Beyond that, experiment proves that geographic location and climate also have impact on bridge deterioration. If the bridge is above river, valley or gully, it will increase the hazard of deterioration. Compare to valley or gully, the bridge over the river has higher hazard. In the aspect of climate, high temperature and snowfall will accelerate bridge deterioration. But it doesn't mean that low temperature is good for bridge, the annual minimum temperature also cannot be too low. Furthermore, bridges in the places with large Annual Average precipitation will be in better condition. All in all, in the places where the temperature is moderate and the precipitation is large, the deterioration of bridge will be slower.

6. Discussion and Conclusion

In view of the two problems of the existing methods, we propose a method based on Cox regression model, which can not only comprehensively analyse the influencing factors, but also correctly handle the censored data. Using real world bridge data, the forecasting model is implemented and validated through investigations into the performance of Cox regression model. Empirical results show that the influence degree of factors on score. In addition, bridge management department can forecast the future bridge technical condition based on the assessment of these factors through the forecasting model and make a long-term maintenance planning.

Some remarks are provided as follows. First, other methods can be tried to fill the missing values in raw data to improve the accuracy of parameter estimation. Second, more feature selection methods can be explored. And more factors that may have influence on bridge condition can be added into the study, such as traffic volume. Moreover, in addition to Cox regression model, other approach might be investigated for handling censored data in real world bridge data. This is important for bridge deterioration analysis to get accurate prediction results.

Acknowledgments

The authors would like to thank the Yunnan Communications Investment & Construction Group LTD for providing the data used in this study. This work was supported by the National Natural Science Foundation of China under Grant No.61462092, and the Open Scientific Research Project of Yunnan Transportation Research Institute under Grant No.2013(B)01.

References

[1] Frangopol, D.M., Lin, K.Y., Estes, A.C.: (1997) Life-cycle cost design of deteriorating structures. Journal of structural engineering 123(10), 1390–1401.
[2] TSUDA, Y., KAITO, K., AOKI, K., KOBAYASHI, K.: (2006) Estimating markovian transition probabilities for bridge deterioration forecasting. Structural Engineering/Earthquake Engineering 23(2), 241s–256s.
[3] Huang, Y.H.: (2010) Artificial neural network model of bridge deterioration. Journal of Performance of Constructed Facilities 24(6), 597–602.
[4] Robelin, C.A., Madanat, S.M.: (2007) History-dependent bridge deck maintenance and replacement optimization with markov decision processes. Journal of Infrastructure Systems 13(3), 195–201.
[5] Hyman, W., Hughes, D., Dobson, T.: (1983) The least cost mix of bridge replacement and repair work on wisconsin's state highways over time—a computer simulation. Technical Rep., WisDOT, Madison, Wis.
[6] Zhao, Z., Chen, C.: (2001) Concrete bridge deterioration diagnosis using fuzzy inference system. Advances in Engineering Software 32(4), 317–325.
[7] Rong, K.: (2008) Review of domestic research on censored data. In: Statistics & Information Forum. vol. 10.
[8] Nakat, Z.S., Madanat, S.M.: (2008) Stochastic duration modeling of pavement overlay crack initiation. Journal of Infrastructure Systems 14(3), 185–192.
[9] Miao, W., Yang: (2018) Cox regression analysis based on censored data. Statistics and Management 255(10), 26–29.
[10] Li, L., Li, F., Chen, Z., Sun, L.: (2016) Use of markov chain model based on actual repair status to predict bridge deterioration in shanghai, china. Transportation Research Record 2550(1), 106–114.
[11] Cesare, M.A., Santamarina, C., Turkstra, C., Vanmarcke, E.H.: (1992) Modeling bridge deterioration with markov chains. Journal of Transportation Engineering 118(6), 820–833.
[12] Ross, S.M., Kelly, J.J., Sullivan, R.J., Perry, W.J., Mercer, D., Davis, R.M., Washburn, T.D., Sager, E.V., Boyce, J.B., Bristow, V.L.: (1996) Stochastic processes, vol. 2. Wiley New York.

IOP Conf. Series: Materials Science and Engineering **787** (2020) 012035 doi:10.1088/1757-899X/787/1/012035

[13] Li, L., Sun, L., Ning, G.: (2014) Deterioration prediction of urban bridges on network level using markov-chain model. Mathematical Problems in Engineering 2014.

[14] Morcous, G.: (2006) Performance prediction of bridge deck systems using markov chains. Journal of performance of Constructed Facilities 20(2), 146–155.

[15] Yanev, B., Chen, X.: (1993) Life-cycle performance of new york city bridges. Transportation Research Record 1389, 17.

[16] Chase, S.B., Small, E.P., Nutakor, C.: (1999) An in-depth analysis of the national bridge inventory database utilizing data mining, gis and advanced statistical methods. Transportation Research Circular 498, 1–17.

[17] Therneau, T.M.: (1997) Extending the cox model. In: Proceedings of the First Seattle symposium in biostatistics. pp. 51–84. Springer.

[18] Yang, Y.N., Kumaraswamy, M.M., Pam, H.J., Xie, H.M.: (2013) Integrating semiparametric and parametric models in survival analysis of bridge element deterioration. Journal of Infrastructure Systems 19(2), 176–185.

[19] Veshosky, D., Beidleman, C.R., Buetow, G.W., Demir, M.: (1994) Comparative analysis of bridge superstructure deterioration. Journal of Structural Engineering 120(7), 2123–2136.

[20] Pan, N.F., Lin, T.C., Pan, N.H.: (2009) Estimating bridge performance based on a matrix-driven fuzzy linear regression model. Automation in construction 18(5), 578–586.

[21] Walters, S.J.: (1999) What is a Cox model? Citeseer.

[22] Cox, D.R.: (2018) Analysis of survival data. Chapman and Hall/CRC.

[23] David, C.R., et al.: (1972) Regression models and life tables (with discussion). Journal of the Royal Statistical Society 34(2), 187–220.

[24] Luo, S.: (2013) Method and application of survival analysis. Zhejiang Preventive Medicine 25(5), 29–31.

[25] D'Agostino, R.B., Lee, M.L., Belanger, A.J., Cupples, L.A., Anderson, K., Kannel, W.B.: (1990) Relation of pooled logistic regression to time dependent cox regression analysis: the framingham heart study. Statistics in medicine 9(12), 1501–1515.

[26] Sauerbrei, W., Schumacher, M.: (1992) A bootstrap resampling procedure for model building: application to the cox regression model. Statistics in medicine 11(16), 20932109.

[27] He, X., Chen, J.: (2014) An application of survival analysis in forecasting changes of stock index futures. Quantitative Economic, Technological and Economic Research pp. 116–126.

[28] Ting, L.: (2017) Research on Influencing Factors of Members' Life Span in Virtual Community. Ph.D. thesis, Harbin University of Technology.

[29] Kalbfleisch, J.D., Prentice, R.L.: (2011) The statistical analysis of failure time data, vol. 360. John Wiley & Sons.

[30] Enders, C.K.: (2010) Applied missing data analysis. Guilford press.

[31] Breiman, L.: (2001) Random forests. Machine learning 45(1), 5–32.

IOP Conf. Series: Materials Science and Engineering **787** (2020) 012036 doi:10.1088/1757-899X/787/1/012036

Mathematical model and simulated power performance of a novel O-type engine

Guanyu Chen [1] and Zhengqing Zhang [1]

[1]SWJTU-Leeds Joint School, Southwest Jiaotong University, Chengdu, Sichuan Province, 611756, China

Abstract. Reciprocating engines such as V-type, L-type and W-type engines are used widely in our daily life. However, there is still some room for the improvement of power performance if the configuration of the cylinder and connecting rod system is changed. In this paper, a novel O-type engine is put forward firstly. The structural features of the O-type cylinder sets are described briefly, and then the relevant mathematical model with hypotheses is established to conduct the performance analysis. Secondly, the O-type engine is compared with the general range of power performance of gasoline engine to demonstrate its merits. Then, the power performance indicators, namely the mean indicated pressure (PMI), space occupation efficiency (SOE) and heat efficiency (HE) of the novel engine are analysed and discussed in detail through MATLAB with the verification and validation which are conducted to ensure the reliability of results. And the results show that the power performance of the O-type engine is better than most of gasoline engines; the power performance indicators in the research are not monotonous at different volumes with the bore stroke ratio increasing; the proper range of the volume and the bore stroke ratio is obtained to make the engine achieve a high power performance. All in all, the research is helpful for the development and application of the O-type engine.

1. Introduction

Today, reciprocating engines, including Horizontal Engines, V-type Engines and W-type Engines and so on, have been widely used in the automobile industry [1-2]. They form the internal heat flow circuit, involving air intake, compression step, combustion, stroke and air exhausting, And the force generated during the stroke is finally applied to the components that require the power, transforming chemical energy into mechanical energy.

Daoud and Friedrich focused their attention on a novel compounding mechanism in the multi-cylinder Franchot engine and removed the rotational components, which made the engine more compact, efficient and economical [3]. And a linear controller system which proved that the performance of the engine with a linear system was better than non-linear classic controller system was designed by Hector, Carlos and Muditha [4]. Furthermore, Mahmood and Mohamad researched the beta-type Stirling engine and found that the length of regenerator would affect the energy efficiency differently, since various working gases, helix and hydrogen and the size of regenerator were capable of influencing the thermal and frictional losses [5].

By means of automotive engine bench test, Xu and Hao established a model based on OBD system constructed in BP neural network, which had a relatively good accuracy to estimate the engine torque [6]. Through the transient Strensith Analysis of Engine Block, Hu, Deng et al. provided a reliable basis for the design of dynamic structural strength of the block [7]. For the issues of Global environment and energy, Hong and Ouyang briefly introduced a mean value model of a spark-ignition gasoline engine and proposed an on-line torque estimation algorithm [8]. Based on BASIC language, Wang used the formula, which was used for calculating the power correction system of automobile engines and for calculating saturated vapour pressure, to write a program for calculation of automobile engine power calibration coefficient [9].

Additionally, Ebrahimi found that the dual-Millar cycle had the most work output among other thermodynamic cycles with the same environmental conditions [10]. Notay et al. studied the effects of the degradation of lubricant and discovered that it would increase the thickness of piston ring pack [11]. Tormos et al. built a model to gain the mechanical losses and the consumption of auxiliary energy, and discussed the promising application of the model, which would do great favour to reduce the emission of greenhouse gases [12]. Östman and Toivonen reduced torsional vibration dramatically by using a cylinder balancing method [13]. The investigations of the systems, vibration and many other features in the internal combustion engine have been shown to be prosperous.

Most of the prior studies focus on the improvement of some adverse phenomenon or other aspects such as vibration and ignition system, but the change of cylinder shape and connecting rod system is not focused on, such as O-type engine. In this paper, a brief introduction to the structural features of the O-type engine is displayed in section 2. Some hypotheses are introduced to control variables and simplify the calculations, and all the crucial performances are defined and displayed with the equations in section 3. In section 4.1, the necessary relevant parameters are pre-set before a comparison of PMI and HE between the O-type engine at a certain volume and general gasoline engine is carried out to show the superiority of the novel engine. Then in section 4.2, the results of simulated power performance indicators including PMI, HE and the SOE and the key structure indicators are displayed. The zero-dimension model is used to simulate the combustion process, as it is relatively simple and convenient, compared with some of the other models [14]. In section 5, the verification and validation are conducted before the results are discussed to define the high-performance range. In section 6, some conclusions are drawn that the power performance of the O-type engine could transcend or be comparable to that of general gasoline engine; the power performance indicators in the research rise first and then drop at different volumes with the bore stroke ratio increasing; when the working range k is set to 60°, the proper range of the volume and the bore stroke ratio is acquired to make the engine achieve a high power performance.

2. Brief construction of a novel O-type engine

The operating principle of O-type engine is similar to that of other reciprocating engines, which use combustion to generate energy for pistons to perform reciprocating motion. The main differences are the arrangement of the cylinder and guide bar mechanism shown in Figure 1 and Figure 2.

2.1. Configuration

As shown in Figure 3 and Figure 4 below, the ring rod connects two piston rods in the same plane. Moreover, one end of the guide bar is fixed by bolts and nuts on the ring rod to avoid the relative motion between them. Moreover, the other end is connected to the auxiliary axis with a bearing. On the guide bar, there is a groove whose side wall is tangent to a cylindrical pin on the crank arm. And the groove length is dependent on the condition that the cylindrical pin could travel without any obstacle. Additionally, a bearing is set around the cylindrical pin to lubricate and alleviate the friction. Except for the specific details mentioned above, the other details are similar to other ordinary internal combustion reciprocating engines.

In Figure 3, 1-cylinder, 2-piston, 3-piston rod, 4-ring rod, 5-guide bar, 6-crank arm, 7-crankshaft, 8-auxiliary axis, 9-groove, 10-cylindrical pin.

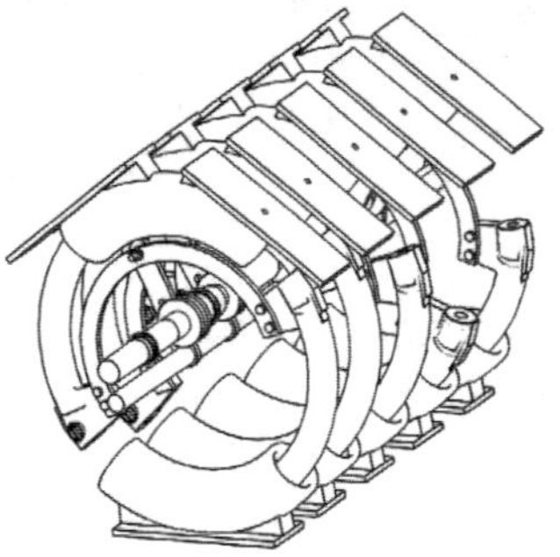

Figure 1. Schematic diagram of the novel O-type engine.

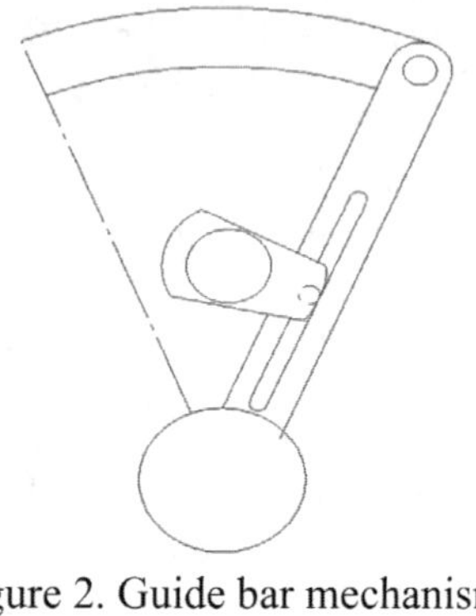

Figure 2. Guide bar mechanism of the O-type engine.

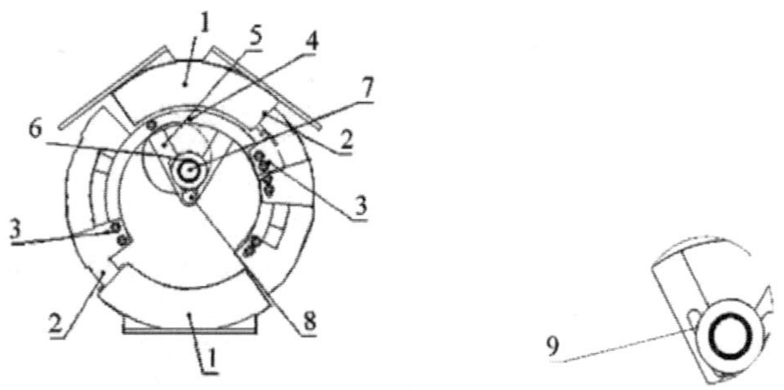

Figure 3. Components of the cylinder sets and the local view.

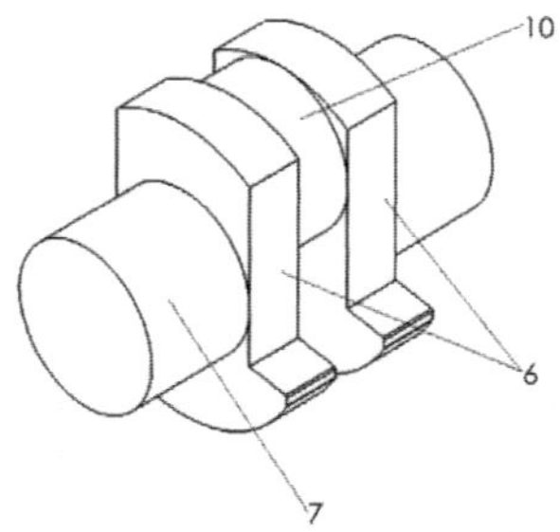

Figure 4. Details of the crankshaft.

2.2. Working Mechanism

The guide-bar mechanism which is applied in the O-type engine, is shown in Figure 2. The guide bar is the driving part while the crankshaft is the driven member. The energy released from gas combustion directly drives the piston and ring rod and hereafter powers the guide bar. The cylindrical pin on the crank arm, with the energy transmitted, traces a circle alongside the inside of the groove on the guide bar to rotate the crankshaft. During the operation, the transmission angle is 90° while the pressure angle is 0°, which suggests the superior force transmission characteristics. In the operation, the guide bar, driven by the piston connecting to the ring rod, sways to the left and right, making the crankshaft rotate.

3. Hypotheses and Methodology

3.1. Hypotheses of the basic parameters

The hypotheses are as follows: the combustion in O- type engines researched is a transient uniform process; the combustion process of V- and O- types of engines researched are the same because we assume the fuel used and other configurative characteristics of the two types of engines are similar.

3.2. Method

PMI indicates the work that is done by the engine in one complete cycle; HE indicates the extent of transformation from the internal energy to the kinetic energy; SOE considering the volume of engine is the value of the mean indicated pressure per unit value of the volume.

3.2.1. For the space occupied by the O-type engine. As shown in Figure 5, due to the fact that all the guide bars only rotate within the range of k degrees, it is suggested that the space produced by the left (360-k) degrees of the circle can be filled to implement other functions. Thus, the real theoretical volume of only one set of cylinder and guide-bar system of the O-type engine (VCGSO), V_o can be shown in the following:

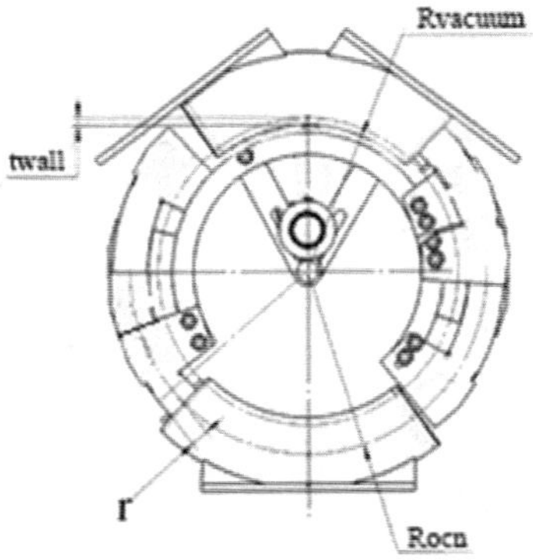

Figure 5. Parameters of the O-type engine.

$$V_o = \frac{kR_{ocn}\pi^2(r+t_{wall})^2}{45} + \frac{k\pi R_{vacuum}{}^2(r+t_{wall})}{180} \quad (1)$$

$$R_{vacuum}=R_{ocn}-r-t_{wall}. \quad (2)$$

where r is the radius of the O-type engine's cylinder, k is the working range of pistons for O-type engine, R_{ocn} is the length from centroid of cylinder to centroid of auxiliary axis, t_{wall} is the thickness of cylinder wall, and R_{vacuum} is the radius from the inside of ring rod to centroid.

3.2.2. Principles for the performances of O- types of engines. The power performance of the reciprocating internal combustion engine could be figured out as follows:

The compression ratio, ε, can be shown as following,

$$\varepsilon = \frac{v_{min} + \frac{k^2\pi^2r^2R_{ocn}}{180}}{v_{min}} \quad (3)$$

where the v_{min} is the volume of combustion chamber.

In Equation (4), S is the stroke of piston.

$$S = \frac{k\pi R_{ocn}}{180} \quad (4)$$

In Equation (5), Ra is the bore-stroke ratio.

$$Ra=\frac{2r}{S} \quad (5)$$

The final expression of the mean indicated pressure, P_{mi}, is shown below,

$$P_{mi} = \frac{P_c}{\varepsilon-1}\left[\frac{v_p}{n_2-1}\left(1-\frac{1}{\varepsilon^{n_2-1}}\right)-\frac{1}{n_1-1}\left(1-\frac{1}{\varepsilon^{n_1-1}}\right)\right] \quad (6)$$

where P_c is the compression terminal pressure, v_p is the pressure rise ratio, n_1 is the average polytropic compressibility index and n_2 is the average variable expansion index.

$$\eta_i = 8.314\frac{\alpha AT_0P_{mi}}{HP_0\emptyset_{in}} \quad (7)$$

where η_i is the heat efficiency (HE), φ_{in} is the aeration efficiency, A is the theoretical air requirement, α is excess air factor, T_0 is the ambient temperature, H is the low calorific value of fuel and P_0 is the ambient pressure. Equation (6), Equation (7) and the derivation procedure refer to Ref. [15].

$$\eta_{op} = \frac{P_{mi}}{V_o} \quad (8)$$

where η_{op} is the space occupation efficiency (SOE).

To calculate the maximum possible value of the crankshaft diameter d_1 which is shown in Figure 6, Equation (10) and Equation (11) are given.

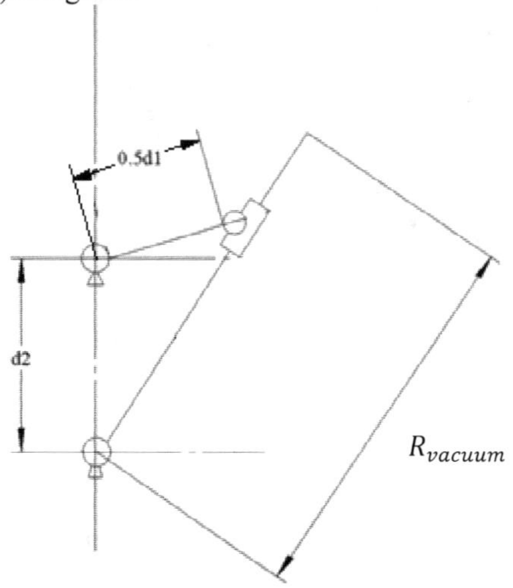

Figure 6. Diagram of the crank mechanism.

$$d_2 = R_{\text{vacuum}} - \frac{d_1}{2} \tag{9}$$

where d_2 is the distance between the auxiliary axis and crank shaft.

$$\frac{d_1}{2\sin(0.5k)} = d_2 \tag{10}$$

$$d_1 = \frac{2\sin(0.5k)\, R_{\text{vacuum}}}{1+\sin(0.5k)} \tag{11}$$

4. Results

4.1. Comparative analysis

The parameters of O-type engines are shown in Table 1 according to usual conditions. Given that the VCGSO is 0.0076 m^3, the PMI and HE are put into comparison with the general range of the indicators of other reciprocating engines to embody the superiority of the O-type engine. The cylinder bore-stroke ratio within a range from 0 to 10 is set on the X-axis, while the PMI and HE are set on the Y-axis in Figure 7 and Figure 8, respectively.

Table1. Parameter defaults of general reciprocating gasoline engine.

Parameters	values	Parameters	values
$v_{\min}$, L	5.465×10^{-2}	T_{out}[a], K	1045
ΔT[b], K	12	a_O[c]	0.004
a_C[c]	0.85	M_{wf}[d]	114
a_H[c]	0.146	H, KJ/Kg	44100
T_0, K	298	A	0.900
P_0, bar	1.013	$\emptyset_{out}$[e]	0.086
n_1	1.36	n_2	1.25
t_{wall}, mm	5		

[a] T_{out} is the exhaust terminal temperature.
[b] ΔT is the intake heating temperature rise.
[c] a_c, a_H and a_o are the fuel composition of carbon, hydrogen, oxygen respectively.
[d] M_{wf} is the relative molecular weight of fuel.
[e] $\emptyset_{out}$ is the residual gas coefficient.

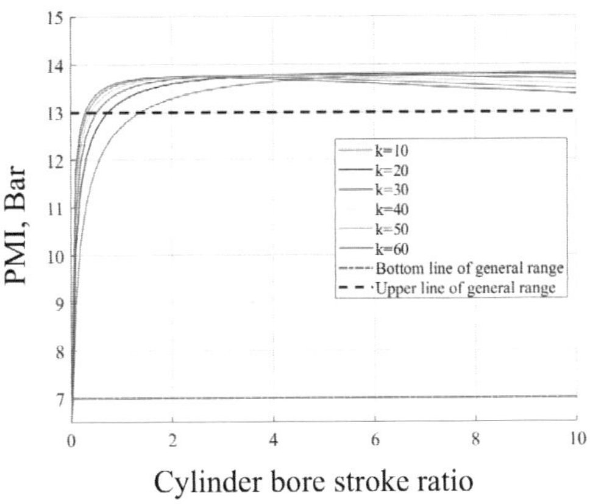

Figure 7. Competitive results of the PMI.

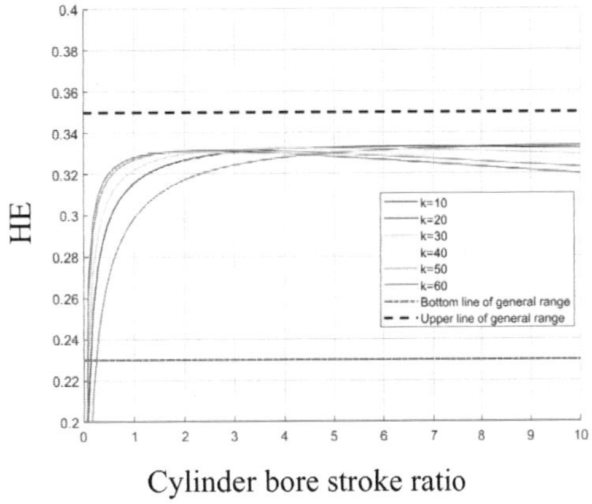

Figure 8. Competitive results of the HE.

From Figure 7 it can be seen that there is a local maximum (approximately 13.7 bar) for all the curves. With k increasing, the local maximum moves to the left, which indicates a smaller cylinder bore stroke ratio. Also, the gradience of curve increment is much steeper than that of the decrement. In addition, most of the values of PMI transcend the general range of PMI of gasoline engine at the relatively large bore stroke ratio, indicating that the O-type engine can do more work per volume of the cylinder in each cycle in this situation. If the working range k is narrow, the bore stroke ratio should become large to exceed the general range of PMI. Nevertheless, at relatively low bore stroke ratio, some of the values of the PMI are at the high position within the general range of PMI, which are also acceptable. From Figure 8 it is self-evident that the trend of HE curve is similar to that of the PMI curve. The maximum value of HE of the O-type engine can reach approximately 0.33, which is almost in the middle of the general range of HE of gasoline engine. When the working range k is large, the values of HE with a wider range of bore stroke ratio are higher than the bottom line of the general range of HE. The more favourable HE is located where the bore stroke ratio is small and the working range k is large simultaneously, and the contrary is the case. It is also true of the PMI.

4.2. Performance analysis

In order to leave enough space for the crank arm, the working range is set to 60° for further calculations and analysis. The range of VCGSO and the bore stroke ratio are between 0.00001 m^3 and 0.01001 m^3 and between 0.00001 and 5.00001 respectively, which are wide enough to do the analysis.

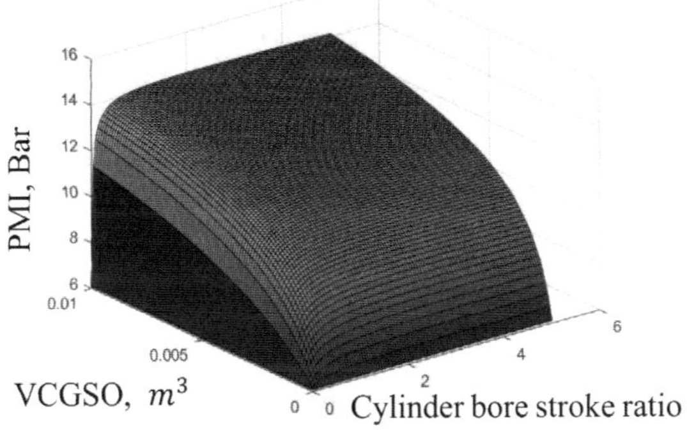

Figure 9. Relationship between PMI, bore stroke ratio and VCGSO.

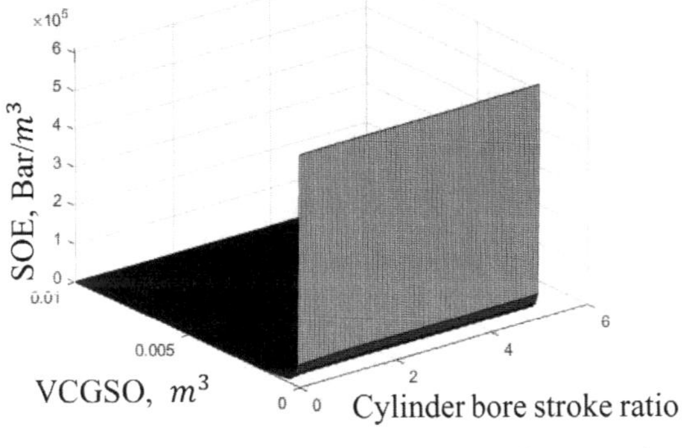

Figure 10. Relationship between SOE, bore stroke ratio and VCGSO.

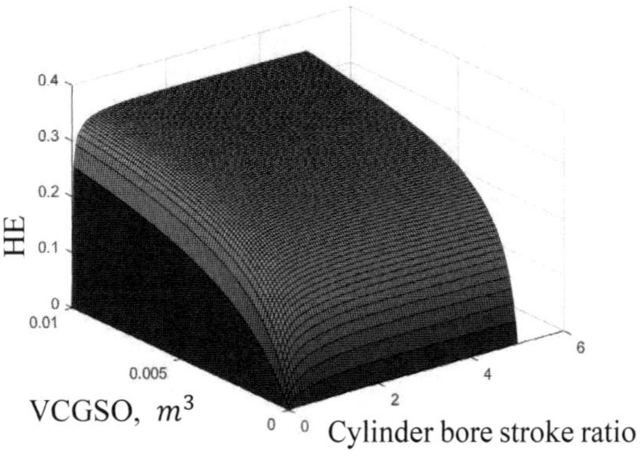

Figure 11. Relationship between HE, bore stroke ratio and VCGSO.

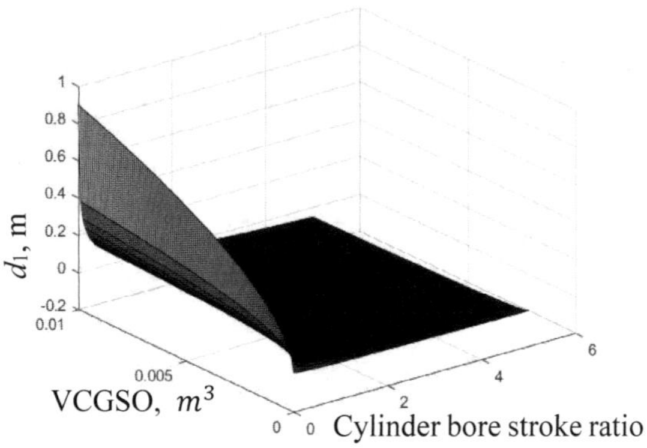

Figure 12. Relationship between d_1, bore stroke ratio and VCGSO.

From Figure 9, it is self-evident that the PMI increases to the maximum along the increasing volume direction, and it increases dramatically firstly and then declines slowly along the direction of the increasing bore stroke ratio. More specifically, the pressure has a global maximum value of about 14.22 bar when the bore stroke ratio is approximately 2, and the volume is near 0.01 m^3. In Figure10 it shows that with the volume declining, the SOE of the O-type engine increases with a more substantial rate when the volume is small. Besides, the SOE rises firstly and then falls along the bore stroke ratio direction, although the change is subtle. From Figure 11 the trends of variation of the HE and PMI are almost identical. However, the HE reaches its maximum value of 0.33 with the values of bore stroke ratio and volume being 2 and near 0.01 respectively. In addition, the HE varies slowly for large volume and bore stroke ratio. In Figure 12 it illustrates that the crankshaft diameter increases more sharply with a smaller bore stroke ratio than with a larger one. For the area where the bore stroke ratio is large, the diameter is not sensitive to the change of the volume and bore stroke ratio. The maximum value of crankshaft diameter is about 0.9 m with the values of bore stroke ratio and volume being near 0 and 0.01 respectively, yet of course this value does not have practical meaning.

5. Discussion

5.1. Verification and validation

Verification is the process that assesses whether a computational model can accurately represent the underlying mathematical model [16]. In this paper, the maximum values are more significant compared to other values, because the highest performance usually relates to the max. values of indicators. And all the results can be calculated with different step sizes in MATLAB. Thus, the max. values of indicators at different step sizes need to be found out and seen whether they are convergent. Note that the step size should be chosen carefully as the results could be the same and unconvincing if one step size is one of the factors of another. The results are shown in Figure13, Figure14, Figure15 and Figure16.

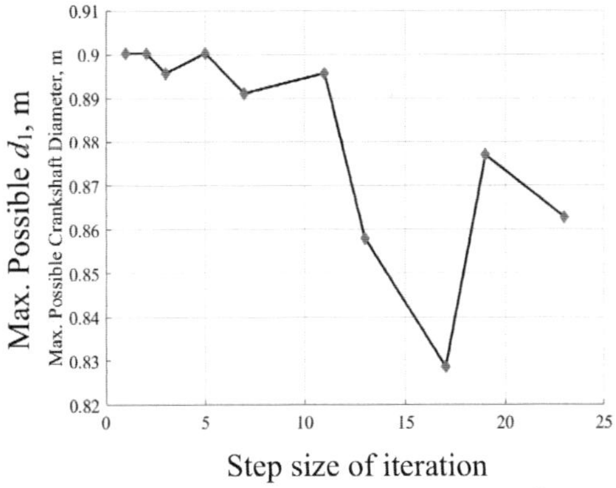

Figure 13. Max. possible crankshaft diameter vs Step size of iteration.

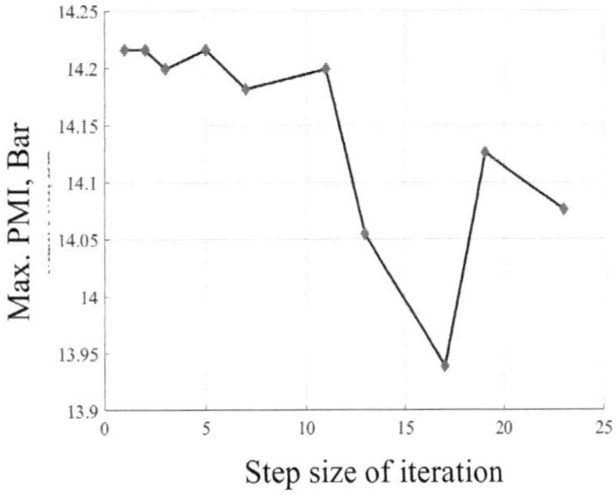

Figure 14. Max. PMI vs Step size of iteration.

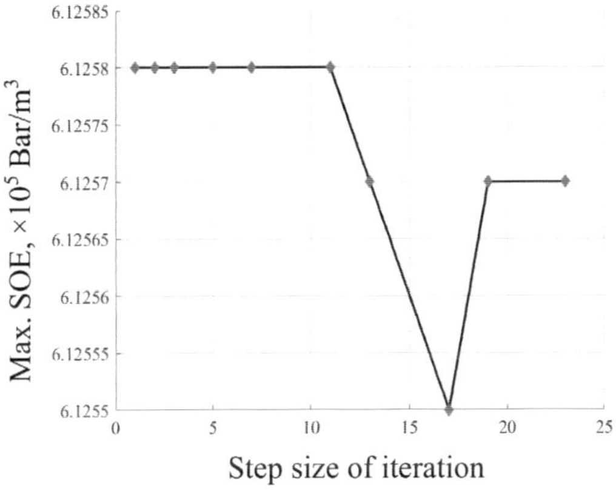

Figure 15. Max. SOE vs Step size of iteration.

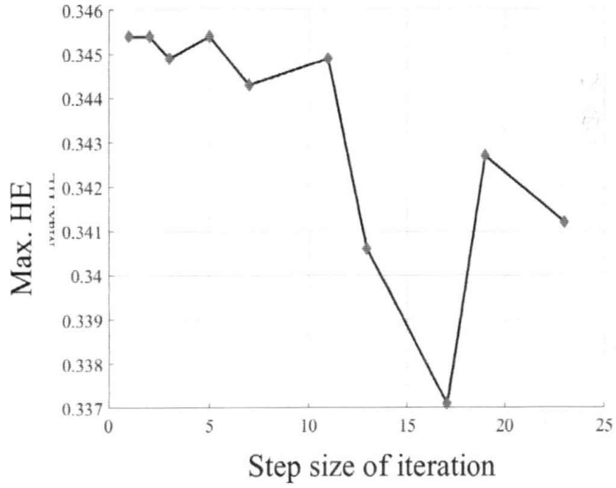

Figure 16. Max. HE vs Step size of iteration.

From Figure 13 to Figure 16, within the range of step size of iteration (SSI) from 1 to 23 the results suggest that the values are becoming convergent within the decreasing SSI. For each indicator of performance, we can take the SSI of 1, which is also what has been taken above to conduct the calculation process. Thus, the accuracy of the solving process can be ensured.

Validation is the process that determines to what extent a model can accurately represent the reality of interest from the perspective of the intended use of the model [16]. The power performance of the O-type engine proposed in this paper is not absurdly high compared to the range of the performance of V-type engine, for the operating principle and configuration of the cylinders of these two types of engines are identical and we only conduct the innovation of the configuration and mechanism. As a result of that, further, the validation of the calculation in combustion process is proved after figuring out the indicators of performance of a specific V-type engine through the same calculation procedure and comparing them to the general range of performance of V-type engine. In Table 2, it can be seen that all of the calculated results are located within the general range of reciprocating gasoline engines. Additionally, in the future we will try to build a prototype to test and compare the experimental results to the computational ones.

Table 2. The configurative parameters and the results of a specific V-type engine.

Configurative Parameters	Values	Results	Values
Bore stroke ratio	1.24	Calculated PMI, bar	11.44
Compression ratio	10	Calculated HE, %	26
Cylinder diameter, mm	92	General range of PMI in reality, bar	7-13
		General range of HE in reality[a], %	23-35

[a] There are only a few exceptions that are beyond the general range of HE, such as Toyota Camry with a HE of 40% and Mazda SKYACTIV X series with a HE of 48% and.

5.2. Interpretation and discussion of the results

In order to confirm the best performance range, the coordination lines of the maximum performance values at different volumes are shown in Figure 17, and maximum performance is introduced which is called the maximum line in Figure 18, Figure 19, Figure 20 and Figure 21.

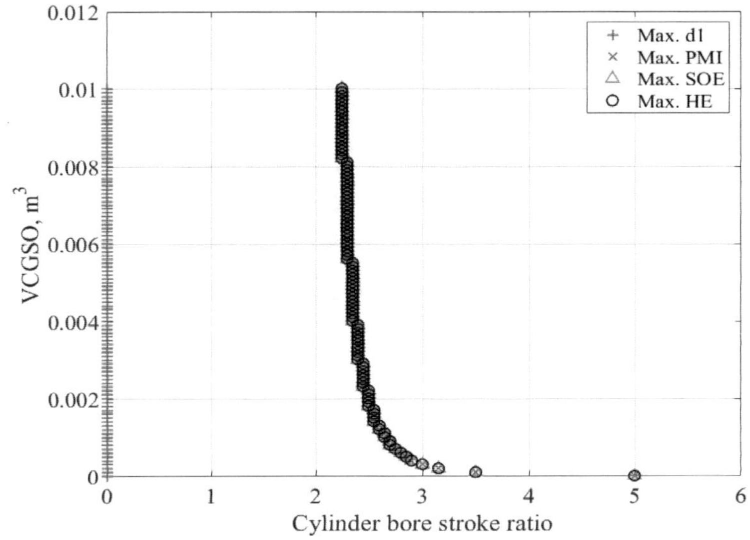

Figure 17 The maximum line of each power performance indicator.

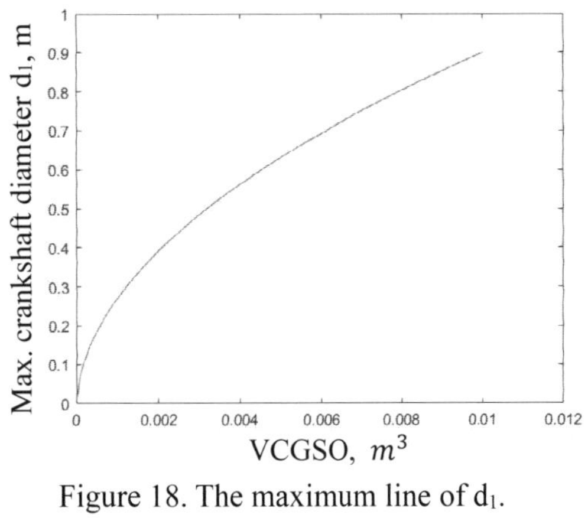

Figure 18. The maximum line of d_1.

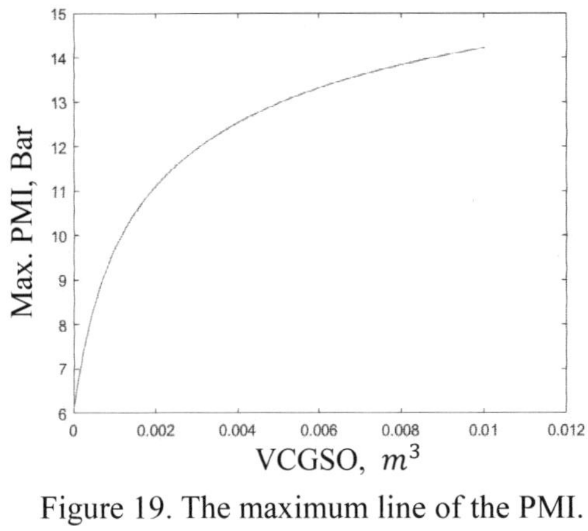

Figure 19. The maximum line of the PMI.

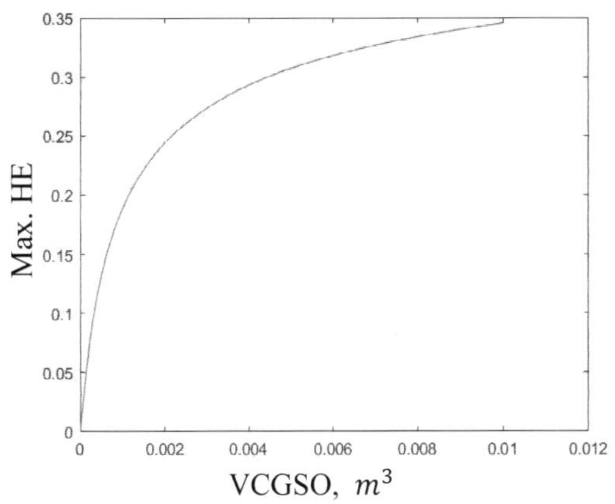

Figure 20. The maximum line of the HE.

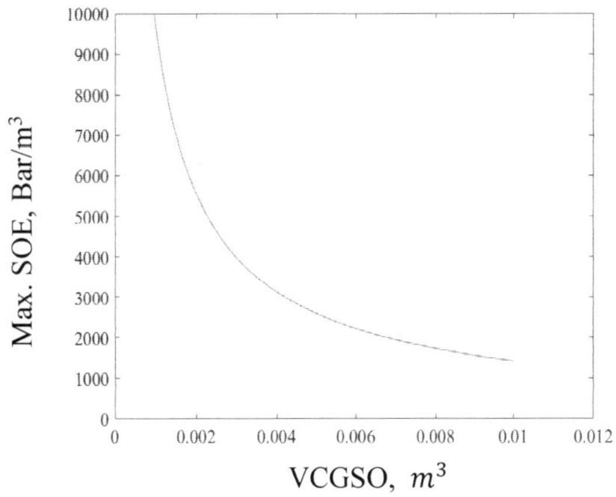

Figure 21. The maximum line of the SOE.

In Figure 17 it illustrates that the maximum PMI line at different volumes has the same coordinates as the maximum HE and SOE line. Nevertheless, the crankshaft diameter has the maximum values when the cylinder bore stroke ratio is near 0. On the maximum line from Figure 18 to Figure 21, the crankshaft diameter, PMI and HE increase with the increasing volume. When the volume is small, they will especially rise sharply. However, the slope is relatively small when the volume is larger than about 0.004 m^3 for the maximum PMI and HE, while for the maximum crankshaft diameter the slope is relatively large all the time. The maximum SOE line drops all the way but especially drops sharply when the volume is less than 0.002 m^3. With a rough estimate of the volume of the cylinder and connecting rod mechanism (0.0076 m^3), the SOE of the specific V-type engine is only 1508.34 bar/m^3. Moreover, the maximum line of the crankshaft diameter is invalid because of the situation where the bore stroke ratio is too near 0, which is inapplicable. Combined with those above and the intention of the superiority over the specific V-type engine, the proper choice of the cylinder bore stroke ratio and VCGSO can be roughly determined.

Then, the limits can be initially set. The bore stroke ratio is theoretically between 2 and 2.5 from Figure 17 as the PME, SOE and He could reach the peak at that time. Before finally defining the proper choices of the bore stroke ratio and VCGSO, some restrictions should be considered. Firstly, the crankshaft diameter is assumed to be more than about 80 mm so as to withstand the torque of the drive shaft, so that the VCGSO should be larger than approximately 0.00013 m^3 from Figure 18. Secondly, PMI should be better than that of the specific V-type engine, so that the volume should be more than about 0.0025 m^3 from Figure 19. Besides, the VCGSO should be smaller than the volume of the cylinder and connecting rod set of the V-type engine if possible. For HE, however, it is sometimes plausible to sacrifice some of it to achieve the higher performance, so that HE is not included within the limits to simplify the calculation. The proper choices of the bore stroke ratio and VCGSO within the limits mentioned above are shown in Figure 22, while the more accurate and convincing choices of the two structural parameters can be determined with other detailed limits in the future.

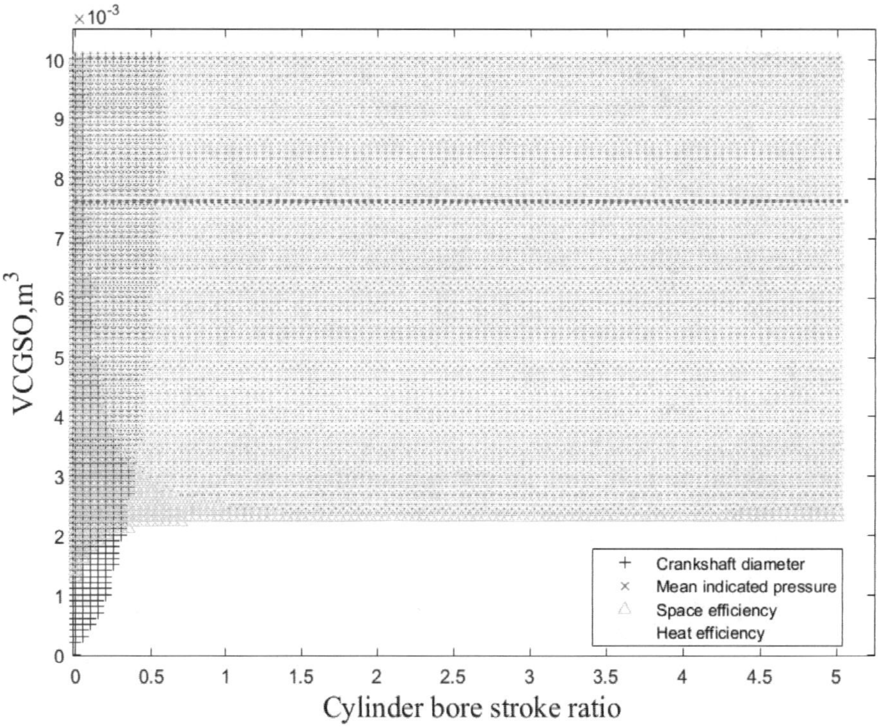

Figure 22. The proper range for the bore stroke and the volume.

In Figure 22 the values of the possible crankshaft diameter in the graph above are the values that can be achieved under the restrictions set above, and so are the values of the possible PMI, the SOE and HE. The overlapped bullet-like area below the dotted red line is the proper range for the choice of bore stroke ratio and volume which makes the performance indicators better than the specific engine.

6. Conclusion

Based on the work and research done by other researchers in this field and in light of the results above, it can be concluded:

- The power performance such as the PMI of the O-type engine could be superior to most of gasoline engines while the HE of the O-type engine is at the middle level within the general range of HE of gasoline engine;
- The PMI, SOE and HE rise first and then fall with an increasing bore stroke ratio at different VCGSO, and PMI and HE increase with increasing VCGSO at different bore stroke ratios;
- In the situation where the working range k is 60°, the proper range of the volume and the bore stroke ratio are obtained---a bullet-like area where the bore stroke ratio is more than about 0.1 and less than about 0.6 and the VCGSO is larger than about 3.1×10^{-3} m^3.

This paper focuses on the power performance of the O-type engine. However, both the equivalent stress and strain conditions on the configurative components when the mechanism is in operation and the effective power output of the O-type engine can be simulated in the future work by using finite element analysis and the calibration coefficient [9] respectively in near future. Furthermore, the other detailed aspects, such as the ignition order, cooling system, exhaust emission, the optimised solution of structure parameters and so on, could be combined with the novel configuration of O-type engine cylinder, and the engine torque could be examined with the combination of the OBD system and the BP neutral work [6] after the first prototype is built in the future. In conclusion, the O-type engine, as a result of the better power performance and relatively small occupying space and so on, is promising and likely to be modified and applied in automobile industry.

IOP Conf. Series: Materials Science and Engineering **787** (2020) 012036 doi:10.1088/1757-899X/787/1/012036

Reference

[1] Pukrabek, W.W. (2004) Engineering Fundamentals of the Internal Combustion Engine. Prentice Hall Publishing, New Jersey.

[2] Heywood, J.B. (2018) Internal Combustion Engine Fundamentals. McGraw-Hill Publishing, New York.

[3] Daoud, J.M., Friedrich, D. (2018) A novel Franchot engine design based on the balanced compounding method. Energy Conversion and Management, 169: 315-325.

[4] Bastida, H., Ugalde-Loo, C.E., Abeysekera, M. (2017) Dynamic Modelling and Control of a Reciprocating Engine. Energy Procedia, 142: 1282-1287.

[5] Chahartaghi, M., Sheykhi, M. (2018) Energy and exergy analyses of beta-type Stirling engine at different working conditions. Energy Conversion and Management, 169: 279-290.

[6] Xu, C., Hao, L.L. (2017) An estimation of the engine torque based on OBD system. http://www.cnki.com.cn/Article/CJFDTotal-SHQC201702005.htm.

[7] Hu, Y.M., Deng, Z.X., Zhu, Z.G., Wang, P., Tan, J., Zhang, J.S., Qing, H.B. (2005) Transient Strensith Analysis of Engine Block. The State Key Laboratory of Mechanical Transmission, 26: 63-67.

[8] Hong, M.N., Ouyang, M.G. (2009) On-board Torque Estimation Base on Mean Value SI Engine Models. Journal of Mechanical Engineering, 45:290-294.

[9] Wang Z.P. (1993) Computerized Calculation of Automobile Engine Power Calibration Coefficient. Journal of Highway and Transportation Research and Development, 10: 47-53.

[10] Ebrahimi, R. (2018) A new design method for maximizing the work output of cycles in reciprocating internal combustion engines. Energy Conversion and Management, 172: 164-172.

[11] Notay, R. S., Priest, M., Fox, M.F. (2019) The influence of lubricant degradation on measured piston ring film thickness in a fired gasoline reciprocating engine. Tribology International, 129: 112-123.

[12] Tormos, B., Martín, J., Carreño, R., Ramírez, L. (2018) A general model to evaluate mechanical losses and auxiliary energy consumption in reciprocating internal combustion engines. Tribology International, 123: 161-179.

[13] Östman, F., Toivonen, H.T. (2008) Active torsional vibration control of reciprocating engines. Control Engineering Practice, 16: 78-88.

[14] Zhong, C.M. (2004) Research on mathematics simulation during the combustion for internal-combustion engine. http://en.cnki.com.cn/Article_en/CJFDTOTAL-KJJJ200406017.htm.

[15] Jiang, D.M. (2002) Advanced principles of internal combustion engine. Xi'an Jiaotong University Press Publishing, Xi'an.

[16] Schwer, L.E. (2006) Guide for Verification and Validation in Computational Solid Mechanics. The American Society of Mechanical Engineers Publishing, New York.

Application of RBF Sliding Mode Control in Pump System for Electric Vehicle Heat Air Conditioning

Taixiong Zheng, Xiaokang Hou and Rongdi Yuan

School of Advanced Manufacturing Engineering, Chongqing University of Posts and Telecommunications, Chongqing 400065, China

E-mail: 935462217@qq.com

Abstract: Aiming at the characteristics of time-varying, time-delay and non-linearity of electric vehicle heat pump air-conditioning system, this paper presents a sliding mode control method based on RBF neural network. The RBF neural network can approximate the nonlinear function with arbitrary precision under certain conditions, and has strong self-learning and self-adaptive ability. Combining the RBF neural network with the sliding mode variable structure control not only has the anti-interference of the sliding mode control and the robustness to the parameter changes, but also effectively reduces the flutter caused by the sliding mode control. The simulation results show that the RBF neural network sliding mode control method can improve the system control accuracy, effectively weaken the chattering, and has good adaptability to the interference conditions.

1. Introduction

The air conditioning system has the functions of heating, cooling, defogging, defrosting, blowing, filtering, etc., and has become an indispensable auxiliary system for automobiles. It regulates the air inside the car to ensure the comfort and safety of the passenger compartment. For pure electric vehicles, the energy source of the whole vehicle is only the power battery. This requires the air conditioning system can not only meet the needs of the passenger cabin comfort, but also to meet the requirements of high efficiency and energy saving. The heat pump air conditioning system is the most suitable air conditioning system for pure electric vehicles because of its high energy efficiency ratio and integration of heating and cooling [1].

The heat pump air conditioning system control has the following characteristics: First, the heat pump air conditioning system itself has hysteresis and non-linearity, and many interference factors, such as ambient temperature, driving speed, personnel intensive, door opening, etc., it is difficult to establish an accurate mathematical model. Secondly, because the pure electric vehicle itself has the shortcomings of short mileage, the energy saving of the air conditioning system is also very important [2].

The PID controller is widely used in the control field because of its simple structure and easy implementation. However, in the face of the complicated and variable working conditions of the heat pump air conditioning system, it is difficult for the PID controller to make adjustments in time, and the control effect is not good. The fuzzy controller can simplify the complexity of the system and is suitable for the control of such nonlinear, time-varying, and hysteretic systems, and does not depend on the accurate mathematical model of the system. X. Wang used the fuzzy theory in the research of automotive automatic air conditioning control. The difference between temperature and set temperature and the temperature difference change rate were used as the dual input of the system. Under different vehicle speed conditions, the fan air volume and compressor were controlled by fuzzy controller to achieve better control results. However, the fuzzy rules and membership functions of fuzzy controllers are all designed by personal experience and lack of systematization [3]. Zheng combined traditional PID control with Smith algorithm and fuzzy control algorithm, proposes Smith-Fuzzy-PID control method, and applies the method in automotive air conditioning system, which reduces system overshoot compared with traditional PID controller. And response time [4]. He proposed a Neuro-Fuzzy control method. This method and the traditional fuzzy controller in the automotive air conditioning temperature control comparison experiment, greatly improve the system's adaptability and sensitivity [5]. Zhang proposed a sliding mode control method, which controls the temperature inside the vehicle through the compressor speed, so that the air conditioning system has good robustness to external temperature changes and wind speed changes, and mitigates chattering by switching gain method [6].

2. Model of electric vehicle heat pump air conditioning system

2.1 heat pump air conditioning working principle

The heat pump air conditioning system is mainly composed of electric compressor, four-way reversing valve, two-way expansion valve, in-vehicle heat exchanger and external heat exchanger. The electric energy drives the compressor to work, so that the working fluid repeatedly undergoes the physical process of evaporation endotherm and condensation heat release, and realizes heat exchange and transfer inside and outside the passenger compartment [7-9].The heat pump air conditioning system controls the flow direction of the working medium through the four-way reversing valve to realize the switching between the heating mode and the cooling mode, and realizes the adjustment of the cooling capacity and the heating amount by controlling the rotation speed of the compressor.

2.2 heat pump air conditioning mathematical model

The thermal load of a car mainly comes from the heat of the air, the heat of the solar radiation, and the heat radiated by the human body. In order to establish an applicable air conditioning model, the temperature distribution of the member cabins of the car is evenly distributed, and the heat balance equation of the member cabins can be written as:

$$\rho V C_p \frac{dT_i}{dt} = \rho V_f C_p (T_f - T_i) + \sum_{j=1}^{4} K_j F_j (T_0 - T_i) + Q \tag{1}$$

Where, ρ is the air density; C_p is the specific heat; T_i is the air temperature inside the vehicle; T_f is the fresh air temperature; V is the air volume inside the vehicle; V_f is the fresh air volume; K is the heat transfer coefficient; F is the heat transfer area; j takes 1, 2, 3, 4, respectively represent the roof, body side, floor, window; Q is the heat released by the people inside the car.

Bringing a specific parameter of a car into Eq.1, the equation of state is written as:

$$\dot{x} = Ax + Bu \tag{2}$$

where, $A = \begin{bmatrix} 0 & 60 \\ 0 & -1 \end{bmatrix}, B = \begin{bmatrix} 0 \\ 1 \end{bmatrix}.$

2.3 heat pump air conditioning model

According to the working principle of heat pump air conditioner, the model of electric vehicle heat pump air conditioning system is built by AMESim/MATLAB software, and joint simulation is carried out to explore the control effect of control method on heat pump air conditioning system and its influence on system power consumption and energy efficiency ratio (COP).

3. controller design

In order to improve the control accuracy and anti-interference of the heat pump air conditioning system, the RBF neural network sliding mode controller is designed here. The control objective is to track the actual temperature of the passenger compartment to the desired temperature to ensure the comfort of the driver.

3.1 RBF neural network sliding mode control law design

Assume the state equation of a nonlinear uncertain :

$$\dot{x} = f(x) + g(x)u \tag{3}$$

Where $x = [x_1, x_2 ... x_n]$ is the system observable variable, u is the control input, f(x) and g(x) is an unknown nonlinear function.

The design goal of the sliding mode controller is to make the initial point of the system, regardless of the position of the state space, the movement of the system can tend to switch faces for a limited time [10-11]. Taking the passenger compartment temperature as the tracking object, taking the temperature feedback to obtain the state error and its derivative, the design sliding surface is:

$$s = c_1 e + c_2 \dot{e} + ... + c_{n-1} e^{(n-2)} + e^{(n-1)} = 0 \tag{4}$$

Where e is the tracking error, the polynomial $P(p) = c_1 + c_2 p + ... c_{n-1} p^{(n-2)} + p^{(n-1)}$ is Hurwitz stable, p is the laplace operator, c_i is the normal number, i=1...n-1, and its value determines the speed of the additional error decay.

The tracking problem of the system can be transformed into the stability problem of the sliding surface. The Lyapunov function is defined as follows:

$$V = \frac{1}{2} s^2 \tag{5}$$

The sliding mode existence condition can be written as:

$$s \dot{s} \leq -\eta |s| \tag{6}$$

Where η is constant. It ensures the stability of the system. Deriving the sliding mode switching function s can be obtained:

$$\dot{s} = c_1 \dot{e} + c_2 \dot{e} + ... + c_{n-1} e^{(n-1)} + e^{(n)} = \sum_{i=1}^{n-1} c_i e^{(i)} + x_1^{n} + y_d^{(n)} = \sum_{i=1}^{n-1} c_i e^{(i)} + f(x) + g(x)u + p + y_d^{(n)} \tag{7}$$

If f(x) and g(x) are known precisely, the ideal sliding mode control law is:

$$u^i = u_{eq}^i + u_{sw}^i = \frac{1}{g(x)} \left[- f(x) + y_d^{(n)} - \sum_{i=1}^{n-1} c_i e^{(i)} - k \operatorname{sgn}(s) \right] \tag{8}$$

Where $k \geq P + M + \eta$.

U^i_{eq} is the ideal equivalent control and U^i_{sw} is the ideal switching control. Since f(x) and g(x) are actually unknown, an ideal sliding mode controller cannot be implemented. Considering the universal approximation theorem of RBF neural network, we use the RBF neural network to adaptively approximate the f(x) and g(x), thus constructing and replacing f(x) and g(x), then the actual slip. The mode control law is:

$$u^i = \frac{1}{\hat{g}(x)} \left[- \hat{f}(x) + y_d^{(n)} - \sum_{i=1}^{n-1} c_i e^{(i)} - k \operatorname{sgn}(s) \right] \tag{9}$$

3.2 Adaptive Law Design

Substituting Eq.8 into Eq.6 yields

$$\begin{aligned}\dot{s} &= \sum_{i=1}^{n-1} c_i e^{(i)} + f(x) + g(x)u + p - y_d^{(n)} \\ &= \sum_{i=1}^{n-1} c_i e^{(i)} + f(x) + \hat{g}(x)u + g(x)u - \hat{g}(x)u + p - y_d^{(n)} \\ &= f(x) - \hat{f}(x) + \left[g(x) - \hat{g}(x)\right]u - k\,\mathrm{sgn}(s) + p \\ &= (w_f^{*T} - w_f^T)\phi(x) + (w_g^{*T} - w_g^T)\phi(x)u - \delta - k\,\mathrm{sgn}(s) \\ &= \eta_f^T\phi(x) + \eta_g^T\phi(x)u - \delta - k\,\mathrm{sgn}(s)\end{aligned} \tag{10}$$

Where $\eta_f = w_f^* - w_f$, $\eta_g = w_g^* - w_g$.

The task of the adaptive law is to adjust the weights w_f and w_g such that the sliding variable s and the parameter errors n_f and ng approach 0, so consider the following Lyapunov function as follows:

$$V = \frac{1}{2}s^2 + \frac{1}{2\rho_1}\eta_f^T\eta_f + \frac{1}{2\rho_2}\eta_g^T\eta_g \tag{11}$$

Where ρ_1 and ρ_2 are constant. The derivative of the Eq.10 can be written as

$$\begin{aligned}\dot{V} &= s(\eta_f^T\phi(x) + \eta_g^T\phi(x)u - \delta - k\,\mathrm{sgn}(s)) + \frac{1}{\rho_1}\eta_f^T\dot{\eta}_f + \frac{1}{\rho_2}\eta_g^T\dot{\eta}_g \\ &= \frac{1}{\rho_1}\eta_f^T(\rho_1 s\phi(x) + \dot{\eta}_f) + \frac{1}{\rho_2}\eta_g^T(\rho_2 s\phi(x)u + \dot{\eta}_g) - s\delta - k|s|\end{aligned} \tag{12}$$

By setting $\dot{\eta}_f = -\rho_1 s\phi(x)$, $\dot{\eta}_g = -\rho_2 s\phi(x)u$.

Eq.11 is rewritten as:

$$\dot{V} = -s\delta - k|s| \tag{13}$$

According to the approximation theory of RBF neural network, as long as there is enough implicit interruption, the adaptive RBF neural network can realize the approximation error is very small, so the adaptive law is designed as:

$$\dot{w}_f = \rho_1 s\phi(x) \tag{14}$$

$$\dot{w}_g = \rho_2 s\phi(x)u \tag{15}$$

4. system simulation and analysis

In order to verify the above design and control methods, this paper analyzes and compares the control effects of RBF neural network sliding mode controller and PI controller under steady state and transient conditions in AMESim/MATLAB joint simulation environment.

Due to changes in the driving conditions of the car, factors such as light, vehicle speed and ambient temperature will affect the operation of the heat pump air conditioner. The control effect of the two control methods is verified by the vehicle speed as the interference amount. The increase of the vehicle speed will increase the heat exchange between the passenger compartment and the external environment, causing the passenger compartment temperature to fluctuate with the change of the vehicle speed. The transient conditions are set to: ambient temperature -10 °C, humidity 40%, light intensity 0, vehicle speed is NEDC working condition, heat pump air conditioning system enters inner circulation heating mode. The simulation sets the passenger cabin target temperature to 20 °C.

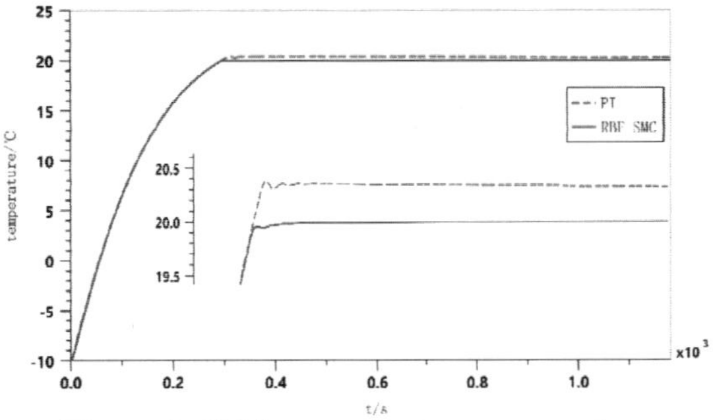

Figure 1. Cabin temperature curve under steady-state conditions

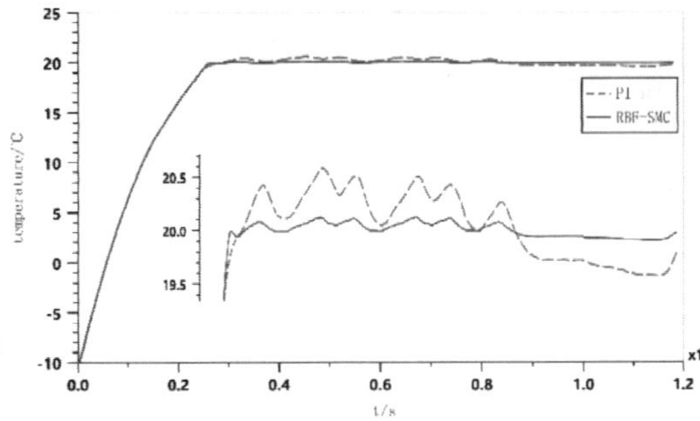

Figure 2. Cabin temperature curve under dynamic condition

The passenger cabin based on the RBF sliding mode controller has higher temperature accuracy and less fluctuation. Compared to PI control, the RBF neural network-based sliding mode controller output can quickly change the corresponding vehicle speed, and has better accuracy and anti-interference for maintaining the passenger compartment temperature.

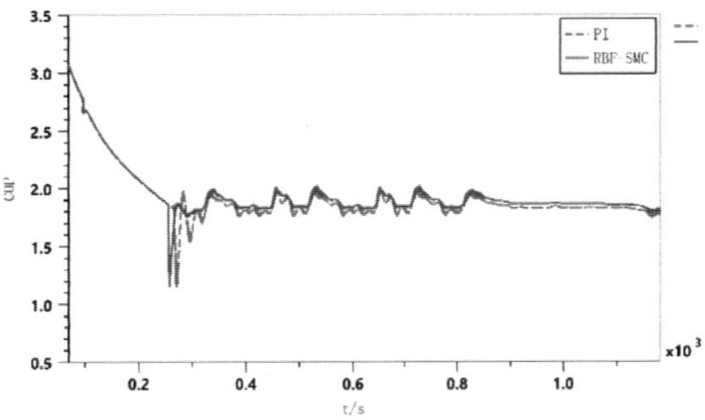

Figure 3. Compressor speed curve

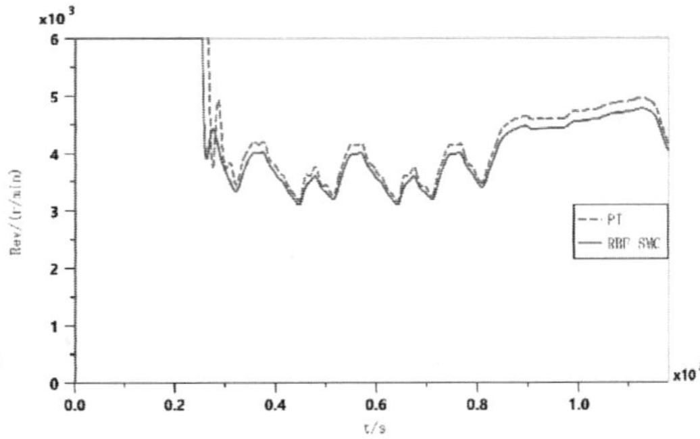

Figure 4. The COP curve of air conditioning system

The RBF sliding mode controller based compressor has a relatively small rate of change in speed, which can effectively reduce the vibration of the compressor and prolong its service life, and the air conditioning system operates at a higher energy efficiency ratio, which is more energy efficient.

5. Conclusion

For the control of nonlinear systems that are difficult to establish accurate mathematical models, traditional control theory and control methods have great limitations. Sliding mode control has the advantages of fast response, insensitivity to system parameter changes and strong anti-interference, and is widely used in the control of nonlinear systems.

Aiming at the problems of low control accuracy and poor anti-interference of heat pump air-conditioning system, this paper designs a sliding mode controller based on RBF neural network. The simulation results show that the RBF neural network sliding mode controller not only improves the control precision and anti-interference of the heat pump air conditioning system, but also effectively reduces the chattering, and smoothly controls the compressor and prolongs the service life.

Acknowledgment

This work was financially supported by Science and Technology Commission of Chongqing Municipality (2018YFB0106100).

IOP Conf. Series: Materials Science and Engineering **787** (2020) 012037 doi:10.1088/1757-899X/787/1/012037

Reference

[1] F. Z. Peng, M. S. Wei, Influence of ambient temperature on the performance of electric vehicle air-conditioning system, Journal of Beijing University of Aeronautics and Astronautics, Vol.40, No.12, 1741-1746, 2014

[2] H. T. Min, Y. B. Cao, Modeling of electric vehicle air conditioning system and its impact on vehicle performance, Journal of Jilin University (Engineering Edition), Vol.39, No.1, 53-57, 2009

[3] X. Wang, H Jiang, Multi-model fuzzy PID controller design for aircraft air-conditioned vehicles, Control Engineering, Vol.23, No.4, 453-457, 2016

[4] Y. Zheng, Experimental study on optimization control of automotive air conditioning temperature system, Computer Simulation, No. 3, 149-154, 2016

[5] B. He, R. Liang, J. Wu, A temperature controlled system for car air condition based on neuro-fuzzy, International Conference on Multimedia Information Networking and Security, No2, 564-567, 2009

[6] Q. Zhang, Sliding mode control of an automotive air conditioning system, American Control Conference, 5748-5753, 2013

[7] Q. F. Peng , H Zhao , X J Chen , Design and experimental study of novel heat pump air conditioning system for electric vehicles, Automotive Engineering, No37, 1467-1432, 2015

[8] Z. Zhang, J. Wang, The solutions to electric vehicle air conditioning systems, Renewable and Sustainable Energy Reviews, No91, 443-463, 2018

[9] Q. F. Peng, H Zhao, Design and Experimental Study of New Heat Pump Air Conditioning System for Electric Vehicles, Automotive Engineering, Vol.37, No.6, 1467-1470, 2015

[10] J. M. Wang, X. M. Dong, Hypersonic vehicle RBF neural network sliding mode variable structure control, Journal of Electric Machines and Control, Vol. 20, No. 05, 103-110, 2016

[11] B. Qin, H. Zhou, Variable-speed sliding mode control of wind turbine based on RBF network, Journal of Electrical Engineering and Technology, Vol.28, No. 05, 37-41, 2013

Discussion on construction technology of Deep Foundation Pit Support in Construction Engineering

Zhengwei Feng [1], Yaohui Shen[1*]

[1]Department of Engineering Management, Xiamen University Tan Kah Kee College, Zhangzhou, Fujian Province, 363105, China

[*]Corresponding author's e-mail: yhshen@xujc.com

Abstract. With the expansion of the construction scale of urban building basement, the technical requirements of excavation and support of deep foundation pit are higher. Based on an engineering example, this paper expounds the construction scheme of deep foundation pit excavation and support in a residential area project. It introduces the construction methods of soil nailing wall support in the north side of the foundation pit, retaining row pile support in the east side and inter pile wall engineering construction. It analyzes the method of controlling measurement error in the process of steel casing burying, and the technical problem that the reinforcement cage does not jam in the process of large diameter long pile construction. In this paper, the key points of quality control in the construction of deep foundation pit support are discussed and summarized to ensure the smooth progress of the project. The research results provide references for other similar projects.

1. General situation of the project

1.1. The project information

The project is located in Xinluo District of Longyan City, Lianzhuang Road is in the east, one planning road is in the north and the phase II project is in the west. It consists of 3 buildings i.e. Building # 17, Building # 18 and Building # 12 #. Each of these buildings has two underground floors. The ground floors of each of the building are 26 floors for Building # 12 and 27 floors for Building # 17 and Building # 18. The height of each of these buildings is about 78 meters, and these buildings cover an area area of 1455.73 square meters, with a total construction area of 39668.8 square meters. These buildings are all class 1 high-rise buildings, ± 0.000 is the Yellow Sea elevation 343.70m. The main building adopts the raft foundation, and the pure basement part adopts the pier foundation. The structural type of the project is frame shear wall structure and the excavated earthwork depth of the foundation pit is about between 5.9 -7.5 meters. The foundation pits fall into the category of deep foundation pit.

1.2. The geological conditions of the project

The project site is located in the southeast of Longyan City, the geomorphology belongs to the class I accumulation terrace in rivers. The south west side is the geomorphology of the foothill slope accumulation, and the terrain is relatively flat. The site was originally a piece of root soil and a site of demolished old houses. Because of the demands for construction , it has been back filled and leveled, with a ground elevation between -5.2 meters and 3.7 meters. For the deep foundation pit here, the soil mass of the pit wall is mainly composed of miscellaneous fill, root soil, silty clay, cobbles, silty clay, gravelly silty clay and the like. The underground water level is -4.50 meters. When the foundation pit was excavated, the measure which has been adopted was using the water collection wells for water drainage.

2. The technical scheme for construction

After considering the safety, the feasibility of the construction technology, the economic and the construction period, etc., according to the actual environment of the site, the supporting scheme for the deep foundation pit were stipulated as follows: the north side of the foundation pit was to be supported by the soil nailing wall; the west side was an open space, the conditions for grading of a slope were available, so an open slope was to be excavated and the surface of the slope was to be provided with a protective layer through spraying concrete, the row piles supports were to be adopted which was close to Lianzhuang Road.

The amount of earth excavation of this project was about 50,000 cubic meters, and the area of the foundation pit was large. The supporting structure was the rotary excavating pile and the soil nailing wall support. The conditions for excavation in parts and layers were available. It was considered to carry out the vertical soil excavation due to the enclosure pile support in east side; as to the other sides, the excavations were to be performed in layers. For the earth excavation, three backhoe excavators (1 cubic meter) were to be used; for transportation of the earthwork, 20 sets of 10t earthwork vehicles were to be adopted, which was to be subject to adjustment in real time depending on the excavation progress and the road transportation conditions.

The deep foundation pit supporting structure stage: In this project, the row pile support,the inter-pile wall support and the soil nailing wall support were adopted. The section of the crown beam was provided along the periphery of the foundation pit, and the trench was excavated to the working surface at three stages. The first stage was to construct the bored in situ piles. The construction of the crown beam was carried out after the construction of the enclosure row piles was completed. The elevation of the crown beam surface was -2.0m, the bottom elevation of the crown beam was excavated to -2.6m. The crown beam construction, pile driving head chiseling and side slope trimming shall be carried out together, and the side slope outside the foundation pit shall be sloping according to the design requirements, after completion of the crown beam, the intercepting drain was constructed and spraying of concrete was carried out. The excavation was carried out in parts and and then the crown beam was constructed with the flow operation method. When the construction of the crown beam was completed, this part was excavated in layers to the bottom elevation of the foundation pit (-9.2m). A slope was set above and below the ditch. The temporary protective rails and warning signs were provided at the top of the slope. In the excavation and construction of the earthwork, attention was also paid to the protection of the supporting structure [1].

The temporary slope at the inner side of the foundation pit adopted the two-stage grading method. Since the foundation pit was a rectangle, in order to facilitate the excavation of soil reduce the effect on the soil excavation, which was caused by the slope, the soil excavation slop was arranged at the north side of the foundation pit. The ramp was of an internal ramp, with a gradient between 1:4 and 5, and a ramp width between 7 and 8 meters (taking into considering parallel driving of two vehicle).

First of all, the construction of the in situ cast concrete piles was carried out, and the earthwork excavation of Building 18# and Building 17# was carried out immediately after the construction of the in situ cast concrete piles was completed. In the process of the earthwork excavation, anchoring of the pipelines as well as the soil nailing wall support were carried out. The transportation road, which was located on the east side of the foundation pit, was excavated in layers and in the backward direction from the west side of the foundation pit to the east side. The excavation of the reserved road on the north side was carried out the latest.

The construction sequence for the foundation pit engineering was as follows: leveling of the site→ survey and setting out of the control points in the construction→construction of the drainage ditch on the top of the slope→casting of the concrete piles→excavation of earthwork→ construction of the crown beam→construction of the steel pipes and the soil nailing→setting up of the meshs and spraying of concrete was carried out→excavation to the bottom of the pit→ construction of the cushion layer on the pit bottom.

3. The technologies used in construction of the foundation pit support

The location of the project provided it with a complex terrain condition, the west side was an open space, so the excavation was carried out by means of an open slope, and the meshs were set up for spraying of concrete; the north side was supported by the soil nailing wall; the east side, which was close to Lianzhuang Road, was supported with the row piles and enclosure wall. In this paper, the supporting methods used for the north side and the east side were mainly studied.

3.1. The soil nailing wall support on the north side

The soil nailing wall method is a technology in which, the soil nails which are made of reinforcing steel are driven into the original soil, then the steel meshes are laid on the surface of the side slope and the concrete is sprayed, so that the soil nails, the reinforcing steel meshes and the sprayed concrete can be formed as a whole so as to maintain the stability of the side slope and ensure the stability of the slope surface after the excavation of the foundation pit [2]. The anchor rods used as the soil nailing in the project adopt HRB400 C20 reinforced bars, which are set up at an interval of 1500 mm, the lengths of which are 4500 mm, 6000 mm and 9000 mm, and the diameter of the construction hole is 110mm; the arrangement of the reinforcing ribs is 4B16, the strength grade of the concrete sprayed on the surface layer is C20, the sprayed thickness is 100mm, and the 8@200x200 is made as the reinforcing mesh, and the anchoring force of the supporting anchor rod in the construction process is less than 90kN. The length of the water discharge hole is 800 mm, the diameter of the pipe is 50 mm, the interval is 1500 mm.

The slope were drilled before the anchor rod construction was carried out. The depth of the holes was greater than the length of the soil nail reinforcement by + 0.5m. Then the holes were cleaned and the anchor rods were inserted in them and then grouting was carried out. The tilting angle error of the completed hole must be within the range of $\pm 1^{\circ}$, and the depth error of the hole must be within the range of ± 100 mm. The cement paste with the strength of M10 was used for grouting, the water to cement ratio was 0.5. The triethanolamine early strength agent of a concentration of 0.05% was added. The low-pressure grouting process with the grouting pressure of 0.5-1.0MPa was adopted. The grouting process was carried out continuously. The grouting would not be stopped until the slurry had overflown from the hole, or the grouting pressure had reached the design requirements and had held for 3 minutes. The grouting pipe was tied to the soil nail body. A grout stopping plug, an exhaust pipe and the like was provided around the position of the hole. The distance between the grouting pipe and the hole bottom was 150 mm.

3.2. The construction technology of the enclosure piles on the east side

The enclosure pile supports for the foundation pit on the east side adopted the in situ cast concrete piles, each having a diameter of 1 meter, at an interval of 2 meters. The height of pile top is -5.60 meters, the pile length is 10.1 meters, the number of piles is 35, concrete grade is C30, the main rib in the section reinforcement, which is subjected to the longitudinal force is 16C20, the arrangement of the reinforcement ribs is CC20@2000, and the arrangement of the spiral ribs is A10@120. The schematic diagram of the enclosure piles and drainage is as shown in Figure 1.

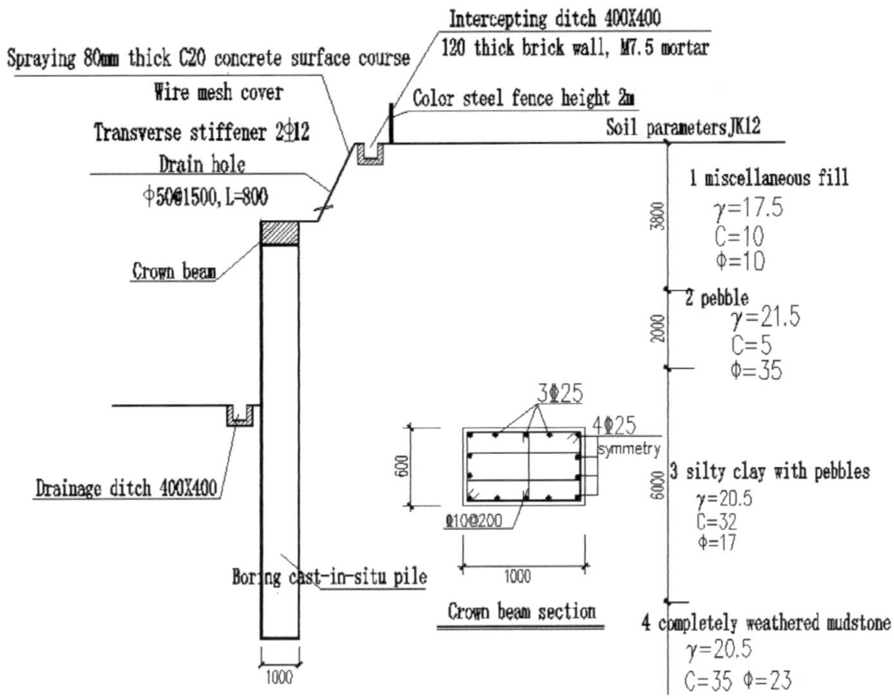

Figure 1. *A schematic diagram of the enclosure pile sand the drainage on the east side.*

3.2.1. Installation of steel guard cylinders: For the enclosure piles in the project, the holes were formed by using a rotary drilling rig. The steel casing is made of a 5mm steel plate, the diameter of the pile is 1000 mm, the inner diameter of the pile casing is 1030mm, the length is 4m. The top, the middle part and the bottom of the steel casing are welded with a reinforcing ring with a thickness of 5 mm and a height of about 15 cm. The welding of the joints in the steel casing should meet the welding requirements, which should be able to resist to pull and pressure and water leakage.

The steel casing can be laid by hammering or pressurization, which should be place at a place which is 0.5m above the hard and dense soil. After determining the center of the pile, it is possible to lay the steel casing d, and the specific position of the steel casing should meet the requirements for deviations, the deviation between the center of the top surface of the steel casing and the designed pile position should not be more than 5 cm, and the tilting angle deviation should be less than 1%. The drilling can be carried out only after the measurement is correct [4].

3.2.2. Construction of the large diameter long pile: Since the diameter of the main rib in the steel bar cage is relatively larger, the steel bar cage is fabricated by adopting a building platform. For the main reinforcement rib which is to be hoisted, the method for reinforcement is to arranged steel bars on the cross section in a square crossing manner. During the transportation and hoisting, two cranes should be employed, and a single crane is to be used, both of its main hook and auxiliary hook should be used to carry out simultaneous hoisting, To prevent the reinforcement cage frame from being deformed due to collisions or its own weight, welding of the main reinforcement rib should be carried out by two welders at the same time so as to ensure the verticality of the reinforcement cage after the butt joint is made meets the requirements. Before lowering the reinforcement cage, an reinforcement cage made with short steel bars, with a slightly larger diameter and carrying weights shall be lowered in advance to check the quality of the holes, and the phenomenon of a seized cage should not occur at the time when the reinforcement cage is being lowered [5].

- As to the bored pile with a larger diameter, it is possible to protect its wall with the high grade mud, that is, the PHP mud, which is made by incorporating the polyacrylamide into the ordinary mud and is known as the PHP mud. Generally, the content of polyacrylamide is 0.003% of the amount of the mud. Compared with the common clay which is made with the natural pulp, the PHP mud is made by adding PHP colloid into the base slurry which is made using bentonite. It has the advantages such as: the water loss is small, the mud skin is thin, compact and flexible, and the effect of wall protection is better.
- In order to avoid the collapse of the hole or seizure of the drilling rig during the drilling operation, attention should be paid to controlling and adjusting the consistency of mud. When clay layer is encountered during the drilling operation, clean water should be injected to reduce the mud consistency. Clay should be added to a sand layer and a gravel layer in order to improve the mud consistency and ensure safety during the drilling operation.
- In case a gravel layer is encountered in the drilling operation and the drilling speed becomes slower, first a small drill of 1 meter should be used for drilling up to the designed elevation. Then the hole should be drilled once again with a drill of 1.5 meters. This method can greatly improve the working efficiency.

3.2.3. Construction of the inter-pile wall: In order to ensure the safety of the foundation pit support and ensure the stability of the soil mass between the piles, a thick concrete wall with 80mm thickness should be formed between the supporting piles by spraying C20 concrete during the earthwork excavation. The steel mesh of Φ 8@200mm×200mm shall be set up within the supporting piles.

- The soil between the piles should be repaired. It is required to repair the soil at the same time the excavation is carried out. It ensures the stability of the side wall under natural conditions.
- After the repair is completed, the mesh shall be hung. Two Φ 16@2000 reinforcement steel bars should be pressed longitudinally above the mesh, and the reinforcement shall be secured on the pile.
- The mesh can be prefabricated and fixed to its position between piles with the short steel bars (soil nails).

4. The key points in quality control of the deep foundation pit support

The foundation plan should be based on, the axis and geometric size of the foundation pit should be checked in real time during excavation. The position and elevation of each control pile should be checked regularly during the excavation process.

The large diameter hole piles are protected by high-grade mud, namely PHP mud. PHP mud is made by adding polyacrylamide into the ordinary mud. Its content shall be strictly controlled and generally it accounts for 0.003% of the mud amount.

In order to ensure the safety of the foundation pit, the steel bars and the structures. It masters the working conditions of the support system in detail to meet the requirements of information construction. This supporting work was being monitored according to the design requirements. So it was possible to know any changes in the foundation pit in time, and non safety accident has occurred.

5. Conclusion

The deep foundation pit work of the residential district project adopts the soil nailing support and the enclosure pile support technology. This support work is close to the highway and the terrain is complex, the construction technology of the soil nailing wall and the supporting wall row pile is introduced. The key points in the quality control technology used for the deep foundation pit support are put forward for the purpose of reducing the construction risk of the project, which can be used for reference to the construction of similar foundation pit engineering project.

IOP Conf. Series: Materials Science and Engineering **787** (2020) 012038 doi:10.1088/1757-899X/787/1/012038

References

[1] Wang C.D, Xia X. (2003) Application of shotcrete anchor support in deep foundation pit engineering. Journal of rock mechanics and engineering, 167: 83-86.
[2] Feng Z.W. (2016) Discussion on the Construction scheme of shotcrete-bolt support for the slope and foundation pit. Building Technology Development, 140: 62-63.
[3] Li Y, Dong W.H, Huang X.F. (2013) Reinforcement and Reconstruction technology of existing foundation pit support structure under reverse operation. Construction technology, 204: 55-58.
[4] Ma Y.A. (2015) Analysis of pile foundation construction technology of highway bridge. Sichuan cement, 53: 98-100.
[5] Tan X.S. (2018) Research on design and construction of deep foundation pit support in geotechnical engineering. Architectural technology development, 68: 121-122.

AUTHOR INDEX